Equilibrium Between Phases of Matter

Equilibrium Between Phases of Matter

Phenomenology and Thermodynamics

H.A.J. Oonk
Utrecht University, The Netherlands

and

M.T. Calvet
University of Barcelona, Spain

 Springer

A C.I.P. Catalogue record for this book is available from the Library of Congress.

ISBN 978-90-481-7542-0
ISBN 978-1-4020-6408-1 (e-book)

Published by Springer,
P.O. Box 17, 3300 AA Dordrecht, The Netherlands.

www.springer.com

Printed on acid-free paper

TABLE OF CONTENTS

LEVEL 2:   *phase theory: the thermodynamics of equilibrium between phases*

SOLUTIONS OF EXERCISES

INTRODUCTION

About the book, the project

Equilibrium Between Phases of Matter – Phenomenology and Thermodynamics is a textbook, in which the phenomenology, the thermodynamic theory, and the practical use of phase diagrams are presented in three levels that diverge in nature – in particular as regards the role of thermodynamics. The book has been written from a chemical and geological teaching background. Each of the three levels of the book is representative of a particular course in a curriculum.

Level 0: an introduction to phase diagrams

The philosophy behind the ground level is that most of the characteristics of equilibrium between phases can be understood without the use of thermodynamics, realizing that, in a common-sense manner, the experimental observations on equilibria and spontaneous changes, and elementary notions about interactions, indicate the way to go. In spite of all this, the central figure in level zero, right from the beginning, is the *chemical potential* – a concept firmly rooted in thermodynamics. Equilibrium conditions in terms of chemical potentials, and the variables necessary to define a system in equilibrium are, are the basic elements of the *system formulation*.

The first three sections deal with the characteristics of a system in equilibrium; the variables necessary to define the system; the conditions for equilibrium and the rules – the phase rule and the lever rule. Pure substances and their forms are the subject of the fourth section. In the fifth section the phenomenology of binary phase diagrams is developed in terms of the nature of the interaction between the entities of the components; and, subsequently ternary phase diagrams are developed from the diagrams of the binary subsystems. Much of the use of binary phase diagrams is based on the fact that phases in equilibrium have different compositions – sixth section. The last section is devoted to homogeneous and heterogeneous chemical equilibria.

Level 1: an introduction to thermodynamics and phase theory

The intermediate level is an introduction to classical thermodynamics - culminating in the *Gibbs energy* as the arbiter in equilibrium matters - followed by the thermodynamic treatment of equilibrium between phases for a number of elementary cases.

The thermodynamics part starts with differential expressions; work, heat, and energy; heat capacity and enthalpy. After the ideal gas and its expansion and compression; and a section on chemical energy; entropy and Gibbs energy make their appearance – in particular their role in spontaneous changes and matters of equilibrium. In the last sections the principles of equilibrium are applied to pure substances – where *molar Gibbs energy* has the status of *chemical potential* – and to homogeneous and heterogeneous chemical equilibria.

Level 2: phase theory: the thermodynamics of equilibrium between phases

In the upper level the step is made to *mixtures* with their 'puzzling' *partial molar properties*, revealing the identity between *chemical potential* and *partial Gibbs energy*. After the introduction of *ideal mixtures*, a *'magic formula'* for the Gibbs energy of mixing is used to explain the phenomenon of demixing, and to derive the properties of the *ideal dilute solution*. Ideality and non-ideality, in a certain sense, give rise to the existence of two sub-levels. In the first sub-level the function recipes of the chemical potentials are such that explicit relationships can be formulated for systems where the phases are pure substances and/or ideal mixtures or ideal dilute solutions. The second sub-level, the last part of the work is devoted to systems where the phases are non-ideal mixtures, whether or not, in combination with phases of fixed composition – and where the leading role of *partial* Gibbs energy, read the chemical potential, is taken over by the *integral* Gibbs energy.

A priori and a posteriori

To assist the reader in assessing her/his own level of understanding the ins and outs of equilibrium between phases, each of the sections is provided with an *a priori*, in which the subject is introduced, and an *a posteriori*, in which the main conclusions are summarized.
Besides, owing to the level structure of the work, there is a certain, small overlap between parts of different levels. Or, in other terms, some overlap has not been kept out of the authors their way.

Exercises

Exercises are given at the end of each section, and the solutions of most of them at the end of the book. Their main function, of course, is to interest and stimulate the reader or a group of students in the class room. Occasionally, exercises are used as a vehicle to introduce issues that are not treated in the main text.

Follow-up

Equilibrium Between Phases of Matter – Phenomenology and Thermodynamics, is the first volume of a two-volume project. The second volume, by M.H.G. Jacobs and H.A.J. Oonk, *Equilibrium Between Phases of Matter – Thermodynamic Analysis and Prediction*, more than the first, has a postgraduate and professional character – and is written from a research background in *materials science* and *geophysics*.

In terms of thermodynamic properties, the distinction between the two volumes is in the role of the temperature and pressure derivatives of the Gibbs energy. In the first volume a dominating role is played by the Gibbs energy itself and its first derivatives, which are entropy and volume. The three together account for most of the phenomenology of equilibrium between phases – but not all of it; and one can think of retrograde phenomena. And when it comes to the real thermophysical properties of a given system, and the prediction of its behaviour under circumstances far from ambient temperature and pressure, one is completely lost without the incorporation/availability of the second and higher-order derivatives.

About languages

Writing a text in a foreign language is a delicate undertaking – in the sense that one cannot express oneself as clearly and linguistically correctly as one would like to. It is also extra time-consuming: the time needed to consult dictionaries and books of synonyms is commensurate with the time needed for the real work.

The manuscript of the senior author's 1981 book *Phase Theory* was read and corrected by Dr Philip Spencer, who was rather satisfied with the English. On the other hand, an American reviewer of the book, found the English somewhat clumsy.

The authors of this book have been working together for more than twenty years as participants in the REALM (Réseau Européen sur les Alliages Moléculaires) – a European network on molecular alloys, as may be clear. French is the REALM's official language – in the family circle one of the authors speaks Catalan, and the other Dutch.

Anyhow, dear reader, the message is clear. Please, enjoy the science and wink at the linguistic shortcomings.

Not only communication in terms of real languages, but also the communication between thermodynamicists from different branches of materials science has its typical difficulties. Just to give an idea, in certain fields a prominent role is played by activities and fugacities, whereas in other fields these terms virtually are non-existent.

And, of course, the relationship a physical chemist has with thermodynamics is quite different from the one built up by a geologist. A chemist creates his own systems out of pure chemicals, and having the phases of his own choice. The geologist, on the other hand, focuses his attention to stones created by nature and often having a multitude of phases, which, sometimes, reflect the history of the material.

The first volume of *Equilibrium Between Phases of Matter* has been written in the tradition of the Dutch School - in the footsteps of Josiah Willard Gibbs's disciples J.D. van der Waals, H.W. Bakhuis Roozeboom, F.A.H. Schreinemakers, J.J. van Laar, J.L. Meijering, and others. In this tradition a prominent role is given to *chemical potentials* – and, when it comes to deal with deviation from ideal behaviour, the use of *activities* and *fugacities* is avoided.

Acknowledgements

We should like to take the opportunity of expressing our gratitude to Mercedes Aguilar, Jan den Boesterd, Ingrid van Rooijen, and Aloys Lurvink for fine artwork; to Petra van Steenbergen, Hermine Vloemans, and Tamsin Kent for most valuable suggestions related to publishing; and to Koos Blok for thorough bibliographic research.
We are grateful to our students, who, through their remarks and probing questions, have kept us alert – thermodynamics is tricky business after all.

Dedication

We dedicate this book to our spouses and the other members of our families, and to our friends, who all, every now and then, remind us of the fact that *equilibrium between faces* is also a matter of great importance.

Utrecht, The Netherlands,   *Harry A.J. Oonk*
Barcelona, Spain   *M. Teresa Calvet*

Spring 2007

LIST OF FREQUENTLY USED SYMBOLS

Latin letters

A	Helmholtz energy; magnitude parameter of $AB\Theta$ model
A	first component in binary and ternary system
α	activity
a, b, c	system-dependent parameters
B	general for substance; second component in binary and ternary system
B	asymmetry parameter in $AB\Theta$ model
C	third component in ternary system
c	number of components
d	ordinary differential
C_P	heat capacity at constant pressure
C_V	heat capacity at constant volume
E	electromotive force
e	electric charge
f	variance, number of degrees of freedom; fugacity; activity coefficient
G	Gibbs (free) energy
g	acceleration of free fall; parameter in excess Gibbs energy
H	enthalpy
h	parameter in excess enthalpy; altitude
$\ln$	natural logarithm
LN	defined as $LN(X) = (1-X)\ln(1-X) + X\ln X$
M	set of variables necessary to define an equilibrium system, and number of elements in it; molar mass
m	mass; molality
N	set of equilibrium conditions, and number of elements in it
N_{Av}	Avogadro's number
n	amount of substance
P	pressure
p	number of phases
Q, q	heat
R	gas constant
S	entropy
s	parameter in excess entropy
T	thermodynamic temperature
t	Celsius temperature
U	energy
V	volume
W, w	work
W	number of configurations
X	variable in general; mole fraction; mole fraction of second component

Y	variable in general; mole fraction of third component
Z	variable in general; general for thermodynamic quantity like energy and entropy

Greek letters

α	cubic expansion coefficient
$\alpha, \beta, \gamma, \delta$	denote phases
β	pressure coefficient
γ	activity coefficient
Δ	operator for difference and change
δ	operator for virtual change, except for special use in § 212
Θ	class- or system-dependent temperature parameter of $AB\Theta$ model
κ	compressibility
μ	chemical potential
ν	stoichiometric coefficient
Π	osmotic pressure
ρ	density
φ	osmotic coefficient
Ω	interaction parameter in magic formula
ω	parameter in excess Gibbs energy
∂/∂	partial differential coefficient

Superscripts

E	refers to excess quantity
id	refers to ideal-mixing behaviour
$\alpha, \beta, \gamma, \delta$	to refer to phases
liq	for liquid
sol	for solid
vap	for vapour
o	for standard state; for transition temperature and equilibrium pressure of pure component in a binary or ternary system
*	for a pure-substance quantity like entropy

Subscripts

A, B, C	for property of substance/component in system
c	for critical point
e	to refer to equilibrium
f	for formation from the elements, attached to Δ
fox	for formation from the oxides, attached to Δ
m	for molar quantity; from § 203 on the subscript is dropped: only molar quantities are being used

LEVEL 0

AN INTRODUCTION TO PHASE DIAGRAMS

§ 001 EQUILIBRIUM

Three simple experiments are examined to find out what the characteristics are of equilibrium between phases.

first experiment

Two identical vessels made of copper and having a square cross-section are filled with water. One of the vessels is filled with water from the cold tap; the other with water from the hot tap. The vessels are put against one another; inside a box made of polystyrene foam, see Figure 1.

By reading the thermometers it is observed that, when time goes by, the temperatures of the two parts of the set-up become equal and remain equal. Inside the box, then, there is a uniform temperature: the whole inside the box is in a state of *thermal equilibrium*.

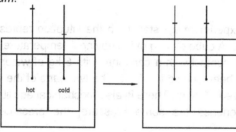

FIG. 1. On the way to thermal equilibrium. Thermal equilibrium is characterized by equality of temperature, equality of thermal potential

In learned terms: when two bodies with different temperatures are put into thermal contact, there will be a flow of heat - from the body with the higher to the body with the lower temperature - such that the temperatures of the two become equal. Obviously the flow of heat comes to a stop when the temperatures have become equal - when the thermal potentials, so to say, have become equal.

second experiment

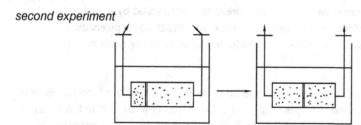

FIG. 2. On the way to mechanical equilibrium. Mechanical equilibrium is characterized by equality of pressure, equality of mechanical potential

In Figure 2 the features of the second experiment are sketched. A horizontal cylinder with two compartments separated by a piston and at both sides provided with a manometer is immersed in a thermostat. The thermostat - a water bath with heating and cooling facilities - guarantees thermal equilibrium at a selected temperature. The two compartments contain equal amounts of air.

In the left-hand situation the piston is locked; the manometers indicate different pressures. When it is set free, the piston moves to the right - to the middle of the cylinder. In that situation, represented by the right-hand side of Figure 2, the manometers indicate the same pressure.

In this case there is a flow of space, from the right-hand to the left-hand compartment. The flow is stopped when the mechanical potentials, the pressures of the two compartments have become equal. Thereafter there is *mechanical equilibrium*.

third experiment

For the third experiment we start from the situation represented by the left-hand side of Figure 3. A cube of 1 kg of ice having a temperature of –5 °C is placed aside and in thermal contact with a container with 5 kg of water, which by careful experimentation has been cooled to –5 °C. The two parts of the whole, i.e. ice and water, are in thermal equilibrium. There is also mechanical equilibrium inasmuch as ice and water experience the same pressure, the pressure exerted by the atmosphere.

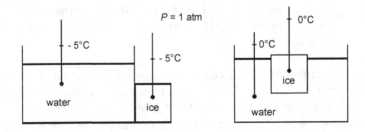

FIG. 3.   On the way to equilibrium between the solid and liquid forms of the substance water. The equilibrium is characterized by equality of chemical potential. Under 1 atm pressure the chemical potentials of the substance water in the solid and liquid forms are equal at 0 °C (only)

In spite of the equality of temperature and the equality of pressure, the investigator will observe changes when the cube of ice is transferred to the water in the container; right-hand side of Figure 3. It is observed that the temperatures of both ice and water rise, and come to a stop at 0 °C. In that situation there will be no more changes: the temperatures of ice and water remain at 0 °C.

We might ask ourselves if there is, again, a flow of something from one part of the system to the other. In order to answer this question, let's assume that during the whole experiment there is no flow of heat to and from the system as a whole - accounted for by thermal insulation. In that case then, the heat needed to increase the temperature of both ice and water from –5 °C to 0 °C has to come from the system itself. Knowing that heat is needed to melt ice, we may conclude that heat will be produced when an amount of water is transferred from the liquid to the solid form - and that is exactly what is going to happen when the cube of ice is transferred to the water.

In terms of potentials and by analogy with the first two experiments we may make the following observations. First, there is - apparently - a potential for the *transfer* of a given *substance*. That potential is the *chemical potential*, and its symbol μ has to be extended with a subscript to indicate the substance; in the case of the substance water: μ_{H_2O}. Next, the properties of the substance water are such that at 1 atm and –5 °C the chemical potential of H_2O in the solid form is not equal to the chemical potential of H_2O in the liquid form:

$$\mu_{H_2O}^{sol}(1\,atm, -5\,^{\circ}C) \neq \mu_{H_2O}^{liq}(1\,atm, -5\,^{\circ}C). \tag{1}$$

As a result of this inequality there will be a flow of matter - the substance H_2O - until the chemical potentials of the solid and liquid parts of the system have become equal. At 1 atm pressure this is at 0 °C:

$$\mu_{H_2O}^{sol}(1\,atm, 0\,^{\circ}C) = \mu_{H_2O}^{liq}(1\,atm, 0\,^{\circ}C). \tag{2}$$

In fact, as we will see later on (→108), there is a flow of matter form the part of the system where the potential is higher to the part of the system where the potential is lower, see Figure 4. Above 0 °C the situation is the other way round: if it were possible to repeat the third experiment with water and ice initially having a temperature of +5 °C, one would observe a fall of the temperature to 0 °C and an increase of the amount of liquid (water) at the expense of the amount of solid (ice).

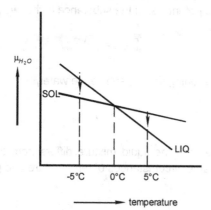

FIG. 4. The chemical potentials of solid and liquid water in the vicinity of 0 °C and under a pressure of 1 atm

The equilibrium between ice and water is an example of *heterogeneous equilibrium* where different and distinguishable homogeneous parts coexist with one another. The two different homogeneous parts are (referred to as) *phases*, a solid phase and a liquid phase. By the way, when several cubes of ice are floating in water there is still only one solid phase.

equilibrium between two phases - both containing alcohol and water

In the situation sketched by Figure 5, a vessel provided with a manometer and containing alcohol and water is immersed in a thermostat. There is an amount of liquid; and the space above the liquid is occupied by vapour. In this case there is equilibrium between a liquid phase, which is a mixture of alcohol and water, and a vapour phase, which is also a mixture of alcohol and water.

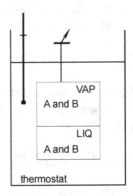

FIG. 5.   Isothermal equilibrium between a liquid and a vapour phase. Each of the two phases is a mixture of two substances A and B, such as alcohol and water

In this situation of equilibrium there is no net transfer of alcohol form the liquid to the vapour phase: the potentials, of the substance in the two phases are equal:

$$\mu_{alcohol}^{liq} = \mu_{alcohol}^{vap}.\tag{3}$$

On the same lines of reasoning, for the substance water,

$$\mu_{water}^{liq} = \mu_{water}^{vap}.\tag{4}$$

NB: the potential of water in the liquid mixture differs from the potential of pure liquid water under the same circumstances of temperature and pressure!

generalizing

For a system in equilibrium, such that the substances A, B, C... are present in the phases α, β, γ...,

- there is a uniform temperature and a uniform pressure;

- for each substance there is uniformity of its chemical potential throughout the system:

$$\mu_A^\alpha = \mu_A^\beta = \mu_A^\gamma \ldots$$
$$\mu_B^\alpha = \mu_B^\beta = \mu_B^\gamma \ldots \qquad (5)$$
$$\mu_C^\alpha = \mu_C^\beta = \mu_C^\gamma \ldots$$

These facts are the *characteristics of equilibrium* and at the same time they are a set of *conditions for equilibrium*: without uniformity of temperature, pressure and chemical potentials there is no equilibrium.

In what follows, the conditions of uniformity of temperature and uniformity of pressure will be referred to as the *a priori equilibrium conditions*.

the first phase diagram

In general terms, and returning to the equilibrium between ice and water, the chemical potentials $\mu_{H_2O}^{sol}$ and $\mu_{H_2O}^{liq}$ are functions of T and P and, when these two functions are subjected to the condition

$$\mu_{H_2O}^{sol}(T, P) = \mu_{H_2O}^{liq}(T, P), \qquad (6)$$

an equation appears in the variables T and P. The solution of this equation corresponds to a curve in the PT plane, see Figure 6. In other words, when a condition, Equation (6), is imposed on the two variables T and P, one of the two will get the status of dependent variable. If the investigator wants to fix the pressure to a certain value, say 10 atm, (s)he has to know that the system will adapt the temperature such that Equation (6) is satisfied.

On the same lines of reasoning, it is obvious that there will be another curve in the PT plane - in the PT *phase diagram* - representing the equilibrium between liquid and vapour, i.e. representing the solution of

$$\mu_{H_2O}^{liq}(T, P) = \mu_{H_2O}^{vap}(T, P). \qquad (7)$$

The intersection of the curve - the *boiling curve* - with the solid-liquid equilibrium curve - the *melting curve* - is a point in the PT plane (at about 0.006 atm and +0.01 °C), which is called *triple point*.

At the *PT* conditions of the triple point, three phases - solid, liquid and vapour - coexist, are in equilibrium with one another. It is obvious that at the triple point the chemical potentials of water in the solid and vapour forms are equal:

$$\mu_{H_2O}^{sol} = \mu_{H_2O}^{vap} ; \tag{8}$$

and from this observation it naturally follows that from the triple point a curve goes out that represents the equilibrium between solid and vapour: the curve which is the solution of the equation

$$\mu_{H_2O}^{sol} (T, P) = \mu_{H_2O}^{vap} (T, P). \tag{9}$$

Finally, the three *two-phase equilibrium curves* divide the *PT* plane into three fields, three *stability fields*, marked by the capitals S, L, and V. For *PT* conditions of the S field the solid form ice never will change spontaneously into liquid or vapour. The other way round, for conditions of the S field, vapour and liquid change spontaneously into solid (be it that they sometimes take their time to do so - which enables us to use liquid water of −5 °C for the third experiment).

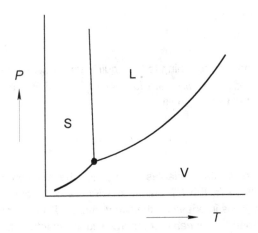

FIG. 6.   Schematic drawing of the phase diagram of the substance H_2O
around the triple point

The three experiments, which have been considered, correspond to spontaneous changes, which proceed all by themselves and lead to a state of equilibrium. Equilibrium between phases is characterized by equality - uniformity - of temperature, of pressure and of chemical potentials.

EXERCISES

1. *vessels with water in thermal contact*

A vessel containing 10 kg of water having a temperature of 50 °C and another vessel containing 5 kg of water of 25 °C are put into thermal contact and then thermally isolated from the surroundings.
   - What will be the temperature of the water in the vessels when thermal equilibrium has been reached?

   NB The heat involved in changing the temperature of the vessels when empty may be neglected.

2. *sulphuric acid and water in thermal contact*

A vessel containing 10 kg of water having a temperature of 50 °C and another vessel containing 5 kg of sulphuric acid of 25 °C are put into thermal contact and then thermally isolated from the surroundings. From a certain moment on the thermometers in the two liquids indicate the same temperature, which is 46.4 °C.
   - What is the significance of this observation?

   NB see foregoing exercise.

3. *sulphuric acid poured into water*

In the experiments considered in the foregoing two exercises, two vessels with liquids having different temperatures are put into thermal contact as a result of which a uniform temperature is reached.
   - Will, in each of the two cases, the same final temperature be obtained - i.e. 41.67 °C and 46.4 °C - if the liquids are poured into one another (sulphuric acid into water and not the other way round!) instead of just put into thermal contact?

§ 002 VARIABLES

It is examined what kind of, and which variables are needed to characterize the state of a system where phases are in equilibrium.

intensive variables

Let's suppose that you are invited to investigate the characteristics of a system in equilibrium - defined as to its chemical composition and the number and nature of the phases - and to write a report on the results. Obviously, and before you are going to start the experiments, you must have a clear idea of the actions you must take, and can take - given the experimental set-up and its dimensions and the restrictions imposed by the definition of the system.
As an example let's take the *equilibrium between liquid and vapour* in a *binary system*, i.e. a system composed of two substances A and B. The experimental set-up, which you are going to use, is sketched in Figure 1.
In your report, of course, you give the dimensions of the vessel, a description of the thermostat, the characteristics and the degree of perfection of the manometer and thermometer and you mention the provenance of the chemicals and their purities. Next, you state clearly how the experiment is carried out, you mention the amounts of A and B needed and you carefully indicate how the phases are analyzed after equilibrium has been ascertained.

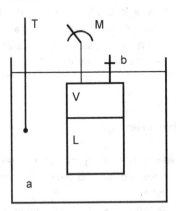

FIG. 1.   Sketch of set-up for studying the equilibrium between liquid (L) and vapour (V). a. thermostat bath; b. passage; T. thermometer; M. manometer

In your quality as experimentalist, and as follows from Figure 1, you have command over the temperature of the thermostat and the amounts and relative amounts of A and B in the vessel. The absolute amounts of A and B you need - at a selected temperature - depend on the dimensions of the vessel. For example, if your vessel has a volume of 400 cm^3 and you need 10 drops of liquid A to saturate the space with vapour of A (the 11th remaining liquid) then you know that if the volume were 800 cm^3 you would need 20 drops to saturate the space. And if you use 20 drops in the case of 400 cm^3, you increase the amount of the liquid phase. Clearly, in all of the three situations the concentration of the vapour in the vessel is the same, and thereby the pressure indicated by the manometer. By changing the relative amounts of A and B you can change the overall percentages of the two components in the system. The division of the components over the two phases, after your choice of the amounts of A and B, is a spontaneous event, dictated by the intrinsic properties of the system.

In summarizing, the experimentalist has the freedom to set the temperature of the thermostat and to make his own choice of the contents of the vessel. Thereafter, the system dictates the pressure and also the compositions of the two phases in terms of percentages. What is left for the experimentalist, is to read the manometer and to analyze the phases as to their composition.

As a result, an equilibrium system can be characterized or rather is characterized by a set of *intensive variables*, i.e. variables that are independent of the extension of the system. These variables are temperature, pressure and the smallest set of variables necessary to define the chemical composition of each of the phases. The important consequence of this observation: what is found in the laboratory can be realized on the scale of an industrial plant.

temperature and temperature scales

The *temperature scales* that generally are used are the *thermodynamic temperature scale*, or *absolute temperature scale*, and the *Celsius temperature scale*, named after Anders Celsius (1701-1744).

Both scales make use of *fixed points* - and equilibria between phases are used in both cases to define the fixed points. The *zero point* of the Celsius scale is the *normal freezing point* of water: the temperature at which the solid and the liquid forms of the substance water are in equilibrium at a pressure of 1 atm (see below for atm).

The *degree Celsius* (°C) is 0.01 times the temperature difference between the *normal boiling point* and the normal freezing point of water. In other words 100 °C - the second fixed point of the Celsius scale - is the temperature at which the liquid and gaseous forms of the substance water are in equilibrium at 1 atm. It may be remarked that, from a point of view of careful calibration, the two fixed points have the disadvantage that they need the adjustment of pressure (a disadvantage which is absent for the triple point (←001)).

After Celsius's time the existence was discovered of a *natural zero point* of temperature: the lowest value temperature can have.

The *absolute zero of temperature* is at about −273.15 °C. The existence of a natural zero point implies that just one fixed point is needed to define the *unit of temperature*. The unit of thermodynamic temperature is the *kelvin* (symbol K) and the fixed point is the triple point of water (about +0.01 on the Celsius scale). The unit kelvin, therefore, has been defined as

$$1 K = \frac{\text{triple point temperature of water}}{273.16}. \qquad (2)$$

With this definition the kelvin is virtually equal to the original degree Celsius. The modern degree Celsius (°C) is exactly equal to the kelvin; and the modern zero point of the Celsius scale is laid at 273.15 K exactly. As a result, the modern Celsius temperature (*t*) and the thermodynamic temperature can be converted into one another by means of

$$\frac{t}{°C} \equiv \frac{T}{K} - 273.15. \qquad (3)$$

The *thermodynamic temperature* concept is based on the *second law of thermodynamics*, which goes back to 1824 to the work of Sadi Carnot (1796-1832) and the concept of *entropy* (→106). Entropy was exactly defined in 1851 by William Thompson, to whom was conferred the title of Lord Kelvin (1824-1907).
The thermodynamic temperature is equal to the *ideal-gas temperature*, which is defined by the *ideal-gas equation*

$$P \cdot V = n \cdot R \cdot T. \qquad (4)$$

According to this equation the product of pressure (*P*) and volume (*V*) of a given amount (*n*) of ideal gas is proportional to its thermodynamic temperature (*T*). The value of the constant *R*, the *gas constant*, which is used throughout this text, is the value recommended by CODATA, the International Committee on Data for Science and Technology:

$$R = 8.314472 \text{ kg} \cdot \text{m}^2 \cdot \text{s}^{-2} \cdot \text{K}^{-1} \cdot \text{mol}^{-1}. \qquad (5)$$

The ideal-gas temperature and the *ideal-gas thermometer* have played an important part in thermometry. In modern thermometry use is made of a number of fixed points that define the *International Temperature Scale* (ITS). The ITS is adjusted from time to time (apart of course from the triple point temperature of water). Just to mention a few fixed points of The International Temperature Scale of 1990 (ITS-90): the triple point of e-H_2 (hydrogen at the equilibrium composition of the ortho- and para-molecular forms) at 13.8033 K; the triple point of Ar at 83.8058 K; triple point of Hg at 234.3156 K; the normal freezing points of In at 429.7485 K, and Cu at 1084.62 K (Preston-Thomas 1990).

SI units

The units appearing in Equation (5) are the units of five of the seven basic physical quantities of the International System of Units (SI), agreed in 1960 by the 11th "Conférence Générale des Poids et Mesures" as extension and perfection of the Metric System. The complete set of basic physical quantities and their units is displayed in Table 1.

Some special attention has to be given to *amount of substance* and its unit mole. One mole is the amount of substance of a system which contains as many as *elementary entities* as there are carbon atoms in 0.012 kg of carbon-12. The number of carbon atoms in 0.012 kg of carbon-12 is referred to as the Avogadro constant, named after Amadeo Avogadro (1776-1856), and of which the CODATA value is

$$N_{Av} = 6.0221415 \times 10^{23} \text{ mol}^{-1}. \tag{6}$$

Table 1:   The SI basic physical quantities and their units

physical quantity	SI unit	symbol for unit
length	metre	m
mass	kilogram	kg
time	second	s
electric current	ampere	A
thermodynamic temperature	kelvin	K
luminous intensity	candela	cd
amount of substance	mole	mol

It must be realized that the concept of elementary entity may give rise to some confusion: elementary entity is not necessarily synonymous with physically existing particle. For a system which is oxygen gas at low pressure, it is obvious to take an O_2 molecule as the elementary entity, so that one mole of the substance has a mass of 0.031999 kg. Moreover, one can be rather sure that the physically existing particles are indeed O_2 molecules. For a system which is liquid silica one has every right to say that its elementary entity is represented by the formula SiO_2 and that, as a consequence, one mole has a mass of 0.060085 kg. In reality liquid SiO_2 does not contain separate SiO_2 molecules: it has a polymeric nature with rather undefined particles.

In the following, one mole of a given substance is defined by its chemical formula and, in order to avoid confusion, eventually with the corresponding mass. As an example, let's take the mineral dolomite. If we are going to represent it by the formula $CaMg(CO_3)_2$, then one mole of the substance has a mass of 0.184403 kg. If, on the other hand, we are going to represent it by $Ca_{0.5}Mg_{0.5}CO_3$ (e.g. to compare its properties with calcite, $CaCO_3$, and magnesite, $MgCO_3$), then we will be dealing with a molar mass of 0.092202 kg·mol^{-1}.

pressure and its units

The physical quantity *pressure* is a derived quantity: it is force divided by area. Force and area are derived quantities by themselves and their units, accordingly, are derived units: *newton* ($N = kg \cdot m \cdot s^{-2}$) and square meter ($m^2$), respectively. Therefore, the SI unit of pressure is $Kg \cdot m^{-1} \cdot s^{-2}$; its name is *pascal* and its symbol Pa. Other units that will be used are the *bar*, the *atmosphere* and the *torr* and their exact (modern) definitions are given in Table 2.

The torr is named after Evangelista Torricelli (1608-1647) who in 1643 showed that the pressure exerted by the atmosphere of the earth corresponds to the pressure exerted by a vertical column of mercury having a length of 760 mm. Torricelli predicted that the length of the column of mercury, read the pressure of the air, would be lower at higher altitudes. This prediction was experimentally verified in 1648 by Blaise Pascal (1623-1662) who, to that end, assisted by his brother-in-law Périer, brought his barometer into action at the top of the Puy de Dôme, near Clermont-Ferrand, in France.

Table 2:   Survey of pressure units used in this text

name of unit	symbol for unit	definition of unit
pascal	Pa	$Kg \cdot m^{-1} \cdot s^{-2}$
bar	bar	10^5 Pa
atmosphere	atm	101325 Pa
torr	Torr	(101325/760) Pa

barometric formula

The pressure exerted by the air at altitude h is given by the *barometric formula*

$$P = P^0 \, e^{-M \cdot g \cdot h / RT}, \tag{7}$$

where M stands for the (mean) molar mass of the air and g for the acceleration of free fall (whose value is about $9.81 \ m \cdot s^{-2}$).

Obviously, the change of pressure with altitude has to do with gravity - just as the pressure exerted by a vertical column of liquid (Torricelli's tube with mercury) which is given by

$$P = \rho \cdot g \cdot h, \tag{8}$$

and referred to as hydrostatic pressure. In this equation the symbol ρ represents the density of the liquid, g again the acceleration of free fall and h the height of the column.

Strictly speaking, equality of pressure to denote (mechanical) equilibrium is only true inside a horizontal plane. This fact, which is the *principal law of hydrostatics*, goes back to Simon Stevin (1548-1620).

In our treatment of equilibria between phases we will neglect the change of pressure within a system-in-equilibrium due to gravity. In other terms, the systems are considered to be so small that the influence of gravity is negligible.

composition variables

The most unambiguous way to state the composition of a mixture of two or more substances is by means of *weight percentages* or weight *fractions* (or *mass percentages* and *mass fractions*, if you like): there is *conservation of mass* - the masses of the substances do not change on mixing, and neither with temperature or pressure. From a thermodynamic point of view, however, a composition description in terms of molecular entities is preferred; and for that reason we, will make use of *mole fractions* (in spite of the observations made above as to the possible difference between elementary, read molecular entity and physically existing particles).

In the case of a mixture of two substances A and B we proceed as follows. First we (we ourselves) define the *molar masses* M_A and M_B of A and B. Next, for a mixture in which A has a mass m_A and B a mass m_B, we define the mole fraction of B by

$$X_B = \frac{\dfrac{m_B}{M_B}}{\dfrac{m_A}{M_A} + \dfrac{m_B}{M_B}}. \qquad (9)$$

The mole fraction of A, X_A, is defined on the same lines, and obviously the sum of the two has to satisfy the equality

$$X_A + X_B = 1. \qquad (10)$$

For a mixture of two substances, therefore, only one mole fraction quantity is needed to specify its (relative) composition. We invariably will use the *mole fraction of the second component*, the mole fraction of the substance B as the (independent) composition variable; and use for it the symbol X. In graphical representations the mole fraction axis is a line of unit length. Usually the mole fraction axis is the horizontal axis. If so, the vertical axis $X = 0$ displays a property or properties of the first component, substance A, and the vertical axis $X = 1$ has the same function for the second component. An example is given by Figure 2; the pure component property on the two vertical axes is the *equilibrium vapour pressure* over the pure liquid.

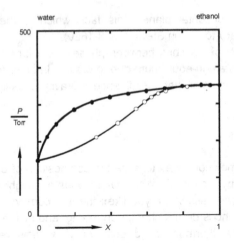

water   ethanol

FIG. 2.   Liquid + vapour equilibrium in the system water + ethanol at 60 °C. Filled circles represent liquid phase; open circles vapour phase

Generally, for a mixture of the number c of components one has for the i-th component

$$X_i = \frac{n_i}{\sum_{i=1}^{c} n_i},$$ (11)

where n_i is the number of moles of the i-th component in the mixture. There are $(c - 1)$ independent mole fractions, because

$$\sum_{i=1}^{c} X_i = 1.$$ (12)

In other words, for a mixture of c components there are $(c - 1)$ mole fraction values needed to define its composition.

the example of liquid + vapour equilibrium

      In conclusion, and returning to the example of the liquid + vapour equilibrium in the system A + B, the state of the system in equilibrium is fully described by its temperature (T), pressure (P), B's mole fraction in the liquid phase (X^{liq}) and B's mole fraction in the vapour phase (X^{vap}). Therefore, if you are invited to write a report on the experimental determination of the (liquid + vapour) equilibrium for a combination of two substances, then the essential part of it is a table with the experimental data quartets $[T, P, X^{liq}, X^{vap}]$. A concrete example is Table 3 for the system water + ethanol.

Table 3: Liquid + vapour equilibrium data for the system {(1–X) water + X ethanol}, (Schuberth 1980)

T/K	P/Torr	X^{liq}	X^{vap}
333.15	149.4	0.000	0.000
333.15	212.5	0.050	0.328
333.15	249.9	0.100	0.449
333.15	287.7	0.200	0.547
333.15	305.6	0.300	0.593
333.15	317.4	0.400	0.628
333.15	327.3	0.500	0.664
333.15	336.1	0.600	0.707
333.15	343.6	0.700	0.760
333.15	349.0	0.800	0.824
333.15	351.7	0.900	0.903
333.15	351.8	0.950	0.949
333.15	351.0	1.000	1.000

Note that all data quartets displayed in Table 3 are for the same temperature and that, as a result, the data set is an isothermal section of the equilibrium system considered. The isothermal section is represented by Figure 2, the *phase diagram* such that the horizontal mole fraction axis is used for X^{liq} as well as for X^{vap}.

The state of a system in equilibrium is defined by a set of intensive variables. These variables are temperature, pressure and the number of mole fractions necessary and sufficient to define the composition of each of the phases.

EXERCISES

1. *the air's pressure at the Puy de Dôme*

What are the pressures indicated by barometers in the center of Clermont-Ferrand (altitude 401 m) and at the top of the Puy de Dôme (altitude 1465 m) when the pressure at sea level is 760 Torr and the temperature 15 °C?.
Take 29 g·mole^{-1} for the air's molar mass.

§ (002)

2. *pressure at top and bottom*

An amount of argon (Ar, molar mass = 39.95 g·mole^{-1}) is contained in a vertical cylinder, with a piston at the upper side and 25 cm above the bottom of the vessel. The pressure exerted by the argon molecules on the piston is exactly 101325 Pa.
- Calculate the pressure exerted by the argon molecules on the bottom of the vessel; the temperature is 25 °C.

3. *Fahrenheit's temperature scale*

The temperature scale named after Daniel Gabriel Fahrenheit (1686-1736) was based on the two fixed points
i. the temperature of the cryogen mixture of water, ice and common salt (0 °F);
ii. the temperature of the human body (96 °F)
The Fahrenheit temperatures of the normal freezing point and the normal boiling point of water are 32 °F and 212 °F, respectively.
- Give the formula for the interconversion of Celsius and Fahrenheit temperatures and calculate the Celsius temperatures of Fahrenheit's fixed points.

4. *two phases and their amounts of two substances*

C and A are substances that are liquid at room temperature. When added together they give rise to two layers of liquid - a state of equilibrium between two liquid phases L_I and L_{II}.
In an experiment 10 mol A and 10 mol C are added together and it is established that L_I is composed of 6 mol A and 2 mol C and that, consequently, L_{II} is composed of 4 mol A and 8 mol C.
- Complete the table for three other experiments with other amounts of C and A, all expressed in mole.

added together $n(A)$ $n(C)$		in phase L_I $n(A)$ $n(C)$		in phase L_{II} $n(A)$ $n(C)$	
10	10	6	2	4	8
5	5				
6	4				
2	8				

5. *ethanol and water saturate a space*

Assess the amounts of ethanol and water needed to saturate at 60 °C a space having a volume of 50 dm^3 such that the pressure exerted by the vapour - which is a mixture of ethanol and water - is 300 Torr.
It may be assumed that the vapour obeys the ideal-gas equation (←Table 3; Figure 2).

Obviously one cannot give an arbitrary value to each of the variables necessary to define the state of a system in equilibrium. It is examined how the number of independent variables can be found, once the equilibrium system has been defined.

independent and dependent variables

In the foregoing section we considered the equilibrium between liquid and vapour in a system of two substances. The state of the system in equilibrium, as we found, is fully defined by four variables, four intensive properties, which are temperature and pressure and for each of the two phases one composition variable. For the latter we took the mole fraction of the second component: X^{liq} for the liquid phase and X^{vap} for the vapour phase. At the end of the section we observed that, at constant temperature, the mole fractions of the liquid phase in equilibrium with vapour are represented by a curve in the PX plane. In other words, the mole fraction of a liquid phase in equilibrium with vapour can be represented as a function of pressure

$$X^{liq} = X^{liq}(P) \quad T \text{ const. .} \tag{1}$$

Similarly, there is a relation between pressure and the mole fraction of the vapour phase:

$$X^{vap} = X^{vap}(P) \quad T \text{ const. .} \tag{2}$$

Moreover, if one likes, one can replace the first of these expressions by

$$P = P(X^{liq}) \quad T \text{ const. ,} \tag{3}$$

saying that the pressure indicated by the manometer is a function of the mole fraction of the liquid phase. And substitution of Equation (3) into Equation (2) gives rise to

$$X^{vap} = X^{vap}(X^{liq}) \quad T \text{ const. .} \tag{4}$$

The important observation we can make is, that for the equilibrium case considered, out of the three variables P, X^{liq} and X^{vap} only one can be freely chosen and that, thereafter, the remaining two are fixed. We are free to say that we want to realize that system such that $X^{liq} = 0.2$. Then we should know that in the case of Figure 002:2 the pressure can only have the value of 287.7 Torr and the mole fraction of the vapour phase the value of 0.547.

Generally, that is to say taking into account the influence of temperature as well, one can write

$$X^{liq} = X^{liq}(T, P) \tag{5}$$

$$X^{vap} = X^{vap}(T, P) \tag{6}$$

or, if one likes

$$P = P(T, X^{liq}) \tag{7}$$

$$X^{vap} = X^{vap}(T, X^{liq}). \tag{8}$$

Out of the four variables (T, P, X^{liq}, X^{vap}), two variables, say T and X^{liq} can be chosen - by the investigator - and after that choice, the remaining two are adjusted by the system itself.

In order to find out what really is going on with the variables involved in the liquid-vapour equilibrium, let's start again. Start again by taking a liquid mixture of the two substances, and, separated from it, a vapour mixture. The liquid mixture and the vapour mixture can be given different compositions and temperatures and put under different pressures: two times three variables that can be freely chosen. If we, next, bring the two together such that there will be equilibrium between a liquid and a vapour phase, we will observe that things are going to happen with the 2 x 3 = 6 values selected for the variables. First of all the *a priori equilibrium conditions* come into action, as a result of which the two phases will have a uniform temperature and a uniform pressure. At this 'moment' the number of variables involved has become four: four variables are needed to define the state of the system in equilibrium. These four variables constitute - what we are going to call - the *set M of variables*:

$$M = M\left[T, P, X^{liq}, X^{vap}\right]. \tag{9}$$

Next, there is a set of (other) conditions that come into action: the *conditions of uniform chemical potentials* - the chemical potential of A in the liquid has to be equal to the chemical potential of A in the vapour, and the same holds true for B's potentials. The equilibrium conditions in terms of chemical potentials constitute the *set N of thermodynamic equilibrium conditions*:

$$N = N\left[\mu_A^{liq} = \mu_A^{vap}; \mu_B^{liq} = \mu_B^{vap}\right]. \tag{10}$$

The influence of the two conditions is such - and that is purely mathematical - that four minus two variables remain as independent variables. In other terms and in harmony with the observations made above, in the equilibrium situation the values of X^{liq} and X^{vap} are fixed after the choice of T and P.

In the following we will use the symbols M and N, not only to refer to the two sets, but also for the numbers of elements in the two sets M and N, respectively. The number of independent variables of the system in equilibrium is found as

$$f = M\left[T, P, X^{liq}, X^{vap}\right] - N\left[\mu_A^{liq} = \mu_A^{vap}; \mu_B^{liq} = \mu_B^{vap}\right] = 4 - 2 = 2. \qquad (11)$$

The number f of independent variables is called the *number of degrees of freedom* or the *variance* of the equilibrium system. The equilibrium between liquid and vapour in a system of two substances is bivariant: has two degrees of freedom. If for these two degrees of freedom temperature and pressure are taken, then the type of mathematical solution of the set of N equations is

$$\begin{cases} X^{liq} = X^{liq}(T, P) \\ X^{vap} = X^{vap}(T, P) \end{cases}. \qquad (12)$$

It means that the solution of the set of N equations corresponds to two surfaces in *PTX* space which bear a certain relation to one another. One of the two surfaces represents the liquid phases and the other the vapour phases. Each pair of phases in equilibrium is represented by a point on the *liquidus surface* and a point on the *vaporus surface*, such that the line connecting the two points is parallel to the X axis. Figure 002:2 is an isothermal section of the two surfaces.

Generally, the chemical potentials figuring in the set N of thermodynamic equilibrium conditions are functions of the variables figuring in the M set. Therefore, in order to be able to solve the set of equations, one has to know - from case to case! - how the chemical potentials depend on the variables. One has to know, in other words what the *function recipes* are of the chemical potentials.

At this place it may be right to say a few words about the philosophy behind the construction of this work, in particular in relation to thermodynamics, the theoretical language of equilibria between phases. In level 1, the next part of the work, a start is made with thermo: to the extent that equilibria can be treated in which all phases are pure substances. In level 2 the thermodynamics of mixtures is introduced. In the first part of level 2, equilibria are considered for idealized cases such that the set of equilibrium equations can be solved in an analytical manner, read, such that explicit formulae can be derived for the M-f dependent variables in terms of the f independent variables. In the last three chapters of level 2, the systems considered are non-ideal, and, in order not to lose clarity, the complications related to the use of chemical potentials will be circumvented as much as possible. For the time being, that is to say in level 0, the approach will be of a phenomenological nature: without the use of thermodynamics; however, invariably starting from the thermo oriented sets M and N. The approach of this level is visualized in Scheme 1; again the binary liquid + vapour equilibrium and this time once more under isothermal conditions. An important role is reserved for expressions like the one represented by Equation (11) - for these expressions we will use the term *system formulation*.

The system
     substances: A, B
     equilibrium: A(liq) = A(vap); B(liq) = B(vap)
     constraints: T is constant

     the variables: $M = M\,[P, X^{liq}, X^{vap}]$

     the conditions: $N = N\,[\mu^{liq} = \mu^{vap}; \ \mu^{liq} = \mu^{vap}]$

     the variance $f = M - N = 3 - 2 = 1$

     type of solution $\begin{cases} X^{liq} = X^{liq}\,(T, P): & \text{curve in } PX \text{ plane} \\ X^{vap} = X^{vap}\,(T, P): & \text{curve in } PX \text{ plane} \end{cases}$

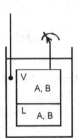

SCHEME 1.   The isothermal binary liquid + vapour equilibrium

the phase rule

For the general equilibrium case of c components (substances) divided over the number p of phases, the number of degrees of freedom is, simply, given by

$$f = c - p + 2. \tag{13}$$

This relation, the so-called *Phase Rule*, is named after Josiah Willard Gibbs (1839-1903), the father of Phase Theory.

    The phase rule can be derived, in a straightforward manner, by means of the M minus N approach - Scheme 2 may be of help.

	phases $\rightarrow$	1	2	3	p
components $\downarrow$		α	β	γ	
1	A	x	x	x	$(p-1)$
2	B	x	x	x	$(p-1)$
3	C	x	x	x	$(p-1)$
c	number of mole fractions:	$(c-1)$	$(c-1)$	$(c-1)$	$\uparrow$
					number of conditions

SCHEME 2.   On the derivation of the Phase Rule

In each phase there are $(c - 1)$ independent mole fractions, as a result, for p phases,

$$M = M[T, P \text{ and } p \text{ times } (c - 1) \text{ mole fractions}]. \tag{14}$$

Next, the chemical potential of the substance A in phase α has to be equal to A's potential in each of the $(p - 1)$ other phases. The number of independent equilibrium conditions, the number of signs of equality is $(p - 1)$

$$\mu_A^\alpha = \mu_A^\beta = \mu_A^\gamma = \cdots\cdots = \mu_A^{\text{phase } p}. \tag{15}$$

As a result for all c substances

$$N = N[c \text{ times } (p - 1) \text{ signs of equality}]. \tag{16}$$

The number of degrees of freedom now follows as

$$f = M - N = 2 + p(c - 1) - c(p - 1) = c + p - 2. \tag{17}$$

isothermal and/or isobaric conditions

Most generally, the liquid + vapour equilibrium in the system A + B has two degrees of freedom, as is shown by the (M - N) equation, Equation (11), and in agreement, of course, with the phase rule

$$f = c - p + 2 = 2 - 2 + 2 = 2. \tag{18}$$

In practice liquid+vapour equilibria are mostly studied either under *isothermal* or under *isobaric conditions*, which means that one of the two conditions has been consumed 'a priori'. If one likes one can say that for isobaric or isothermal conditions the phase rule reduces to

$$f = c - p + 1 \quad (T \text{ or } P \text{ constant}). \tag{19}$$

And similarly for both isothermal and isobaric conditions

$$f = c - p \quad (T \text{ and } P \text{ constant}). \tag{20}$$

It is not unlikely that the three different equations will give rise to some confusion. At any rate it is clear that the *system formulation*

$$f = M\left[P, X^{liq}, X^{vap}\right] - N\left[\mu_A^{liq} = \mu_A^{vap}; \mu_B^{liq} = \mu_B^{vap}\right] = 1 \tag{21}$$

for the isothermal binary $(c = 2)$ liquid + vapour equilibrium, does not give rise to any guesswork whatsoever.

the lever rule

It may be emphasized that in the treatment as it is presented above it invariably is assumed that the equilibrium system has been realized and that, for that situation, the variables M and the conditions N are defined and the number of degrees of freedom are found. This is quite easy and quite straightforward. In practice, on the other hand, a series of obstacles has to be removed. The first of these is related to the amount of substance to be used.

It is clear that 1 drop of water is not enough to realize, say at 60 °C, the equilibrium between liquid and vapour in a space of 1 m³.
It is also clear that it is impossible to realize the equilibrium between a vapour phase of (water + ethanol) having an ethanol mole fraction of 0.40 and a liquid phase of the two substances having an ethanol mole fraction of 0.08 by bringing equal amounts of the two substances in the experimental space. Indeed, that equilibrium can only be realized if the *overall mole fraction* of ethanol (X^o) is between 0.08 and 0.40, i.e. if $0.08 < X^o < 0.40$.

Generally, when $(1 - X^o)$ mol A and X^o mol B divide themselves over two phases α and β, having mole fractions values of X^α and X^β respectively, then the amounts of the two phases, $n(\alpha)$ and $n(\beta)$ respectively, satisfy the rule ($\rightarrow$Exc 3):

$$n(\alpha):n(\beta) = (X^\beta - X^o):(X^o - X^\beta) . \qquad (22)$$

This rule is referred to as the *lever rule* or *centre of gravity principle*.

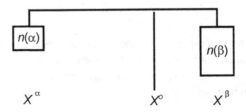

intellectual devices

In the treatment of phase equilibria, two popular intellectual devices are the *cylinder-with-piston* and the *vessel-with-manometer*. In the case of the first, the piston is weightless and able to move freely, without friction. The pressure exerted from the outside on the piston invariably is equal to the pressure inside. The experimentalist has the freedom to subject the system to the temperature and the pressure of his/her own choice. In the case of the second device, which is a vessel to which a manometer is connected ($\leftarrow$Figure 002:1), the experimentalist has the freedom to adjust the temperature of the system. At the imposed temperature, the pressure is a property fixed by the equilibrium condition(s).

If an equilibrium system is defined by M variables and there are N equilibrium conditions, then the number of independent variables - the number of degrees of freedom - is f = M − N. For a system in which c substances give rise to an equilibrium between p phases that number is given by f = c − p + 2. This is the Phase Rule, named after Josiah Willard Gibbs.

EXERCISES

1. *three variables subjected to two conditions*

A certain system is characterized by the set of three variables M [X, Y, Z], which are subjected to the two conditions N [$f^I = f^{II}$; $g^I = g^{II}$], where f^I and f^{II} and g^I and g^{II} are functions of X, Y and Z:

$$f^I = X + Y + Z$$
$$f^{II} = 2X + 2Y + 2Z$$
$$g^I = 3X + 2Y + 2Z$$
$$g^{II} = X - Y - 2Z$$

- Which relations do exist between the variables?

2. *phase diagram or not?*

Can Figure a) be a phase diagram of a binary system at isobaric conditions? And what is your opinion about Figure b) as a phase diagram?

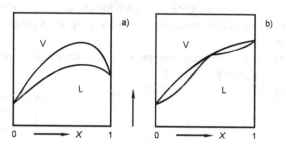

3. *derivation of lever rule*

Show that when a mixture of two substances A and B, of which the overall mole fraction of B is equal to X^o, separates into a phase α of composition X^α and a phase β of composition X^β, the amount of substance in phase α and the amount of substance in phase β are related as

$$n(\alpha):n(\beta)=(X^\beta - X^o):(X^o - X^\alpha)$$

4.   *a system formulation*

An empty space is filled, to about 50% of its volume, with equal amounts of three moderately volatile liquids A, B and C, which are pure substances that are partially miscible and give rise to three liquid phases.
- Determine the variance of this system at isothermal conditions, and so by first enumerating the M variables and the N equilibrium conditions.

5.   *amounts of three phases out of three substances*

Starting with 1 mol A, 1 mol B and 1 mol C, the equilibrium is realized between the three phases α, β and γ. The mole fractions of B and C in each of the phases are given in the scheme below. For each of the phases, calculate its amount of substance.

phase	X_B	X_C
α	0.2	0.2
β	0.5	0.3
γ	0.1	0.6

6.   *the experimental advantage of a small vapour phase*

A liquid mixture of water and ethanol, containing 20 mole percent of the latter, is brought a vessel-with-manometer. Next, the device is immersed in a thermostat adjusted at 60 °C. At this temperature the equilibrium is realized between liquid and vapour. It is estimated that the vapour phase occupies only 1 percent of the volume of the space.
- What is the pressure indicated by the manometer and what is the composition of the vapour?

Data are in § 002.

7. *naphthalene is added little by little to toluene*

In a continuous series of isothermal ($T = T_a$) vessel-with-manometer experiments 0.5 mole of naphthalene, which is solid at T_a, is added, little by little, to 0.5 mole of toluene. At the start of the series there is equilibrium between a liquid and a vapour phase, both composed of pure toluene. It is observed that, after the addition of 0.25 mol, naphthalene does not dissolve any more (the liquid has become saturated with naphthalene). In the final situation there is equilibrium between three phases: vapour (which is still pure toluene); liquid (saturated solution of naphthalene in toluene); and solid (pure naphthalene).

- Using straight lines, make a diagram in which the course of the pressure indicated by the manometer is plotted as a function of the overall mole fraction of naphthalene in the system (which increases from $X° = 0$ to $X° = 0.5$).

The following two (mutually related) general properties should be taken into account: 1) A's chemical potential, in a homogeneous mixture with B, decreases when B's mole fraction increases; and 2) the chemical potential of a gaseous substance increases with increasing pressure.

8. *does an empty place matter?*

In this section's derivation of the Phase Rule, it is assumed that each component is present in each of the phases - scheme 2.

- Is that necessary? Or, in other words, will the result $f = c - p + 2$ be influenced if one or more crosses are deleted from the scheme?

Pure substances are materials that are characterized by a molecule when they are molecular materials like n-butane (molecule $CH_3CH_2CH_2CH_3$) and in all other cases by a chemical formula like NaCl for sodium chloride, common salt. Pure substances can take different forms, all having their own stability conditions in terms of temperature and pressure.

phases and forms

The three opening phrases, written by Gibbs under the heading "On Coexistent Phases of Matter" (p. 96 in the 1906 edition of The Scientific Papers of J. Willard Gibbs, Vol I, Thermodynamics) read

"In considering the different homogeneous bodies which can be formed out of any set of component substances, it will be convenient to have a term which shall refer solely to the composition and thermodynamic state of any such body without regard to its quantity or form. We may call such bodies as differ in composition or state different phases of the matter considered, regarding all bodies which differ only in quantity and form as different examples of the same phase. Phases which can exist together, the dividing surfaces being plane, in an equilibrium which does not depend upon passive resistances to change, we shall call coexistent"

And (p. 97), after the appearance of $n + 2 - r$ (our $c + 2 - p$; ←003)

"Hence, if $r = n + 2$, no variation in the phases (remaining coexistent) is possible. It does not seem possible that r can ever exceed $n + 2$. An example of $n = 1$ and $r = 3$ is seen in the coexistent solid, liquid, and gaseous forms of any substance of invariable composition".

In the case of the first quotation the matter considered (any set of component substances) is (brought to equilibrium in an experimental set-up under certain conditions and subsequently) analyzed as for its homogeneous parts, which correspond to one or more phases (numbered I, II, III, ..., or labelled α, β, γ, ...). In the case of the second quotation the matter considered is a substance of invariable composition, and the phases are specified by their forms.

Following Gibbs, we will use term *'phase'* in the general context of equilibrium between phases of matter. The word phase is present in compound words, having a general nature. Compounds like phase diagram, Phase Rule, two-phase equilibrium curve, single-phase field, three-phase equilibrium.

The term *'form'* will be used primarily in the context of materials science. The determination of the form is the first step to the identification of a homogeneous material (phase). A form is characterized by a certain molar volume - which is a continuous function of temperature and pressure - by a certain crystal structure (when it is crystalline solid), and so on. And whenever, within a given form, two or more substances can form homogeneous mixtures of variable composition, the properties of the mixtures like molar volume will be continuous functions of the composition variables. Two phases in equilibrium can have the same form; and if so, their compositions necessarily are different.

the PT phase diagram

The basic structure element of the *PT phase diagram* for a pure substance is composed of three *two-phase equilibrium curves* emanating from the *triple point*, Figure 1.

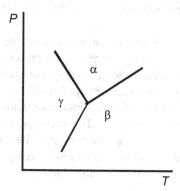

FIG. 1.   Basic structure element of a pure substance's phase diagram

A two-phase equilibrium curve has a *double function*: the ($\alpha + \beta$) two-phase equilibrium curve i) represents the *PT* circumstances for which equilibrium can be realized between the α- and the β- forms; and ii) divides the *PT* plane into two parts, such that at the α side the form β can change spontaneously into α, whereas α never will change into β (at the β-side these things are the other way round).

volume changes and heat effects

In terms of pressure, and for the situation sketched in Figure 1, α can be referred to as the *high-pressure form* and β as the *low-pressure form* of the two. The change of the material (at the two-phase equilibrium curve) from the high-pressure form to the low-pressure form is accompanied by a *change in volume*. This change, no matter how small, is always positive:

$$V^\alpha < V^\beta; \quad \text{or} \quad \Delta_\alpha^\beta V \ (= V^\beta - V^\alpha) > 0. \tag{1}$$

In terms of temperature, β is the *high-temperature form* and α the *low-temperature form* of the two. The change from the low-temperature to the high-temperature form is accompanied by a *heat effect*, $Q^{\alpha \to \beta}$. This effect, no matter how small, is always positive and it means that heat is needed to change α into β.

$$Q^{\alpha \to \beta} > 0 \qquad\qquad (2)$$

In the case of Figure 1, β is the high-temperature form and at the same time the low-pressure form: the slope of the two-phase equilibrium curve is positive. For the substance water, on the other hand, liquid water is the high-temperature form and at the same time, the high-pressure form: the slope of the two-phase equilibrium curve (the melting curve) is negative; see Figure 001:6. The change from ice to water at 0 °C, 1 atm is accompanied by a decrease in volume of $(-19.63 + 18.00 =) -1.63$ cm^3·mol^{-1} and a heat effect of 6008 J·mol^{-1}.

As a matter of fact, the thermal counterpart of ΔV is not the heat effect Q itself, but a quantity related to it. The volume change ΔV is a property which is expressed in m^3·mol^{-1}, and m^3·mol^{-1} is equal to joule-per-mole-per-pascal; J·mol^{-1}·Pa^{-1}. It means that when ΔV is multiplied by the *mechanical potential*, the pressure P, a property is obtained in J·mol^{-1} - like the heat effect Q. Analogously, the thermal counterpart of ΔV is a property expressed in J·mol^{-1}·K^{-1}. This property is obtained when the heat effect (of the change α → β) is divided by, the *thermal potential*, the thermodynamic temperature at which the change takes place. It is the *change in molar entropy* ΔS (see below; →106, in particular).

Summarizing, in a *PT* phase diagram of a pure substance the *single-phase fields* are positioned such that i) on increasing pressure the changes are in decreasing volume, and ii) on increasing temperature the changes are in increasing entropy.

Clapeyron's equation (→110)

The volume change, $\Delta_\alpha^\beta V$, and the heat effect, $Q^{\alpha \to \beta}$, not only (by their signs) determine the sign of the slope of the equilibrium curve, they also, by their magnitudes, determine the magnitude of the slope:

$$\frac{dP}{dT} = \frac{Q^{\alpha \to \beta}}{T \cdot \Delta_\alpha^\beta V} \cdot \qquad\qquad (3a)$$

In this equation, T is the thermodynamic temperature, for which the slope is taken. The relationship is referred to as *Clapeyron's equation*, after Emile Clapeyron (1799-1864). Replacing the quotient of heat effect and temperature by the change in entropy, the equation changes into

$$dP/dT = \Delta S/\Delta V .$$ (3b)

The Clapeyron equation is of general validity; its thermodynamic derivation is given in §110. From Equation 3, a special equation is obtained for the case that β is gaseous, and α solid or liquid, and moreover, if β is taken as an *ideal gas*, and α's volume is neglected. For this special case ΔV reduces to RT/P, and with $(1/P) dP = d \ln P$ the equation changes into

$$\frac{d \ln P}{dT} = \frac{Q^{\alpha \to vap}}{RT^2} .$$ (4)

As a next step, with $d(1/T) = - (1/T^2) \, dT$,

$$\frac{d \ln P}{d\left(\frac{1}{T}\right)} = - \frac{Q^{\alpha \to vap}}{R} .$$ (5)

This 'special equation' suggests a linear relationship between ln *P* and 1/*T*. In reality the relationship is only roughly linear - owing to the fact that *Q*, along the equilibrium curve, becomes smaller with in creasing temperature. In any case, it has become common practice to represent vapour pressure data - the pressure of vapour in equilibrium with solid or liquid - in a ln*P* vs 1/*T* diagram. Such a representation of vapour pressure data is referred to as *Clausius-Clapeyron plot*; see Figure 2.

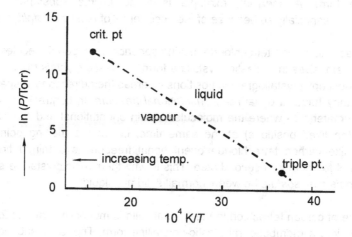

FIG. 2.   Clausius-Clapeyron plot of the boiling curve for water; from below the triple point up to its end point, the *critical point* where the distinction between liquid and vapour comes to an end (→206)

metastability

One of the experiments described in § 001 was carried out – under atmospheric pressure – with an amount of water having a temperature of –5 °C; i.e. experimental circumstances under which liquid water may change spontaneously into ice. Circumstances that correspond to a point in the phase diagram lying in the single-phase field for ice.

In itself, water having a temperature of –5 °C is not less stable than water having a temperature of + 5°C. The only thing is that at –5 °C liquid water may change spontaneously into ice, whereas at + 5°C it never will do. For these reasons liquid water at 1 atm and –5 °C is said to be *metastable*.

It is (even) possible to experimentally study the equilibrium between *supercooled water* and gaseous water (→Exc 3). It is customary to refer to this equilibrium between phases as an example of *metastable equilibrium*. Accordingly, the extension (beyond the triple point) of the (liquid + vapour) equilibrium curve is referred to as *metastable extension*.

Metastability is a fascinating and, at the same time, a complicating phenomenon. Complicating, because spontaneity (irreversibility) does not go well with controllability (reversibility). Diamond and graphite are two forms of the substance carbon. One is inclined to think that diamond is the stable one of the two forms, but it is not (→109).

polymorphism

In the realm of crystalline materials *polymorphism* is a common phenomenon - it is the fact that many substances give rise to more than one crystalline form. A speaking example is found in the substance carbon tetrachloride, especially so because of the occurrence of *plastic crystalline forms*.

Substances, like carbon tetrachloride, having spherically shaped molecules, often manifest themselves in a plastic crystalline form. In *plastic crystals* the individual molecules occupy crystallographic positions whereas their directions in space are arbitrary: they have (a dynamical) *orientational freedom*. In contrast to 'normal' crystalline materials - where the molecules obtain orientational and *translational freedom* (no fixed positions) at the same time, i.e. at the melting point - for materials like carbon tetrachloride orientational freedom is obtained first and translational freedom in a second step. This is why the plastic crystalline state is referred to as a *mesostate* between 'normal' solid and liquid.

In the case of carbon tetrachloride the normal solid is monoclinic, and at 225.4 K it changes into a rhombohedral plastic-crystalline form. The latter subsequently changes into liquid at 250.3 K. The curious thing about carbon tetrachloride is that it can take a second plastic crystalline form - face-centred cubic (fcc) - which, invariably, is metastable. The form can be obtained by crystallization from the liquid, and, when it is heated, it melts at 244.8 K; see Figure 3.

The rhombohedral plastic crystalline form and the low-temperature monoclinic form stand to each other in, what is called, and *enantiomorphic relationship*: the two crystalline forms can exist at the same time, in equilibrium with one another; a phenomenon referred to as *enantiomorphism*. The relationship between the two plastic crystalline forms, on the other hand, is a *monotropic relationship*: the two crystalline forms never coexist - the only thing is that the invariably metastable fcc form can change spontaneously into the other; the phenomenon is referred to as *monotropism*.

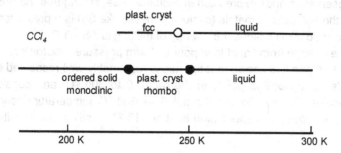

FIG. 3.   Carbon tetrachloride's polymorphic relationships. Top: metastable sequence; bottom: stable sequence

Occasionally, a form, which is stable in a certain span of temperature, re-enters as the stable form in another span. *Re-entrant behaviour* is rather frequently observed for forms belonging to the mesostate of *liquid crystals* (van Hecke 1985). Re-entrant behaviour becomes a possibility when the heat effect of the transition is so small that it can change sign (before the form changes into liquid, or isotropic liquid in the case of liquid crystals).

The scheme shown in Figure 4 pertains to iron, whose re-entrant form is the body-centred cubic one. At 1184 K an amount of heat of 900 J·mol⁻¹ is needed to change bcc iron into fcc – at 1665 K, 837 J·mol⁻¹ is needed to realize the opposite change!

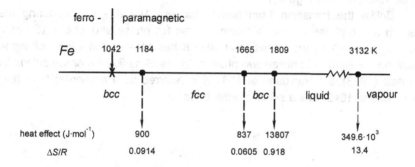

FIG. 4.   Sequence of iron's forms. Heat effects of the transitions; increases in entropy divided by the gas constant

In the scheme, Figure 4, the row labelled $\Delta S/R$ stands for the change in entropy (see above) divided by the gas constant. ΔS and R are expressed in $J{\cdot}K^{-1}{\cdot}mol^{-1}$, their quotient being a *dimensionless quantity*. The advantage of using (here) ΔS over using Q, the heat effect, is that, for certain classes of change, the former, unlike the latter, has a more or less constant value. As an example, for metallic sodium, melting at 370.98 K, the *heat of melting* is 2603 $J{\cdot}mol^{-1}$; as distinct from 1809 K and 13807 $J{\cdot}mol^{-1}$ for iron. Notwithstanding these large differences, the values of ΔS are quite comparable: 0.84 R for Na and 0.92 R for Fe. This observation finds expression in *Richards' rule*, stating that the *entropy of melting* of (the outspoken) metals (excluding metals like Sb) is represented by $\Delta S \approx R$. For the ionic alkali halides the entropy of melting is $\Delta S \approx 3\,R$.

For the change from liquid to vapour at 1 atm pressure *Trouton's rule* states that $\Delta S \approx 11\,R$; and it is known that the rule is reasonably well respected by *non-polar liquids* having boiling points in the range 0 to 300 °C, and consisting of small molecules. For Fe, far outside the 0 $\rightarrow$ 300 °C temperature range (see Figure 4), the *entropy of vaporization* is about 13 R - still in the vicinity of the rule.

a lambda type of transition

In the scheme for iron, the cross on the temperature line represents the *Curie temperature* (1042 K): the temperature at which iron changes from a ferromagnetic to a paramagnetic material. The magnetic susceptibility of a ferromagnetic material decreases with temperature, and very rapidly so in the vicinity of the Curie point - where it falls down to a paramagnetic level (virtually zero). At the Curie temperature there is neither a change in crystal structure, which remains bcc, nor an (isothermal) heat effect.

The heat effect involved in the change is spread out over a range in temperature. It manifests itself in the form of an extra heat capacity ($\rightarrow$103), increasing with temperature, and falling down at the Curie point. The *heat-capacity plot* resembles the Greek letter lambda - it explains the use of the term *lambda transition*, see Figure 5.

Unlike the transition from bcc to fcc at 1184 K - where during the transition a bcc phase is in equilibrium with an fcc phase and where ΔS has a finite, non-zero value - the transition at 1042 K has $\Delta S = 0$, and it has nothing to do with an equilibrium between two phases. In the *classification of transitions* by Ehrenfest (1933), the transition at 1184 K is referred as a *first-order transition*, and the one at 1042 K as a *second-order transition*.

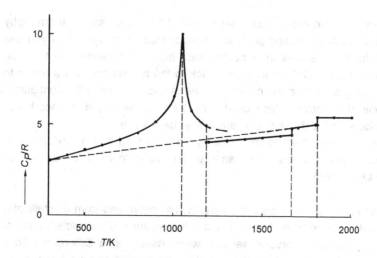

FIG. 5.   Heat-capacities-divided-by-the-gas-constant of iron's forms (Barin
1989); see also Figure 4

a glass transition

The molecule which is pictured, Figure 6, has a rather planar structure: the carbon atoms 1, 2, 3, 5 and 6 and the oxygen atom are about in the same plane (the plane of the paper); carbon atom 4 is below the plane, and the isopropylidene group originates from it in an upward direction. The molecule is a *chiral molecule* (cannot be superimposed on its mirror image) and, therefore, it is characteristic of a substance which is *optically active* (rotates the plane of polarization of polarized light); that substance is laevorotatory carvone. The mirror image of the molecule (C atom 4 above the plane and the isopropylidene group downward) is characteristic of the substance dextrorotatory carvone.

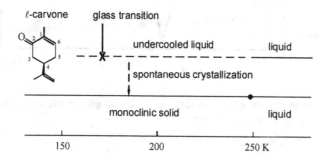

FIG. 6.   Laevorotatory carvone – its forms and its behaviour

Besides, *ℓ*-carvone and *d*-carvone have the same thermo-physical properties, such as melting point and heat of melting (imagine that you can see the individual molecules in a beaker with boiling *ℓ*-carvone, the thermometer indicating *t* = 231 °C; then turn your back to the beaker and use a mirror to see what is going on in it; in the mirror you see molecules having the configuration of *d*-carvone; the thermometer indicates *t* = 231 °C, the only difference being that the numbers on the thermometer are written backwards).
Nota bene, these observations certainly do not mean that a mixture of *d*- and *ℓ*- carvone will manifest the same phase behaviour as *d*- or *ℓ*- carvone individually (→005).

Liquid carvone can be supercooled easily, i.e. prevented from crystallization at, or below its melting point. Figure 7 is a *thermogram* which depicts the behaviour of a supercooled *ℓ*- carvone sample, when heated in a Differential Scanning Calorimeter (DSC). In a DSC, the sample to be studied and a reference (e.g. an empty sample pan) are mounted on a metal block that can be heated at a certain rate. In simple terms, the thermogram represents the heat to be added or withdrawn from the sample, in order to keep it at the same temperature as the reference. The events, registered in *ℓ*- carvone's thermogram in order of increasing temperature, are the so-called *glass transition* (~171 K); a crystallization process (~193 K to 208 K); followed by a *recrystallization* phenomenon (~210 K to 218 K); and, finally, the melting of the sample (~248 K).

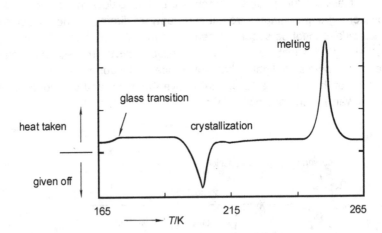

FIG. 7.   Thermogram of a superccoled liquid sample of laevorotatory
           carvone (Gallis et al. 1996)

Like the magnetic transition in the bcc form of iron, the glass transition in a supercooled liquid can be regarded as a continuous, non-isothermal transition within a given form. Unlike the former, the latter has, for a given material, not a fixed position on the temperature scale: the glass transition temperature depends on the cooling conditions to which the liquid is subjected. Below the glass transition temperature the positions of the molecules are fixed: any molecular movement is suppressed by the viscosity, which has become too high (Papon et al. 2002).

Each of the stable forms taken by a pure substance has its own stability field in the PT phase diagram. Forms can appear under conditions outside their stability fields, in which case they are said to be metastable. Frequently forms are seen that are metastable whatever the circumstances.

EXERCISES

1. *the position of phase symbols*

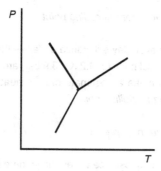

The figure is part of the phase diagram of a pure substance, including the forms, α, β and γ, of which the molar volumes increase in the order β, γ, α.
- Provide each of the single-phase fields with the correct (phase) symbol.
- Arrange the (form) symbols in order of increasing molar entropy.

2. *zero Celsius and zero Celsius*

The rounded difference in temperature between the triple point and normal freezing point of water is 0.01 K. This 10 mK (millikelvin) difference has been used to redefine the zero point of the *Celsius temperature scale* (←002).
- To appreciate this, apply Clapeyron's equation to calculate, as an integer in mK, the difference in temperature between the two point's.

The necessary data are given in the text.

3. *water's triple point pressure*

The table gives for the pure substance water the *equilibrium vapour pressure* over solid (ice) as well as over metastable liquid for six temperatures from $-5\ °C$ to $0\ °C$.

$t\ /°C$	P(solid) /Torr	P(liquid) /Torr
-5	3.013	3.163
-4	3.280	3.410
-3	3.568	3.673
-2	3.880	3.956
-1	4.217	4.258
0	4.579	4.579

- In terms of Equation (5), think up a method in which linear least squares is used to calculate - from the given set of data - water's *triple-point* coordinates.
- Next, carry out the calculation - temperature in two decimal places and pressure as an integer in Pa.

4. *carbon dioxide's metastable normal boiling point*

Carbon dioxide is not an everyday substance as far as the positions of its triple point (216.8 K; 5.1 bar) and critical point (304.2 K; 73.9 bar) are concerned.
- Use the data to make an estimate of i) its heat of vaporization, and ii) its *(metastable) normal boiling point*.

5. *the substance water under high pressure*

The triple point data for the substance water shown here (see Tonkov 1992) involve, apart from the liquid, seven solid forms/phases - indicated by Roman numerals.
- Use the information to construct the *PT* phase diagram; mark the single-phase fields with the appropriate symbol.
- Estimate the pressure which is needed to solidify water at 100 °C.
- Estimate the coordinates of the metastable triple point (II + V + ℓ).

triple point	$t\ /°C$	$P\ /GPa$
I + III + ℓ	-22	0.207
I + II + III	-34.7	0.213
III + IV + ℓ	-17	0.346
II + III + V	-24.3	0.344
V + VI + ℓ	0	0.625
V + VII + ℓ	81.6	2.15
VI + VII + VIII	-3	2.12
II + V + VI	-60	0.6
NB. GPa = gigapascal = 10^9 Pa		

6. *a rule to be respected by metastable extensions*

At the $(\alpha + \beta + \gamma)$ triple point, the *metastable extension* of a two-phase equilibrium curve has to run into the field bounded by the stable parts of the other two equilibrium curves.

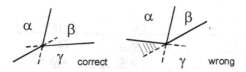

The negation of this statement can be reduced to an absurdity, i.e. is giving rise to contradictory conclusions.

- Demonstrate this for the shaded field, remembering that at the α side of the $(\alpha + \beta)$ equilibrium curve β can change spontaneously into α.

7. *a phase diagram acts as a thermobarometer*

The PT phase diagram of aluminium silicate (Al_2SiO_5) with its three solid forms sillimanite (I), andalusite (II) and kyanite (III) plays an important part in geology – *thermobarometry* in particular. The diagram is given in Tonkov's (1992) compilation, the equations for the (I + II) and (II + III) equilibrium lines being $t\,/^\circ C = -415\,P\,/GPa + 848$, and $t\,/^\circ C = 920\,P\,/GPa + 114$, respectively. According to Althaus (1969), for the change from I to III at 600°C, the heat effect is -9.08 kJ $\cdot mol^{-1}$ (heat given off) and the volume effect -5.51 cm^3 $\cdot mol^{-1}$ (decrease).

- Calculate the coordinates of the (I + II + III) triple point.
- For $300 \leq t\,/^\circ C \leq 800$ and $0 \leq P\,/GPa \leq 1.0$, construct the complete phase diagram, and mark the single-phase fields with the right phase symbol.

8. *superposition of stable and metastable*

For a "monotropic" substance, construct a PT phase diagram, such that the metastable phase relationships (dashed lining) are superimposed on the stable ones (solid lining) and the phases/forms involved are α (stable solid), β (metastable solid), liquid and vapour.

9. *Antoine's equation for vapour pressures*

Five data pairs are given for the equilibrium vapour pressure over liquid 1-aminopropane ($CH_3 \cdot CH_2 \cdot CH_2 \cdot NH_2$), taken from the twelve pairs by Osborn and Douslin (1968).

nr	$t/°C$	$P/Torr$
1	22.973	289.13
2	32.564	433.56
3	47.229	760.00
4	62.235	1268.0
5	77.587	2026

- Use the pairs 1, 3, and 5 to calculate the constants A, B, and C of the "Antoine equation"

- $\ln (P/Torr) = A - \dfrac{B}{\{(T/K) - C\}}$

- Examine the significance of the result by (re)calculation of the vapour pressures for the five experimental temperatures.

10. *supercritical fluid*

A *'supercritical fluid'* is characterized, let's say, by the following (twofold) property.
"When at constant pressure (or temperature) its temperature (pressure) is continuously lowered, it will not undergo a sharp transition - at the liquid + vapour equilibrium curve - from vapour (liquid) to liquid (vapour)".
- Draw a (S + L + V) *PT* phase diagram, and, in it, shade the field that corresponds to supercritical conditions.

11. *iron: the heat effect of magnetic change*

From Figure 5, estimate the heat effect involved in iron's change from ferromagnetic to paramagnetic.
- How much heat is needed to heat one mole of iron from 1500 K to 2000 K?

12. *boiling water altimeter*

Demonstrate that *one degree per 300 m* amounts to a reasonable estimate of the lowering of the boiling point of water resulting from an increase in altitude. First, use Figure 2 for water's heat of vaporization.

§ 005 BINARY AND TERNARY SYSTEMS

The multiplicity of phase diagrams shown by binary systems is reviewed, starting from the isobaric equilibrium between two forms, liquid and vapour, in each of which the component substances are completely miscible. Subsequently, two limiting cases of ternary phase behaviour are derived from the phase diagrams of the binary subsystems.

A and B are completely miscible in α and β

To start with, in the binary system $\{(1 - X) \text{ mol A} + X \text{ mol B}\}$, the case is considered where A and B mix in all proportions in each of the two forms α and β, β being the high-temperature form. For the equilibrium between an α-phase and a β-phase the *system formulation* is

$$f = M \left[T, P, X^\alpha, X^\beta \right] - N \left[\mu_A^\alpha = \mu_A^\beta; \mu_B^\alpha = \mu_B^\beta \right] = 4 - 2 = 2 : \qquad (1)$$

There are two degrees of freedom: the mathematical solution of the combination of equations and variables can be expressed as

$$X^\alpha = X^\alpha(T,P); X^\beta = X^\beta (T,P) , \qquad (2)$$

and it corresponds to two surfaces in PTX space. A surface for the α-phases and another one for the β-phases. For every point on the α surface there is a companion point on the β surface, such that the line connecting the two points is parallel to the X axis.

For the greater part, binary equilibria are studied under either *isobaric circumstances* (constant pressure) or *isothermal circumstances* (constant temperature). One could say that the investigator, before starting the experimental work, has 'consumed' one of the two degrees of freedom. As a result, the outcome of the investigation is depicted in either a TX or a PX phase diagram.

As a first example, Figure 1 represents the isobaric equilibrium between α = liquid and β = vapour in the system A = 1-decanol and B = 1-dodecanol, under the pressure of P = 20 Torr.

Before going on, it is worthwhile to say a few words about the way in which the *interaction between molecules*, A and B, is reflected in phase diagrams. The cigar-type of phase diagram, Figure 1, to start with, is typical of systems where the interaction between the two molecular species has a neutral character. 'Neutral' is to say that for the A molecules it makes no difference whether they are surrounded by other A molecules or by B molecules – and vice versa for the B molecules.

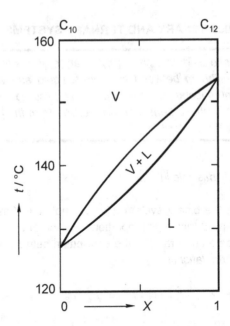

FIG. 1.   Isobaric equilibrium between liquid and vapour in the system
(1-decanol + 1-dodecanol) at P = 20 Torr (Rose et al. 1958).
Liquidus and vaporus divide the TX plane into three regions: the
two single-phase fields L and V, and the two-phase field (V + L)

A non-neutral behaviour is displayed, as an example, by liquid mixtures of
chloroform (C) and acetone (A). The isothermal and isobaric (liquid + vapour)
phase diagrams for this combination of substances are shown in Figure 2. From
the two phase diagrams it follows that the two (C and A) together prefer the liquid
state over the gaseous state:

- from the TX diagram:
  on heating (at increasing temperature) the molecules remain together in the
  liquid state 'for a longer time' than would be the case for neutral behaviour;
- from the PX diagram:
  the amount of matter in the gaseous state is lower than it would be for neutral
  behaviour, as follows from the lower (vapour) pressure.

Therefore, and ignoring the interaction of the molecules in the gaseous state, the
two diagrams are evidence of an extra interaction, having an attractive character.

Incidentally, an attractive extra interaction is the exception rather than the rule; at
least for liquid mixtures of molecular substances (see the compilation by Ohe
1989). In most of the cases molecular substances give rise to TX (liquid + vapour)
diagrams that either have a minimum – evidence of a repulsive extra interaction –
or look like Figure 1.

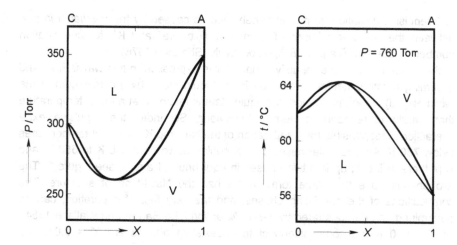

FIG. 2.   For the (liquid + vapour) equilibrium in the system {chloroform
            (C) + acetone (A)}, a minimum in the isothermal diagram
            (t = 35.2 °C, left, Apelblat et al. 1980)), goes together with a
            maximum in the isobaric diagram (right, Reinders and de Minjer
            1940)

A repulsive extra interaction is also generally found for solid mixtures, i.e. mixed
crystals. A beautiful, speaking example is the combination of sodium chloride
(NaCl) and potassium chloride (KCl), whose phase diagram is shown in Figure 3.

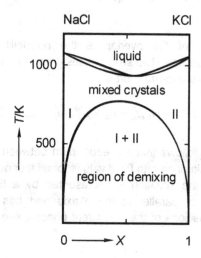

FIG. 3.   The TX phase diagram of the NaCl + KCl system, at 1 atm
            pressure

Crystalline NaCl and KCl both have the so-called *NaCl-type of structure*. In
mixed crystals of the two, the Cl$^-$ ions take all the sites for the negative ions, while
the Na$^+$ and K$^+$ ions are randomly distributed over the sites for the positive ions.

The repulsive deviation from neutral behaviour is caused by the *mismatch in size* between the Na$^+$ and K$^+$ ions. The *ionic radii* of Na$^+$ and K$^+$, for coordination number VI, are 1.02 Å and 1.38 Å, respectively (Shannon 1976).

In the phase diagram the repulsive character is expressed in two ways. First, and assuming a rather neutral interaction in the liquid state, by the minimum in the (solid +liquid) two-phase region: the liquid state is taken up at a lower temperature than would correspond to neutral behaviour. Secondly, the repulsive extra interaction is responsible for a separation of the Na$^+$ and K$^+$ ions⁻ - at temperatures below 780 K. An equimolar mixed crystal cooled down from 800 K to 650 K, and kept there, will split up into two phases in equilibrium, I and II, see Figure 3. The two phases have the same form, which has the NaCl-type of structure. The compositions of the coexisting phases, and the very fact of separation, can be determined and demonstrated by *X-ray diffraction* (see Barrett and Wallace 1954). At T = 650 K the compositions of the coexisting phases are X' = 0.07, and X'' = 0.80.

The two-phase field in the lower part of Figure 3 is referred to as *region of demixing,* or *miscibility gap.*

limited to negligible miscibility of A and B in α

For the combination of NaCl and RbCl (instead of KCl) the mismatch between Na$^+$ and Rb$^+$ (radius 1.52 Å) is so high, that the region of demixing is no longer separated from the (solid + liquid) two-phase region. A like situation is sketched in Figure 4.

The consequence of the overlap is the possibility of a *three-phase equilibrium*: equilibrium between two solid phases (I and II) and a liquid phase (liq); corresponding to the *system formulation*

$$f = M\left[T, X', X'', X^{liq}\right] - N\left[\mu_A^{I} = \mu_A^{II} = \mu_A^{liq};\ \mu_B^{I} = \mu_B^{II} = \mu_B^{liq}\right] = 4 - 4 = 0. \qquad (3)$$

From this expression it follows that the equilibrium between the three phases is *invariant*: it is a unique situation with fixed values for all four quantities of the set M. In the phase diagram the situation is represented by a line, the *three-phase equilibrium line,* which is parallel to the X-axis and has three points on it, representing the compositions of the coexistent phases, see Figure 4, right-hand side.

For NaCl + RbCl the values of the four variables at the three-phase equilibrium are T = 788 K; X' = 0.008; X'' = 0.94; and X^{liq} = 0.56 (Short and Roy 1964).

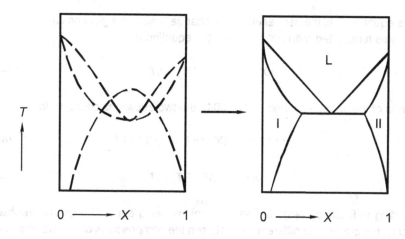

FIG.4.   The (solid + liquid) loop and the solid-state region of demixing
          interfere with one another: the result is a phase diagram with a
          three-phase equilibrium line, which is the lower boundary of two
          (solid + liquid) two-phase regions, and the upper boundary of a
          (solid + solid) two-phase region

For the combination of LiCl and KCl, with Li^+ (0.76 Å) and K^+ (1.38 Å), there is hardly any solid-state miscibility: the TX (solid + liquid) phase diagram is of the simple *eutectic type* (→ Exc 3). The term *eutectic* is from the Greek eutéktos, and it means readily melting: a *eutectic mixture* melts at a temperature lower than the melting points of the pure components. Besides, a three-phase equilibrium, such as in Figure 4 right-hand side, with a liquid phase and two solid phases in the succession solid-liquid-solid, is referred to as a *eutectic three-phase equilibrium*.

Figure 5c is another example of a simple eutectic phase diagram; it is for the combination of dextrorotatory isopropylsuccinic acid (IPSA) and dextrorotatory methylsuccinic acid (MSA); succinic acid is butanedioic acid ($HOOC\cdot CH_2\cdot CH_2\cdot COOH$).

initial slopes (→208)

The TX phase diagrams Figures 1,2b,3,4 all have $(\alpha + \beta)$ *two-phase regions*, limited by two equilibrium curves - one for the compositions of the α-phases, and the other for the β-phases. An *equation for the initial slopes* of the two equilibrium curves is the equation named after Jacobus Henricus Van 't Hoff (1852-1911) - an equation similar to the one named after Clapeyron, Equation (004:4). The equation states that for small $\Delta T = T - T_A^o$, where T_A^o is the temperature at which pure A changes from the form α to the form β, the difference between the initial slopes of the equilibrium curves is given by the expression

$$dX^\beta / dT - dX^\alpha / dT = -(Q^{\alpha \rightarrow \beta} / RT_A^{o2}). \tag{4a}$$

In this equation Q is the heat effect of the change, and R the gas constant. Otherwise formulated, with $\Delta X = X^\beta - X^\alpha$, the equation reads

$$\Delta X = -(Q^{\alpha \to \beta} / RT_A^{o2})\Delta T. \tag{4b}$$

In terms of the *entropy change*, ΔS ($\leftarrow$004), the two equations change into

$$dX^\beta / dT - dX^\alpha / dT = -\Delta S / RT_A^o; \tag{4c}$$

$$\Delta X = (-\Delta S / RT_A^o)\Delta T. \tag{4d}$$

According to Equation (4b), and all other things being equal, the greater the heat effect is, the greater the difference is between the compositions of the two phases.

NB. If A and B do not mix at all in the low-temperature state α, the addition of A to B, and the addition of B to A, as follows from Equation (4c), inevitably will lower the equilibrium temperature. This is the case for the three systems implied in Figure 5.

The *PX* counterparts of the Equations (4), i.e. the equations for the initial slopes in *PX* phase diagrams, are

$$dX^\beta / dP - dX^\alpha / dP = \Delta V / RT; \tag{5a}$$

$$\Delta X = (\Delta V / RT)\Delta P. \tag{5b}$$

For the case that β = ideal gas, and α = liquid or solid, and neglecting the molar volume of α with respect to the molar volume of the gas, the Equations (5) change into

$$dX^{vap} / dP - dX^\alpha / dP = 1/ P_A^o; \tag{6a}$$

$$\Delta X = (1/ P_A^o)\Delta P. \tag{6b}$$

P_A^o represents the equilibrium vapour pressure of pure A.

The last two equations are two of the various expressions of Raoult's Law, the relationship named after François-Marie Raoult (1830-1901).

NB. The Equations (4c,d) and their counterparts Equations (5a,b) are expressions of the close relationship between *entropy* and *volume*. The role played by *entropy changes* in *TX* equilibria is similar to the role of *volume changes* (be it with opposite sign) in *PX* equilibria. The two come together for equilibria where P and T are the leading variables - an example is given by Clapeyron's equation, Equation (004:3b).

stationary points

The *difference* between the initial slopes - of the equilibrium curves in the phase diagrams shown above - is a *pure-substance property*. In the case of Figure 2 right-hand side, the difference between the slopes at the chloroform side is determined by the properties of its change from liquid to vapour - heat of vaporization and boiling point. That difference does not change if the second component, acetone is replaced by another substance, say methanol. Unlike their difference, the *individual* slopes themselves are determined by the combination of the components of the system. The combination chloroform+acetone displays a *TX* phase diagram with a maximum - at both sides of the system liquidus and vaporus have an upward start. Chloroform+methanol, on the other hand, have a *TX* phase diagram with a minimum - at both sides downward start of liquidus and vaporus.

Generally, for *TX* equilibria between liquid and vapour, and starting from neutral interaction, an *attractive* extra interaction has the effect of lifting up the (liquid+vapour) two-phase region. Plausibly, the smaller the difference is between the boiling points of the components, the greater the probability is of producing, for a given extra attraction, a phase diagram with a maximum. In the case of chloroform+methanol there is a *repulsive* extra interaction; the difference in boiling-point temperatures is just 3 K; the phase diagram has a minimum.

Maxima and minima are *stationary points*: liquid mixtures having the composition of the stationary point change - when boiled - completely into vapour at constant temperature, the temperature of the maximum/minimum ($\rightarrow$006). At the extremum, liquid and vapour have the same composition. The other way round, if, for a given temperature, liquid-and-vapour-in-equilibrium have the same composition, the equilibrium state necessarily is an extremum in the phase diagram.

In the context of this passage, reference should be made to the Konowalow Rules (Konowalow 1881). For *TX* liquid+vapour equilibria the rules are

(I) The boiling point as a function of composition shows a stationary point only if liquid and vapour have the same composition;

(II) The boiling point is raised by the addition of that component whose mole fraction in the vapour phase is lower than the mole fraction in the coexisting liquid phase;

(III) The compositions of the coexisting vapour and liquid phases change in the same sense with temperature.

compound formation

Figure 5b is the phase diagram shown by the combination of laevorotatory and dextrorotatory isopropylsuccinic acid (IPSA). The diagram is typical of a *pair of optical antipodes*, of which the two different *chiral molecules* ($\leftarrow$004) cocrystallize in a (1:1) centrosymmetrical structure. The crystalline (1:1) combination is referred to as *racemate,* or *racemic compound*.

The noun 'compound', in racemic compound, is generally used for an intermediate solid phase having a fixed, or a more or less fixed composition. A *stoichiometric compound* has a fixed composition, whereas the composition of a *non-stoichiometric compound* can vary between certain limits.

An example of a non-stoichiometric compound is wustite ($\approx$ FeO) in the systerm Fe + O. In the (Fe + O) *TX* phase diagram, wustite's single-phase field has the shape of a capital V: at the bottom (t = 560 °C) the composition can be written as $FeO_{1.06}$; at the top (t = 1360 °C) $FeO_{1.03}$, and (t = 1424 °C) $FeO_{1.20}$ (Vasyutinskii 1984).

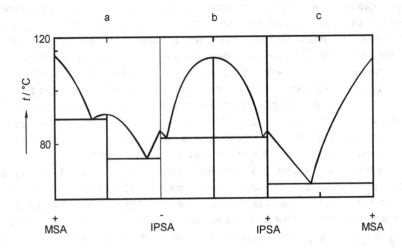

FIG. 5.   The determination of (solid + liquid) phase diagrams is a means to establish the configurative relationships between optically active substances. The compound formed by (-) and (+) IPSA is a racemate. A quasiracemate is formed between (-) IPSA and (+) MSA (Fredga and Miettinen 1947)

isodimorphism, isopolymorphism

Mutual *solid-state miscibility* of two substances A and B in the first place is related to the degree of similarity between the A and B molecules ('molecules' in the sense of building units of the crystal lattice). That similarity regards chemical nature, size and shape. Solid-state miscibility is the exception rather than the rule (Kitaigorodskii 1984).
NaCl and KCl are soluble in one another; in all proportions, however, only above 780 K (see above); they show what is called complete *subsolidus miscibility*. Complete subsolidus miscibility is also found for combinations like gold and palladium (Okamoto and Massalsky 1985) and 1,4-bromochlorobenzene and 1,4-dibromobenzene (Calvet et al. 1991). Naphthalene and 2-fluoronaphthalene

would also show complete subsolidus miscibility, if it were not that they do not crystallize in the same form (van Duijneveldt et al. 1989).

To illustrate these things, we take two solid forms α and β and examine three different situations for their equilibrium with liquid, see Figure 6. In the case of Figure 6a the two substances A and B are isodimorphous in such a manner that there is, for each of the solid forms, a single-phase field extending over the full composition range (there is complete subsolidus miscibility, and so in the form β). The dashed curves in Figure 6a represent the boundaries of the metastable (solid α + liquid) two-phase region. Generally, the position of the metastable two-phase region can be calculated from the positions of the stable regions and the heats of transition of the α to β and β to liquid changes. Exceptionally, a full-range *metastable melting loop* is observed in experimental reality - a beautiful example is shown by 1,1,1-trichloroethane ($Cl_3 C CH_3$) + carbon tetrachloride ($Cl_3 C Cl = C Cl_4$) (Pardo et al. 1999). For the situations depicted by Figures 6b and c, substance A's β-form invariably is metastable, and the same goes for B's α-form. The important thing to note is, that there necessarily are three-phase equilibria (eutectic, Figure 6b, and peritectic, Figure 6c) and that, inevitably, there is incomplete solid miscibility (one speaks of *crossed isodimorphism*: two solid-liquid loops cross one another).

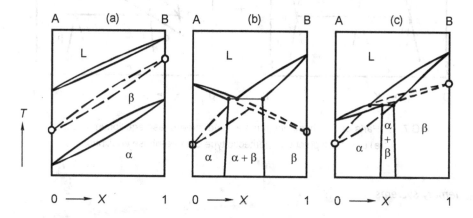

FIG. 6.   Isodimorphism (a) and crossed isodimorphism (b and c). α and β are two crystalline forms in which A's molecules and B's molecules can replace one another. Open circles represent metastable melting points

nomenclature

The two basic types of *three-phase equilibiria* in isobaric binary systems are the *eutectic* - and *peritectic* type, Figure 7.

More specifically, a eutectic type is named

eutectic, for	α = solid I	γ = liquid	β = solid II;
eutectoid, for	α = solid I	γ = solid III	β = solid II;
monotectic, for	α = solid I	γ = liquid I	β = liquid II; and
metatectic, for	α = solid I	γ = solid II	β = liquid.

A peritectic type is named

peritectic, for	γ = liquid	α = solid I	β = solid II
peritectoid, for	γ = solid III	α = solid I	β = solid II

NB! It is imperative that the so-called *metastable extensions,* in Figure 7 indicated by dashes, run inside two-phase fields.

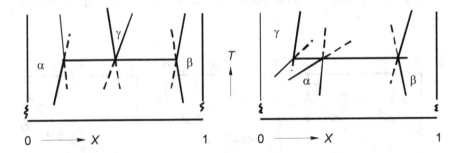

FIG.7.   Parts of *TX* phase diagrams around three-phase equilibrium lines:
a) eutectic type and b) peritectic type of three-phase equilibrium

ternary systems

In ternary systems, defined by

$$\{X_A \text{ mol A} + X_B \text{ mol B} + X_C \text{ mol C}\} \text{ or } \{(1-X-Y)A + XB + YC\}, \qquad (7)$$

compositions are conveniently represented in an equilateral triangle - the *Gibbs triangle,* see Figure 8.

For the isobaric equilibrium between two mixed phases α and β - taken as an example - the system formulation is

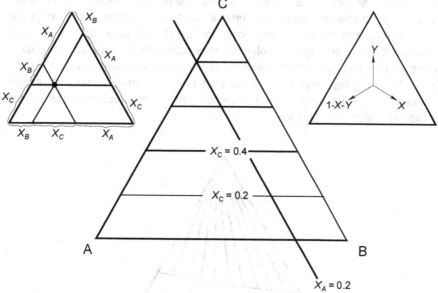

FIG. 8.   Gibbs' composition triangle

$$f = M\left[T, X^{\alpha}, Y^{\alpha}, X^{\beta}, Y^{\beta}\right] - N\left[\mu_A^{\alpha} = \mu_A^{\beta}; \mu_B^{\alpha} = \mu_B^{\beta}; \mu_C^{\alpha} = \mu_C^{\beta}\right] = 2. \qquad (8)$$

There are two degrees of freedom: the values of, e.g., Y^{α}, X^{β}, and Y^{β} are fixed after the choice of T and X^{α}. The solution of the set of equations corresponds to two *surfaces in TXY space*.

For the isobaric equilibrium between three phases, α, β, and γ, the formulation is

$$f = M\left[T, X^{\alpha}, Y^{\alpha}, X^{\beta}, Y^{\beta}, X^{\gamma}, Y^{\gamma}\right]$$
$$- N\left[\mu_A^{\alpha} = \mu_A^{\beta} = \mu_A^{\gamma}; \mu_B^{\alpha} = \mu_B^{\beta} = \mu_B^{\gamma}; \mu_C^{\alpha} = \mu_C^{\beta} = \mu_C^{\gamma}\right] = 1. \qquad (9)$$

After the choice of T, the compositions of all three phases are fixed. In the composition triangle, the three-phase equilibrium for given T corresponds to a triangle whose vertexes represent the compositions of the coexistent phases. The area inside the *invariant triangle* is empty: an overall composition given by a point inside the triangle will split up into the three phases. The fact that the three-phase equilibrium is monovariant implies that for each of the three coexistent phases there is a trajectory in *TXY* space.

An isothermal section of the isobaric equilibrium states, as a result, is represented by a Gibbs triangle that may show one or more *single-phase fields*, *two-phase fields*, *three-phase triangles*, and (only for the temperature of a *four-phase equilibrium*) one *quadrangle*. In isobarothermal representations the use of *tie lines*, connecting the positions of two coexistent phases, is much more common than for binary *TX* or *PX* representations. More than that, for ternary phases of variable composition, the tie lines are an essential part of the representation of the equilibrium states, see Figure 9.

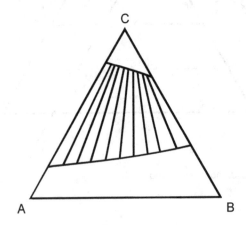

FIG. 9.   The system methylcyclohexane (A) + n-hexane (B) + methanol (C) at 1 atm, 30 °C. Phase diagram showing the boundaries of the region of demixing and *tie lines* connecting the positions of the two coexisting liquid phases (Schuberth 1986)

Like liquidus and vaporus curves in *TX* and *PX* diagrams, liquidus and vaporus surfaces in *TXY* or *PXY* space can have a maximum or a minimum – at which the coexistent phases have the same composition. In addition to the maximum and the minimum, ternary surfaces have the possibility of a third type of stationary point: the *saddle point*, where a minimum in one direction is a maximum in the other. Saddle points are favoured by the combination of two binary subsystems having a minimum (or maximum) with a third subsystem having a maximum (or minimum). A *TXY* example is found in the combination of chloroform, acetone, and methanol (Haase 1950). Two of the subsystems were mentioned above; the third, acetone+methanol, has a minimum.

from binary to ternary

The three *binary subsystems*, (A + B), (A + C), and (B + C), are part of the ternary system (A + B + C), or, in other words, the equilibrium states in the binary

systems are part of the equilibrium states in the ternary systems. The other way round, a substantial part of the equilibrium states in a ternary system are already available in the binary subsystems. Or, somewhat exaggerated, ternary equilibrium can be predicted from the information offered by the binary subsystems. Two examples, Figures 10 and 11, are now discussed, to make these things clear.
In the case of Figure 10, the equilibrium considered is between solid and liquid phases, and the binary subsystems are of the simple eutectic type. For the ternary cross-section the temperature is selected such that it is above the surface for the equilibrium between liquid and solid C. The isothermal section of the other two liquidus surfaces is represented by the curves PR and RQ inside the triangle. The ternary phase diagram now consists of four fields: i) the *single-phase field PRQC*; it corresponds to conditions where all of the material is liquid; ii) the *two-phase field APR*; iii) the *two-phase field BQR*; iv) the *three-phase triangle ABR*. A mixture of overall composition represented by a in the field APR will give rise to equilibrium between pure solid A and liquid mixture represented by point b on PR; the two phases are connected by the *tie line* through a. Overall compositions inside the *invariant triangle ABR* give rise to equilibrium between solid A, solid B, and liquid mixture represented by R.

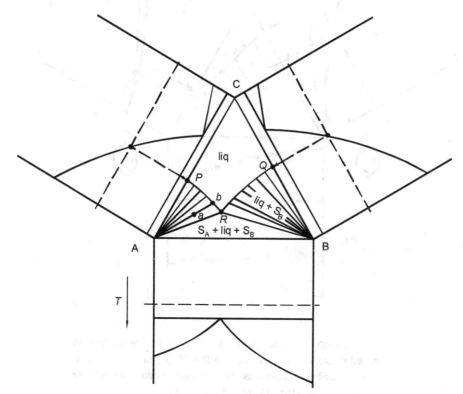

FIG. 10. Isothermal section showing the phase equilibrium relationships of
          a ternary system composed of three eutectic binary subsystems
          under isobaric circumstances

Figure 11 pertains to the equilibrium between (mixed-crystalline) solid and liquid in the system 1,4-dichlorobenzene + 1,4-bromochlorobenzene + 1,4-dibromobenzene. This time the coexisting solid and liquid ternary phases have been calculated, using the information provided by the three binary phase diagrams and the thermochemical properties of the pure components (Moerkens et al. 1983). The result of the calculation, including tie lines, is shown for the isothermal section at 70°C. Note that the difference between the tie lines in Figure 10 and the ones in figure 11 is that, unlike the latter, the former - which emanate from one point - can be drawn without doing any special calculations (one has to be sure, of course, that the solid phase is made of pure B!).

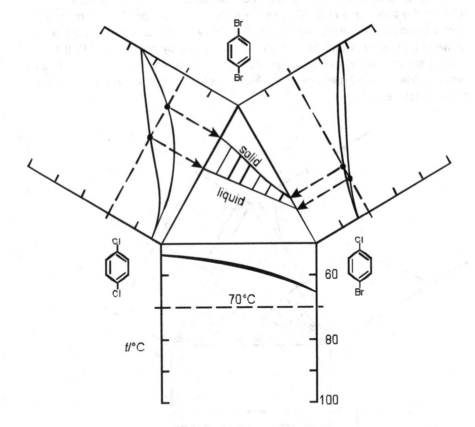

FIG. 11. Ternary system composed of three 1,4-dihalobenzenes forming mixed crystals. Isothermal section (t = 70 °C) of ternary (solid-liquid) equilibrium, calculated from the information available for the binary subsystems Moerkens et al. 1983)

imperfections in phase diagrams

The phase diagrams that are shown in textbooks and scientific publications not all, and not at all, have the same history. A phase diagram shown in a text can range from the outcome of a thermodynamic calculation, via a (subjective) interpretation of experimental observations, to a diagram drawn in a freehand manner.

Calculated phase diagrams that satisfy the laws of thermodynamics always combine correctness with beauty. Diagrams in the other categories occasionally show imperfections that could have been avoided.

Two examples of incorrect phase diagrams are the ones already shown in Exc 003:2. The diagram labelled a) is supposed to represent a vaporus and its companion liquidus. The diagram does not respect the obvious rule that for every point on the vaporus there should be a corresponding point on the liquidus. In the case of the diagram labelled b) the 'obvious rule' is satisfied; however, and for thermodynamic reasons (→Exc 211:10), two phases in equilibrium can have the same composition, but only for a stationary point (first of the Konowalow Rules, see above).

Unintentionally, a phase diagram drawn in a freehand manner may show a (thermodynamic) inconsistency (Nývlt 1977). A *PT* phase diagram for a pure substance may have a region that looks like Figure 12, but not for the case that β = liquid, and γ = vapour. According to the Clapeyron equation, Equation (004:3), a minimum in the boiling curve would imply that, along the boiling curve, and starting from the triple point, the heat of vaporization would change from a negative property to a positive one!

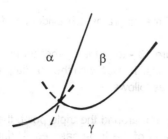

FIG. 12.   Region of a *PT* phase diagram for a pure substance, showing a
two-phase equilibrium curve with a minimum

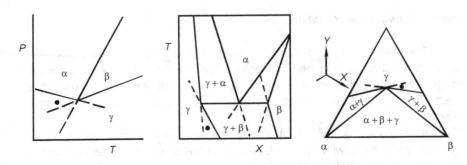

FIG. 13.   Three diagrams incompatible with the principles of stability

The three diagrams shown in Figure 13 all have the same kind of imperfection - expressed by the edges of the γ single-phase fields, with their angle of more than 180°. The bullet, in each of the diagrams is positioned in the γ field, which would mean that γ is stable and should not undergo a spontaneous change.
However:

i) in the *PT* diagram the bullet is at the β side of the (β + γ) two-phase equilibrium curve, and it means that γ might spontaneously change into β;
ii) in the *TX* diagram the bullet is in (the metastable extension of) the (α + γ) two-phase region: the material might split up in a γ part (of another composition) and an α part;
iii) in the *XY* diagram the position of the bullet is such that the material, likewise, might split up in a γ part and an α part.

In other words, all of the three diagrams are evidence of a misapplication of the stability principles.
The three diagrams in Figure 13 - to put it in other terms - do not comply with the rules for metastable extensions. To start with the *PT* diagram for a pure substance, these rules can be detailed as follows.

In a correct *PT* phase diagram, around the triple point, the metastable extensions alternate with the stable parts of the two-phase equilibrium curves (as shown by Figure 12; ←Exc 004:6). In a *TX* or *PX* phase diagram for a binary system, at the three-phase equilibrium, the metastable extensions must fall inside two-phase fields (as shown by Figure 7; →Exc 301:3). In an isobarothermal section for a ternary system, and for the case implied in the *XY* diagram in Figure 13, the two metastable extensions must fall either both in the (α + β + γ) invariant triangle or one in the (α + γ) field and the other in the (β + γ) field (rule named after Franciscus Antonius Hubertus Schreinemakers (1864-1945); (Schreinemakers 1911, see also Hillert 1998).

Valuable guiding principles in understanding phase behaviour are: the notion of interaction between different molecules, the nature of which is ranging from repulsive via neutral to attractive; the knowledge that much of the properties of ternary systems is already present in the binary subsystems; and the necessity to consequently apply the principles of stability.

EXERCISES

1.   *a unary diagram made to look like a binary one*

The two-phase equilibrium curves of a pure substance's *PT* phase diagram can be 'opened up' to show two-phase regions - by replacing *P* by a property which is not necessarily equal for the two phases in equilibrium.

- Draw a (S + L + V) *PT* phase diagram - melting curve with positive slope; boiling curve up to critical point - and translate it into a diagram in which molar volume is plotted against temperature.

2.   *the amounts of the phases during an experiment*

For the phase behaviour depicted in Figure 6c: a homogeneous sample - whose composition is between the compositions L and α at the three-phase equilibrium - is heated under constant pressure "from a position in the α field to a position in the L field".

- Make a scheme, in which for each part of the route, it is indicated - for each of the possible phases - whether the amount of substance of the phase remains unchanged (=), increases (>), decreases (<), or is equal to zero.

The parts of the route are i) through the α field; ii) through the (L + α) field; iii) at the three-phase equilibrium; iv) through the (L + β) field; and v) through the L field.

3.   *phase diagram and cooling curve*

Lithium chloride (LiCl,; melting point 606 °C) and potassium chloride (KCl; m.p. 770 °C) are miscible in all proportions when liquid. Their solid state miscibility is negligible. A liquid mixture, having KCl mole fraction $X = 0.1$, will start to crystallize, when cooled, at $t = 566$ °C. The onset temperatures of crystallization for a number of other composition are ($X = 0.2$; $t = 515$ °C); (0.3; 452); (0.4; 370); (0.5; 443); (0.6; 541); (0.7; 618); (0.8; 685); and (0.9; 733).

- Construct the (solid + liquid) *TX* phase diagram.

In an experiment, heat is withdrawn at a constant rate from a sample having $X = 0.2$. The initial and final temperatures are 600 °C and 200 °C, respectively.

- Make a graphical representation of the sample's temperature as a function of time (*cooling curve*; →006)

4. *a reciprocal system*

The four substances NaCl, KCl, NaBr, and KBr share four structural units which are the two anions and the two cations. A system composed of such four substances is referred to as a reciprocal (salt) system.
- What is the number of independent variables necessary to define the composition of a like system?
- What geometrical figure would you use to represent the compositions of a reciprocal system?

5. *increasing repulsive interaction and the phase diagram*

In the system $\{(1 - X) A + X B\}$ the interaction in the liquid state between A and B is neutral, the (solid + liquid) TX phase diagram is eutectic. The melting points of A and B are 900 K and 450 K, respectively; the eutectic point has $T = 400$ K and $X = 0.90$.
- Make a series of sketch drawings of phase diagrams to demonstrate what happens when the interaction between A and B in the liquid state is going to show an increasing, repulsive deviation from neutral mixing behaviour.

NB. The properties of the pure components remain the same and so do the initial slopes (at $X = 0$, and $X = 1$) of the two liquidus curves (say over the first 10% of the X axis).

6. *overlapping two-phase regions*

The two (on their own correct) two-phase regions, $(\alpha + \beta)$ and $(\beta + L)$, are partly overlapping. As a consequence, the true phase diagram must have an $(\alpha + L)$ two-phase region.
- Guided by the rule for metastable extensions, make a sketch of the true phase diagram (first locate the two $(L + \beta + \alpha)$ three-phase equilibrium situations).

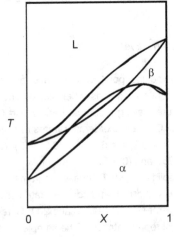

7. *the construction of ternary phase diagrams*

The three optically active substances (-) IPSA (A), (+) IPSA (B), and (+) MSA (C),
acting in Figure 5, are rather alike, and for that reason it can be assumed that the
interaction between their molecules in the liquid state has a nearly neutral character.
As a result, their ternary phase behaviour can be predicted from the binary data with
fair accuracy. The fact is that for neutral mixing behaviour, and for given, constant T,
the equilibrium

solid C plus (liquid mixture containing C)

simply satisfies the relations ship X_C = constant = K_C (likewise for A and B);
and the equilibrium

compound AB plus (liquid mixture containing A and B)

satisfies the relationship $X_A \cdot X_B$ = constant = K_{AB}.

- Use the binary phase diagrams, Figure 5, to construct isothermal sections of
  the ternary system; for i) t = 100 °C; ii) the temperature of the upper eutectic
  in the system (+) MSA + (-) IPSA; and iii) t = 80 °C.

Clue. For each of the solids, stable at the temperature considered, determine the
value of the equilibrium constant - like the above K_C or K_{AB} - and use it to construct
the complete ternary liquidus. Use the ensemble of liquidi to allocate the single- and
two-phase fields, and the invariant triangles.

NB. If an invariant triangle involves a liquid phase, then its "liquid vertex" is the
intersection of two (of the constructed) liquidi.

8. *the appearance of an incongruently melting compound*

In Figure 5a the left-hand liquidus (the A liquidus) intersects the liquidus pertaining to
the 1:1 compound (the AB liquidus) at the left of the equimolar composition – the
point of intersection being a eutectic point. When, in such a case, component A is
given a higher melting point - all other things remaining the same - the situation may
arise that the intersection is a peritectic point, at the right of the equimolar
composition. When that is the case, AB's melting point becomes metastable: on
heating, AB will, at the peritectic temperature, split up in *solid A* and *liquid having the
peritectic composition*. The congruently-melting compound, Figure 5a, has changed
into an incongruently-melting compound.

- Construct the *TX* solid-liquid phase diagram for a system involving an
  incongruently melting compound AB.

9. *ternary compositions having a constant ratio of the mole fractions of two components*

In the ternary composition triangle ABC, the locus of the compositions that have the
same ratio of the mole fractions of A and B is a straight line ending in vertex C.

- Prove the validity of this statement - making use of the properties of similar
  triangles.

Clue. On AB take a point P; on PC take a point Q; similar triangles ABC and RSQ, R
and S being points on AB.

10. *cyclohexane with aniline – mixing and demixing*

{(1–X) mole of cyclohexane + X mole of aniline}, in spite of the fact that aniline (aminobenzene) is a nasty substance, is a superior system to demonstrate phenomena of mixing and demixing. The following description of actions and observations - representative of a classroom experiment - may make this clear.

Eight test tubes provided with screw caps are filled - using a 10 ml measuring cylinder - with varying amounts of cyclohexane and ainiline; see table. After filling, the tubes 0, 6, and 7 show a single liquid; the tubes 2, 3, 4, and 5 show two, clearly separated liquids; and the content of the remaining tube, tube 1, looks milky.

In a next step the tubes 2, 3, 4, and 5 are immersed in water with a temperature of 40 °C, contained in a plastic beaker. As a result of this action the two liquids, in all of the four tubes, change into a clear single liquid. Thereupon, the water and the tubes are allowed to cool to room temperature. Meanwhile the tubes are shaken every now and then; and for each sample the onset temperature of turbidity is registered, see table.

A similar procedure is followed for the tubes 0, 1, 6, and 7. The tubes are immersed in water having a temperature of 25 °C; and the whole is subsequently cooled by means of ice cubes. The samples undergo a change to a two-liquid situation; with the exception of tube 7, whose content is still homogeneous at 1 °C.

Tube	cyclohexane (ml)	aniline (ml)	X	at room temperature	onset of turbidity
0	7.4	0.4	0.06	one liquid	6.0
1	7.9	0.9	0.12	milky	19.5
2	6.9	1.4	0.19	two liquids	26.0
3	5.8	2.8	0.36	two liquids	30.7
4	4.3	3.5	0.49	two liquids	30.5
5	3.0	5.0	0.66	two liquids	27.7
6	1.9	6.3	0.80	one liquid	14.4
7	1.0	7.1	0.89	one liquid	<1.0

- What is the temperature of the room where the experiments are carried out?
- For $t > 0$ °C, make a graphical representation - temperature versus mole fraction - of the binodal curve (the boundary of the region of demixing), and from the curve derive the coordinates of the critical point.
- Sketch a plausible TX phase diagram for the system over the range from –50 °C to 200 °C. In more detail: use the region-of-demixing data; calculate the initial slopes of the two liquidi emanating from the melting points (mp) of the pure components; and consult Exc 212:12 for a rule of thumb, concerning the change from liquid to vapour.

Cyclohexane: 84 g mol^{-1}; 108 cm^3 mol^{-1} (25 °C); mp 6.6 °C; bp 81 °C; heat of melting 2.66 kJ mol^{-1}.

Aniline: 93 g mol^{-1}; 91 cm^3 mol^{-1} (25 °C); mp -6.3 °C; bp 185 °C; heat of melting 10.54 kJ mol^{-1}.

§ 006 DISTRIBUTION AND SEPARATION

An elementary characteristic of equilibrium between phases in systems of two or more components is the fact that the phases as a rule have different compositions. This property holds the possibility that substances can be separated from one another by means of phase changes.

saturation

   When a drop of water is brought in an empty space, say a 200 m³ tank kept at 25°C, it will evaporate rapidly and completely. And free water molecules will be present in every cubic micrometer of the space. Similarly a few sugar crystals will dissolve rapidly and completely in a cup of water. Water is a good solvent for sugar (molecules), and an empty space, or vacuum if you like, is a good 'solvent' for water (molecules).

   The amount of foreign molecules - solute molecules - a solvent can accommodate, however, is limited: at a certain 'moment' the solvent is saturated - and this moment differs from case to case. The moment of saturation, for water in a space, is reached when the chemical potential of gaseous water (μ_{H2O}^{vap}) has become equal to the chemical potential of liquid water (μ_{H2O}^{liq}). Water added after saturation has been reached will not evaporate anymore: there is equilibrium between a liquid and a gaseous phase, governed by the condition

$$\mu_{H_2O}^{liq} = \mu_{H_2O}^{vap}. \tag{1}$$

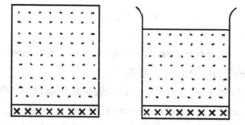

FIG. 1.   Two analogous cases: left, the saturation of an empty space by a volatile substance; right, the saturation of a liquid medium by a soluble substance

And this condition is such that the amount of water in the gaseous phase will correspond to a vapour pressure of 23.756 Torr, for $t = 25$ °C.

For the two crystalline substances naphthalene ($\bigcirc\!\bigcirc$, $C_{10}H_8$; substance A) and 1-chloro-4-iodobenzene (ClC_6H_4I; substance B) the saturation pressures are two orders of magnitude smaller than the one for liquid water: at 25 °C, P_A^o = 9.3 Pa, and P_B^o = 12.7 Pa (van der Linde et al. 1998; Oonk et al. 1998). When brought together in an empty space, the two solids imply three different situations, which are o) the space is saturated neither with A nor B; i) the space is saturated with either A or B; and ii) the space is saturated simultaneously with A and B.

The remarkable thing about the last situation (ii) is, that a manometer device coupled to the space will indicate a pressure that is equal to the sum of P_A^o and P_B^o, which is 22.0 Pa. This observation is evidence of the fact that the two solids act independently: the amount of B that can be taken up by the space is independent of the presence of A, and vice versa. And if so, it is quite easy to construct the PX phase diagram for the (solid + vapour) equilibria in the binary system {$(1 - X)$ mol A + X mol B}.

For the equilibrium between solid A and mixed vapour, the amount of A in the vapour phase is constant. Moreover, that amount can be expressed as $n_A^{vap} = \alpha \cdot P_A^o$, where α is a proportionality constant. The total amount of substance in the vapour phase, as a consequence, is given by $n_A^{vap} + n_B^{vap} = \alpha \cdot P$, where P is the pressure indicated by the manometer. Realizing that the mole fraction variable for the vapour is given by $X^{vap} = n_B^{vap} / (n_A^{vap} + n_B^{vap})$, and $(1 - X^{vap}) = n_A^{vap} / (n_A^{vap} + n_B^{vap})$, it follows that

$$(1 - X^{vap}) \cdot P = P_A^o. \qquad (2)$$

On the same lines, the vaporus for the equilibrium between solid B and mixed vapour is given by

$$X^{vap} \cdot P = P_B^o. \qquad (3)$$

The two vaporus curves, which are arcs, intersect at the *three-phase equilibrium pressure*, and at $X = P_B^o / (P_A^o + P_B^o)$; Figure 2.

The possibility to 'predict' the PX phase diagram, just from P_A^o and P_B^o, does not hold, of course, when A and B are forming mixed crystals, such as is the case e.g. for A = 1,4-dibromobenzene and B = 1-chloro-4-iodobenzene. The isothermal equilibrium between mixed crystals and vapour is analogous to the isothermal equilibrium between liquid mixture and vapour. An example of the latter has been presented in § 002, Figure 002:2 and Table 002:4, for the system water + ethanol. Note that when a space is saturated with vapour from a liquid mixture, the composition of the vapour (and that of the liquid phase) can vary between certain limits; not only the equilibrium conditions have to be respected, but also the *lever rule*.

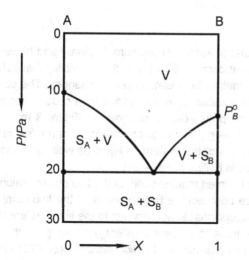

FIG. 2. Upside down (solid + vapour) PX phase diagram, for A = naphthalene and B = 1-chloro-4-iodobenzene at $t = 25\ °C$

distribution

As a next step, the case is considered where a solute B is added to the combination of two immiscible solvents, say the liquid solvents A and C. There are two liquid phases: liquid phase I (A with B dissolved in it); and liquid phase II (C with B). Under isobaric, isothermal circumstances the system formulation, for the equilibrium between the two phases, is

$$f = M\left[X_B^I, X_B^{II}\right] - N\left[\mu_B^I = \mu_B^{II}\right] = 2 - 1 = 1. \tag{4}$$

The system is monovariant, and it implies that the compositions of the two phases, X_B^I and X_B^{II}, are interdependent: the value of X_B^{II} is fixed by the choice of X_B^I, or vice versa.

In its simplest form the relationship between the two is such that X_B^{II} is proportional to X_B^I:

$$\frac{X_B^{II}}{X_B^I} = k. \tag{5}$$

This relationship follows from the thermodynamic laws for *dilute solutions* ($\rightarrow$ 207, 208); it is known as the *Nernst Distribution Law*, after Walter Nernst (1864 - 1941), (Nernst 1891). The proportionality constant k is called *distribution coefficient* - it certainly will be a function of temperature and pressure.

A classical example is the distribution of iodine (I_2) between water (brown colour) and carbon disulphide (CS_2) or carbon tetrachloride (CCl_4, in which it dissolves with a beautiful purple colour).

separation

The very fact that phases in equilibrium, in binary and higher-order systems, generally have different compositions holds the possibility that substances can be separated from one another by means of phase changes. The ease, with which a separation can be effectuated, however, varies from case to case - the sequence of *TX* (liquid + vapour) phase diagrams shown in Figure 3 may make this clear. The succession a → b → c , meant to correspond to a *neutral interaction* (←005) between the components of the system, in liquid as well as in vapour, reflects the influence of difference in boiling point (→210).

It is generally observed that the width of the two-phase region decreases with decreasing difference between the boiling points of the two components. For the cases a and b the vapour phase, with respect to the liquid, is enriched in the lower boiling component - for a to a greater extent than for b. In the limiting case of diagram c - typical of a combination of *optical antipodes* (←005) - liquid and vapour have the same composition, and so over the whole range of compositions. The diagrams d and e have a minimum and a maximum, respectively, where liquid and vapour have the same composition - a special equilibrium state referred to as *azeotrope*. The term azeotrope is a word construction in Greek; to express that there is no change at boiling: constant temperature, unchanging composition.

Phase diagrams, like the ones shown in Figure 3, are graphical representations of the relationships between the intensive variables required to characterize the phases in an equilibrium system. These relationships are generated by the laws of thermodynamics, acting upon the thermodynamic properties of the system - over which the experimentalist has no authority. The experimentalist should call forth all her or his inventiveness in what is left: making use of the possibilities offered by

- the degrees of freedom (the variance) of the system;
- the freedom to select, optimize the relative dimensions of the phases and the dimensions of the whole experimental set-up;
- the kinetic properties of the system, permitting the circumvention of the true equilibrium states;
- the availability of several types of equilibrium: (solid + liquid); (solid + vapour), (liquid + vapour);
- the addition of an extra component;

The remaining part of this section and its exercises should make this clear.

distillation

The simplest distillation set-up is one in which a liquid mixture is heated in a boiler under normal pressure, and, when the liquid boils, the ascending vapour is condensed with cold and the resulting liquid is collected (the distillate). For the case of Figure 3a, a liquid mixture containing 20 mole % of the more volatile component will start to boil at the liquidus and start to produce a distillate containing 67 mole % of the more volatile component.

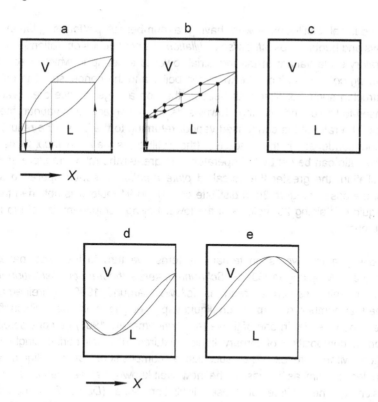

FIG. 3.   A series of TX (liquid + vapour) phase diagrams

Given this fact, two extreme situations can be considered.

In the first of these situations the liquid in the boiler is distilled off to the last drop, and the distillate is collected in its entirety. During the distillation the liquid in the boiler is enriched in the less volatile component (B) and the temperature is rising - such that temperature and composition are following the liquidus curve. Towards the end, the temperature in the boiler is reaching B's boiling point, and the last drop of condensed vapour is virtually pure B. From a point of view of separation the whole operation is useless: the final distillate has the same composition as the original liquid (the operation is meaningful, of course, when it is carried out to free a mixture of volatile liquids from a *non-volatile impurity*).

In the other situation, the boiler contains a large amount of liquid mixture - an infinite amount, to make things easy. And so, from a mixture containing 20 mole % of the more volatile component (A) a large amount of distillate is obtained, and it has 67 mole % of A; Figure 3a. Similarly, and as a second distillation step, from the mixture containing 67 mole % of A, a distillate can be obtained having 95 mole % of A. In the case of Figure 3b at least five distillation steps are needed to realize the enrichment from 20 to 94 mole %.

In an industrial distillation tower, having a number of platforms (plates) with overflows and bubble caps, *stepwise distillation* is a continuous operation.

A laboratory-scale variant of the industrial tower is a set-up in which a so-called *fractionating column* is placed between the boiler and the condenser. The column is constructed such - or filled with beads - that, on a large surface area, vapour-condensed-to-liquid and moving downwards is in contact with vapour moving upwards. The ratio of the condensed vapour returning to the boiler (the reflux) and the vapour condensed in the condenser (the distillate) is the *reflux ratio*. Generally, the reflux ratio can be set by the operator - the greater the ratio, the more efficient the distillation, the greater the so-called *plate number* (a set-up operating such, that, for the case of Figure 3b, a distillate containing 94 mole % is obtained from a boiler liquid containing 20 mole % of the lower boiling component, is said to have plate number 5).

To say a few words on ternary systems, we turn to the 1959 paper by Korvezee and Meijering (1959) on *Schreinemakers's theorem* on *distillation lines*. The five opening sentences read as follows: "Around 1900, Schreinemakers published a number of papers on liquid-vapor equilibria in the Zeitschrift für physikalische Chemie. In one of these (=Schreinemakers 1901) he considered the variation in composition of ternary liquid mixtures in a still during single-stage distillation without reflux. The successive compositions of the liquid were represented by him as curves in the now well-known triangle diagrams. In the discussion of the course of these distillation lines (*DL*'s), Schreinemakers demonstrated that in their end points they must be tangent to one of the sides of the triangle. Consequently at the end of a distillation, when the still contains the least volatile component in a nearly pure state, the last traces of one of the two volatile impurities are much easier to remove than those of the other".

An example of a triangle with distillation lines is shown in Figure 4 - for a simple case where the boiling points of the components decrease from A to C. In Figure 5, the example is shown of a system having a *saddle point* (Reinders and de Minjer 1940).

A notorious system - from the point of view of separation - is the combination of ethanol and water. The *TX* phase diagram of the system has a minimum azeotrope; at about 0.1 K below the boiling point of ethanol and ethanol mole fraction 0.9 (→Exc 6). The presence of the azeotrope is an obstacle for the preparation of pure ethanol by distillation. The problem can be resolved by adding benzene to the azeotropic mixture. Water and benzene are virtually immiscible, whereas benzene mixes in all proportions with ethanol. The ternary system has a minimum boiling temperature, at 64.85 °C, at which two liquid phases are in equilibrium with vapour (Reinders and de Minjer 1940).

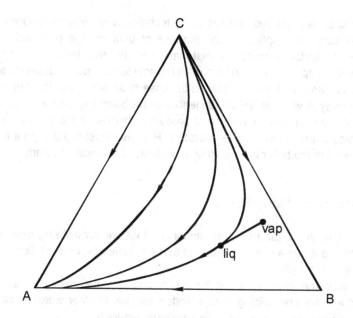

FIG. 4.   Distillation lines (*DL*) in a ternary composition triangle. In each
point of a *DL* there is a *tie line* (liq-vap) tangent to it

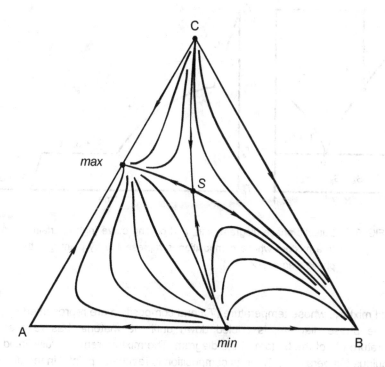

FIG. 5.   Distillation lines; the system has a saddle point (s)

In preparative organic chemistry the techniques of *vacuum distillation* and *steam distillation* are applied for the purpose of purifying the prepared product. Over normal distillation under normal pressure, the two techniques have the advantage that vapour is formed at a relatively low temperature - thereby reducing the risk of decomposing the material. In the case of steam distillation, the product, whose solubility in water is limited, is heated in the boiler together with an amount of water. Vapour containing water and product is formed at the (liquid + liquid + vapour) three-phase equilibrium temperature. Because most of the vapour is water, the water level in the boiler is kept up by a continuous injection of steam.

crystallization from a liquid phase

This time, A, B, (and C) are substances that are crystallizing from a liquid phase, which is either a molten mixture of these substances, or a solution in a non-crystallizing solvent (S).
To start with, the case is taken where A and B have negligible solid-state miscibility, and are crystallizing from a molten mixture. In other words, a case that complies with the simple eutectic phase diagram, Figure 6.

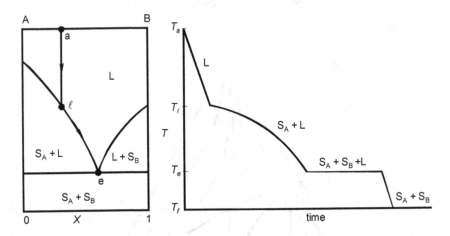

FIG. 6.   Eutectic phase diagram (left); and cooling curve (right) pertaining to a mixture whose composition is represented by point $\underline{a}$ in the phase diagram ($X = 0.3$)

A liquid mixture, whose temperature (T_a) and composition are represented by point $\underline{a}$ in the phase diagram, is cooled down until the material has reached the temperature (T_f) of the bottom of the diagram. The mixture remains fully liquid until the liquidus temperature (T_l) for its composition is reached - point $\underline{\ell}$ in the diagram. From that moment on, solid A crystallizes from the liquid. Meanwhile, the liquid phase 'follows' the liquidus, until the eutectic temperature (T_e) is reached. At T_e, A and B crystallize together; until the full amount of material has solidified. Thereafter, i.e. at the end of the invariant three-phase situation, the temperature can change again, and it drops to its final value, T_f.

The right-hand side of Figure 6 is a so-called *cooling curve*: it represents the temperature of the material as a function of time. Besides, and the other way round, the registration of the temperature of a solidifying sample is a key to the phase diagram. Numerous isobaric solid-liquid phase diagrams have been determined by means of the cooling-curve technique - especially so before the advent of the methods of *differential thermal analysis*.

As regards the separation of A and B, it is obvious that, for the situation sketched in Figure 6, only part of A can be freed from the mixture (5/7, as can be verified, from the coordinates of ℓ and $\underline{e}$, by means of the *lever rule* ($\leftarrow$003)).

A ternary equivalent of the simple eutectic diagram, Figure 6, is shown in Figure 7. The isobaric ternary diagram has three (divariant) *liquidus surfaces*; which, by means of a series of isotherms, display the equilibrium states of the combinations of liquid mixture and solid component. The point of intersection, $\underline{e}$ in the figure, of the three surfaces is the invariant *ternary eutectic*.

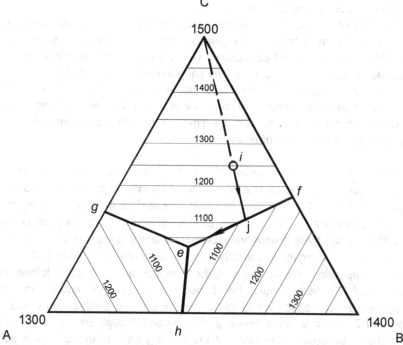

FIG. 7. Ternary eutectic phase diagram showing i) isotherms of the liquidus surfaces for temperatures in hectokelvin; ii) their intersections: the monovariant *ef*, *eg*, and *eh*; and the invariant eutectic point *e*; iii) the composition route followed by the crystallizing liquid mixture whose original composition is represented by point *i*

At the eutectic temperature a liquid mixture, having the eutectic composition, is in equilibrium with all of the three solids. The equilibria between liquid mixture and two solids are monovariant; represented by *ef*, *eg*, and *eh*.

A liquid mixture of A, B, and C, whose composition is given, in Figure 7, by the open circle in the triangle, is cooled down from 1600 K to 800 K. The phase behaviour of the system is easily read from the phase diagram; it is as follows. From 1600 K to 1250 K the whole amount of material is liquid. At 1250 K solid C begins to separate from the liquid. The composition of the remaining liquid follows the line *ij*; and its temperature the liquidus surface. Because from *i* to *j* the liquid phase is losing C, at constant A to B ratio, *ij* is part of a straight line originating from vertex C. At *j* the composition of the liquid phase branches off: the line *je* is followed; and solid B and solid C separate simultaneously from the liquid. Finally, the composition of the liquid has reached point *e*: at constant temperature all of the liquid solidifies - all of the original amount of A, and the remainder of B and C. After complete solidification, the temperature of the material is free to drop to its final value.

The path $i \rightarrow j \rightarrow e$ in Figure 7 is referred to as *crystallization curve* (Geer 1904). The term is applied to crystallization processes under the assumption that no material is removed during cooling.

The absence of solid-state miscibility makes that, during crystallization, the deposited solid phases have unchanging compositions. This circumstance is in favour of the sample's overall equilibrium - which is to say that the role of time, the time needed to reach through and through equilibrium, is rather limited.

For the opposite situation, for the case that the components A and B show complete solid miscibility, time is an important and often determining factor. In the case of the example, which follows, the time taken for the experiment is by far insufficient to reach overall equilibrium.

A liquid mixture of composition X_a - in a system whose phase diagram is shown in Figure 8 - is cooled at a certain constant rate, expressed in amount of heat per unit of time. If *supercooling* is absent, the mixture will start to solidify at T_ℓ (the liquidus temperature for $X = X_a$), producing *mixed-crystalline material* with $X = X_i$. At this stage there is equilibrium between the whole liquid and solid parts of the sample - considering the fact that the initial amount of solid is very small.

During the deposition of solid material, the composition of the liquid follows the liquidus; and if there would be through and through equilibrium, the (homogeneous) solid phase would follow the solidus; and solidification would be complete at T_s. In many cases, however, through and through equilibrium is out of the question - because of the fact that the building units of the crystalline solid (molecules in the case of molecular materials) are immobilized in/by the crystal lattice. At best, there is equilibrium between the whole liquid part of the sample and the surface layer of the solid.

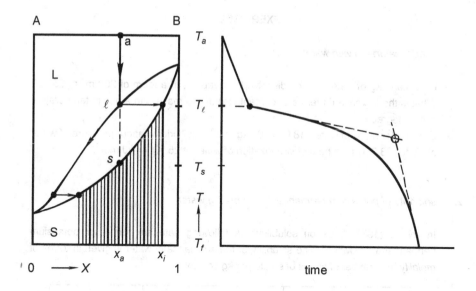

FIG. 8.   Phase diagram for equilibrium between liquid and mixed-
           crystalline solid (left); and cooling curve pertaining to a mixture
           of overall composition X_a

In reality, therefore, the deposited solid will have a zoned character, corresponding
to   a   gradient   in   composition.   And   because   the   first   zones   have
$X > X_a$, the last must have $X < X_a$ (*law of conservation of A and B*, so to say). In
other words, the last zones are deposited at a temperature lower than T_s. In Figure
8's phase diagram the aspect of zoning is represented by the bunch of vertical
lines.

As a remark, it is impracticable to read the *solidus temperature* (for $X = X_a$)
from a cooling curve like the one in Figure 8. Not even the point of intersection of
the tangent lines will provide the correct solidus temperature.

---

*Phase diagrams are a powerful guide to the understanding of separation and
purification processes - even in those cases where through and through equilibrium
is not realized or realizable.*

EXERCISES

1.   *a room saturated with water*

How many kg of water can be dissolved, as vapour, in a room of 200 m^3 at 25 °C?
What is the increase (in terms of percentage) of this mass with a rise in temperature
of one degree?
Data: molar mass of water: $18.015 \cdot 10^{-3}$ kg·mol^{-1}; equilibrium vapour pressure of water
at 25 °C: 23.756 Torr; heat of vaporization of water at 25 °C: 44 kJ·mol^{-1}.

2.   *solubility of potassium permanganate in boiling water*

In Nývlt's (1977) book on solubility the following data are found for potassium
permanganate ($KMnO_4$). The solubility of the substance in water is given in terms of
molality (*m*, number of moles of solute per kg solvent).

t /°C	0	10	20	30	40	50	60
m /(mol·kg⁻)	0.179	0.268	0.401	0.571	0.793	1.065	1.403

- By means of extrapolation, determine the solubility at 100 °C, using a plot of
  i) *m* vs. *t*; ii) ln*m* vs.*t*; iii) ln*m* vs. $(t + 273.15)^{-1}$.

3.   *the space needed to remove an impurity*

How much space is needed to free 1 mole of naphthalene from an impurity of
1 millimole of 1-chloro-4-iodobenzene at 25 °C?

4.   *extraction and clever use of solvent*

An organic solvent S is used to extract an organic substance B from its solution in
water. B's molality (m_B →Exc 2) in the aqueous solution is 0.050 mol·kg^{-1}, and its
distribution coefficient – m_B(in S) : m_B(in water) - is equal to 4.
- Demonstrate, by making a calculation, that it is more effective to extract B
  from water (take 1 kg of water) two times with 0.5 kg of S than one time with
  1 kg of S.

5.   *distillation with a fractionating column*

For each of the five cases shown in Figure 3, a liquid mixture of equimolar
composition is subjected to distillation in a set-up with a fractionating column, having
a very large plate number (say ∞).
- What is the composition of the first drop of distillate?
- What is the final composition of the liquid in the boiler?

6.   *single-step distillation of wine*

Liquidus temperatures as a function of mole fraction are given for the system $\{(1 - X)$ ethanol $+ X$ water$\}$ at normal pressure. The liquid mixture having $X = 0.8$ starts to boil at 83.4 °C (see table) producing vapour with $X = 0.46$ (the point $X = 0.46$; $t = 83.4$ °C is a point of the vaporus).

- With the information given, construct the complete TX liquid + vapour phase equilibrium diagram.
- Is it true that a brandy, having a "strength" of 40% vol. (ethanol mole fraction of 0.171), can be prepared from a 10% vol. (ethanol mole fraction 0.033) wine just by a single-step distillation?

X	0.0	0.1	0.2	0.3	0.4	0.5	0.6	0.7	0.8	0.9	1.0
$t / °C$	78.3	78.2	78.3	78.6	79.1	79.8	80.6	81.8	83.4	86.4	100.0

7.   *vacuum distillation*

According to *Trouton's rule*, the ratio between the heat of vaporization of a substance and its normal boiling point (*nbp*) on the thermodynamic temperature scale is constant – eleven times the gas constant ($\rightarrow$004).

In a lab experiment the substance 2-nitro-p-xylene (*nbp* 513 K) has to be prepared from p-xylene (1,4-dimethylbenzene, *nbp* 411 K).

The by–product of the reaction is 4-methylbenzaldehyde (*nbp* 477 K).

The student is instructed to separate starting material, product and by-product by means of *vacuum distillation*.

- Calculate the boiling points the three substances have at $P = 15$ Torr.

8.   *distillation with steam*

A heterogeneous mixture of water and aniline ($C_6H_5NH_2$; boiling point 185 °C) is subjected to *distillation with steam*. The experiment is ended when the distillate (which has a temperature of 20 °C) contains 1 mol of aniline.

The compositions (mole fraction aniline) at the three-phase equilibrium, which is just below 100 °C, are 0.015 (LI); 0.045 (V) and 0.63 (LII). At 20 °C the compositions of the 'water layer' (LI) and the 'aniline layer' (LII) are 0.0070 and 0.78, respectively.

- Make a sketch drawing of the TX phase diagram.
- How many moles of water are present in the distillate?
- How many moles of (water + aniline) are present in the water layer of the distillate, and how many in the aniline layer?
- Which percentage of the aniline yield will get lost if the water layer of the distillate is rejected, without being extracted?

9.   *a congruently crystallizing compound*

In the system {A + B + H$_2$O}, at constant T and P, three two-phase equilibria make their appearance: (solid A + solution of A and B in water); (solid B + solution); and solid A$_3$B + solution), represented by three *solubility curves* in the triangular phase diagram. The *compound* A$_3$B is a *congruently crystallizing compound*: when it is dissolved in water, and the water is made to evaporate, the compound is the first solid that crystallizes.
-   Sketch the phase diagram for this system.
In an experiment, equal amounts of A and B are dissolved in excess of water. Thereafter the solvent is made to evaporate isothermally until there is no water left.
-   Mark in the phase diagram the route followed by the *mother liquor* (= solution phase).
-   Which solids will be present at the end of the experiment, and what are their relative amounts?

10.   *an incongruently crystallizing hydrate*

Similar to Exc. 9, however with the following differences.
The compound is the *incongruently crystallizing hydrate* H = AB$\cdot^1/_2$H$_2$O: when it is dissolved in water, and the water is made to evaporate, the first solid that will crystallize is either A or B.
During crystallization the deposited solid material is continuously separated from the mother liquor. The experiment is ended the moment the last amount of mother liquor disappears.
-   Sketch a phase diagram for this case.
-   Mark in the diagram the route followed by the mother liquor.
-   What would happen if the solid material were not separated from the mother liquor?

11.   *recrystallization*

In a lab experiment the student is instructed to remove the impurity potassium permanganate (KMnO$_4$) from potassium chlorate (KClO$_3$) by means of crystallization. The crude material has to be dissolved in boiling water, after which the solution has to be cooled to 0°C (using melting ice).
For 100 g of crude material, and considering two cases i) there is 2 g of impurity, and ii) there is 10 g of impurity, answer the following questions.
-   How many kg of water are minimally needed?
-   How many g of pure, solid KClO$_3$ are maximally obtained?
Data.
Solubilities in g per kg of water

| 0 °C | KClO$_3$ | 33.1 | KMnO$_4$ | 28.3 |
| 100 °C | KClO$_3$ | 562 | KMnO$_4$ | see Exc 2 |

§ (006)

12. *a saturation problem*

A and C are two immiscible liquids and B is a solid that is slightly soluble in A as well as in C. In an experiment, carried out at constant T and P, 250g of A and 250 g of C are poured into a beaker, after which B is added in portions of precisely 0.1 g. After the addition of each portion the contents of the beaker are stirred, and allowed to come to equilibrium. After the addition of the tenth portion it appears that liquid phase A is just saturated, and that liquid phase C contains exactly 0.2 g of B.

- What will be B's mass in each of the phases after the addition of the twelfth portion?

13. *back to the analogy – the text around figures 1 and 2*

A hypothetical system {X_A mol A + X_B mol B + X_C mol H_2O}; {T, P constant, and $(X_A + X_B + X_C) = 1$}; has the following properties: A and B are non-interacting solids and their saturated solutions in (liquid) water have $X_A^{liq} = 0.4$ and $X_B^{liq} = 0.2$; in addition, the *amount* of A that can be dissolved in a given amount of water is independent of the presence of *dissolved B,* and vice versa.

- Construct the triangular phase diagram.
- Transform the triangular diagram into a rectangular one, such that the horizontal axis is for $X = X_B / (X_A + X_B)$, and the vertical axis for $Y = X_A^{liq} + X_B^{liq}$.

Up to this point we met with quite a number of substances, but ignored the fact that substances can react with one another. This shortcoming is repaired.

---

the ammonia equilibrium

Again we start off with an example in which an experimental system is on its way to a state of equilibrium. This time we take a cylinder-with-piston. The cylinder is filled with 1.6 mole of hydrogen (H_2) and 0.8 mole of nitrogen, separated from one another by a weak, weightless inner piston; see Figure 1. The set-up is kept at a certain, constant temperature by means of a thermostat, and at a certain, constant pressure by means of a manostat. In this situation the upper piston is 24 cm above the bottom of the cylinder.

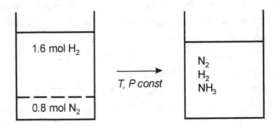

FIG. 1. An amount of hydrogen and amount of nitrogen change spontaneously into a mixture of hydrogen, nitrogen and ammonia to reach a state of chemical equilibrium

At a given moment the inner piston is destroyed and one can observe that, from that moment on, the upper piston goes down - and, after some time, it comes to a stop. The piston is then 20 cm above the bottom, and it remains there: a state of equilibrium has been reached. On its way to equilibrium the volume of the system has reduced to 5/6 of its original value.

According to Avogrado's law (Amadeo Avogadro, 1776-1856), dating from 1811 and saying that, no matter (the name of) the substance, as long as T and P are fixed, equal volumes of gases will contain equal numbers of molecules, the observation comes down to a reduction of moles from 2.4 to 2.0.

The origin of the reduction is in the chemical reaction between nitrogen and hydrogen to form ammonia (NH_3)

$$N_2 + 3 H_2 \rightarrow 2 NH_3 . \tag{1}$$

The original combination of N_2 and H_2 has changed into an equilibrium mixture of N_2, H_2, and NH_3, for which we write

$$N_2 + 3 H_2 \rightleftarrows 2 NH_3 . \tag{2}$$

The two arrows, one to the left and one to the right, underline the dynamic aspect of the equilibrium; in the sense that per unit time equal numbers of NH_3 molecules are formed from /decompose into N_2 and H_2. (As a matter of fact, many chemical reactions need a catalyst to proceed; in the experiment the catalyst has not been forgotten).

As always, the state of equilibrium is governed by one or more conditions in terms of chemical potentials. This time there is just one equilibrium condition:

$$\mu_{N_2} + 3\,\mu_{H_2} = 2\,\mu_{NH_3}; \tag{3}$$

and it stands in a clear relation to the chemical equilibrium equation, Equation (2). The proof of this relationship will follow later on (→202).

As regards the variables that are needed to define the state of equilibrium, we observe that the ammonia equilibrium, apart from T and P, needs two composition variables to define its state. For the latter we may take the mole fractions of H_2 and N_2; denoted by X_{H_2} and X_{N_2}. Of course, the mole fraction of NH_3 in the gaseous mixture follows from

$$X_{NH_3} = 1 - X_{H_2} - X_{N_2}. \tag{4}$$

Four variables, therefore, to define the state of equilibrium, and being subjected to one equilibrium condition: the *system formulation* reads

$$f = M\left[T, P, X_{H_2}, X_{N_2}\right] - N\left[\mu_{N_2} + 3\mu_{H_2} = 2\mu_{NH_3}\right] = 4 - 1 = 3. \tag{5}$$

The chemical potentials of the three substances are functions of the variables of the set M: the equation in N can be solved when it is known how the μ's depend on these variables. The system formulation indicates that there are three degrees of freedom: the numerical values of three of the four variables have to be introduced in the equation, after which the fourth can be calculated. It implies that, any time the values of one or more of the three independent variables are changed, there will be another state of equilibrium. Besides, the state of equilibrium will not change - the equilibrium will not be disturbed - when the complete system at constant T and P is halved: the chemical potentials, just like the variables of the set M, are *intensive properties*.

As a taste of how these things are going to work, we consider the limiting case in which the system respects the ideal-gas laws. In that situation the chemical potential of a pure substance B is given by the recipe

$$\mu_B(T,\ P) = G_B^o\ (T) + RT\ \ln P, \tag{6}$$

where T is the gas constant. According to this expression, B's chemical potential is the sum of a property of B, which is a function of only T, and a term determined by the T and P applied to the system (say, by the investigator).

For B as a component in a gas mixture, an extra term has to be added to the expression - a term which is logarithmic in B's mole fraction:

$$\mu_B = \mu_B(T,P,X_B) = G_B^o(T) + RT \ln P + RT \ln X_B \tag{7a}$$

$$= G_B^o(T) + RT \ln(X_B \cdot P). \tag{7b}$$

The product of X_B and P is often referred to as B's *partial pressure*: the pressure the manometer would indicate if B were alone in the space; symbol P_B. In terms of partial pressure:

$$\mu_B = G_B^o(T) + RT \ln P_B. \tag{7c}$$

After substitution of the recipes for the chemical potentials in the equilibrium condition, Equation (3), we have

$$\begin{aligned} \left\{G_{N_2}^o + RT \ln(X_{N_2} \cdot P)\right\} + 3\left\{G_{H_2}^o(T) + RT \ln(X_{H_2} \cdot P)\right\} \\ = 2\left\{G_{NH_3}^o(T) + RT \ln(X_{NH_3} \cdot P)\right\}, \end{aligned} \tag{8}$$

which, after rearranging, can be written as

$$\ln\left\{\frac{X_{NH_3}^2}{X_{N_2} \cdot X_{H_2}^3 \cdot P^2}\right\} = -\frac{2\, G_{NH_3}^o(T) - G_{N_2}^o(T) - 3\, G_{H_2}^o(T)}{RT}. \tag{9}$$

From this equation it is clear, once more, that there are three degrees of freedom (or rather three degrees of obligation, because): the values of three variables are needed to calculate the fourth, and with that the complete equilibrium state. For example, with T, P, X_{N_2}, the values of X_{H_2} and X_{NH_3} (= $1 - X_{N_2} - X_{H_2}$) follow from Equation (9).

Generally, equilibrium equations like Equation (9) are abbreviated to

$$\ln K = -\frac{\Delta G^o(T)}{RT}. \tag{10}$$

The capital K is the *equilibrium constant* and ΔG^o - as we will come to know later on - the *standard Gibbs energy change* of the chemical reaction involved in the equilibrium. In equilibrium matters, ΔG^o is a key property; it is substance- and reaction dependent.

The change of equilibrium constant with temperature, as follows from Equation (10), is determined by the change with temperature of ($\Delta G^o/T$). The change of ($\Delta G^o/T$) with T is related to the *heat effect* (ΔH^o) of the reaction (→107; Equation (107:26)). The relationship which results is the equation named after Jacobus Henricus van 't Hoff (1852 - 1911):

$$\frac{d \ln K}{dT} = \frac{\Delta H^\circ}{R T^2};$$
(11)

or, in terms of reciprocal temperature $(1/T)$, by substitution of $d(1/T) = -(1/T^2)\,dT$,

$$\frac{d \ln K}{d(1/T)} = -\frac{\Delta H^\circ}{R}.$$
(12)

Upon integration, and neglecting the change of ΔH° with temperature,

$$\ln\left(\frac{K_2}{K_1}\right) = -\frac{\Delta H^\circ}{R}\left(\frac{1}{T_2} - \frac{1}{T_1}\right);$$
(13)

this equation can be used to calculate (to estimate, if you prefer) K's value at T_2 from a known value at T_1.

homogeneous chemical equilibrium in solution

The ammonia equilibrium is a representative example of a gaseous *homogeneous chemical equilibrium*. The adjective 'homogeneous' is to indicate that the equilibrium system is a homogeneous mixture; there is just one phase. A second category of homogeneous chemical equilibria is found in the equilibrium between species dissolved in a (liquid) solvent.

The analogy between the dispersion of molecules in an empty space and the dissolution of species in a solvent (←006) also finds expression in the recipes for the chemical potentials. In Equation (7b), the recipe for B's chemical potential in an ideal-gas mixture, which is repeated here,

$$\mu_B = G_B^\circ + RT \ln\left(X_B \cdot P\right),$$
(14)

the product of X_B and P can be rewritten as $(n_B/n) \cdot (nRT/V) = (n_B/V)RT$; where n is the total amount of matter in the space, and n_B the amount of B. The quotient of n_B and V, which is the concentration of B in the system, is often indicated by $[B]$. In terms of concentration, Equation (14) can be transformed to

$$\mu_B = G_B^{\circ'} + RT \ln[B].$$
(15)

And this formula is the recipe for B's chemical potential in a dilute solution (more precisely, an *ideal dilute solution*, → 207).

As an example, we take the case of a weak acid HA (such as acetic acid) dissolved in water - giving rise to the equilibrium

$$HA \rightleftharpoons H^+ + A^-, \tag{16}$$

where the H^+ ion, the proton is taken up by a water molecule. Neglecting the (weak) electrolytic dissociation of the solvent, the concentrations of H^+ and A^- are equal: $[H^+] = [A^-]$.

Under isothermal, isobaric circumstances the system formulation for the equilibrium is

$$f = M\ \{[HA], [H^+]\} - N\left\{\mu_{H^+} + \mu_{A^-} = \mu_{HA}\right\} = 1. \tag{17}$$

The system is monovariant: $[H^+]$ (= $[A^-]$) is determined by the numerical value of $[HA]$. The relationship between the three concentrations is the solution of the equilibrium condition in N, Equation (17):

$$\ln\{\ [H^+] \cdot [A^-] / [HA]\ \} = \ln K_a = -\frac{\Delta G^{o'}}{RT}. \tag{18}$$

heterogeneous chemical equilibrium

In the case of heterogeneous chemical equilibrium the substances involved are distributed over at least two phases. For the two examples discussed here, we go back to two investigations of historical importance – the investigations carried out by Debray (1867), and Horstmann (1869), and pertaining to the equilibria

$$CaCO_3\ (solid) \rightleftharpoons CaO\ (solid) + CO_2\ (gas),\ and \tag{19}$$

$$NH_4Cl\ (solid) \rightleftharpoons (NH_3 + HCl)\ (gas), \tag{20}$$

respectively.

In the case of the $CaCO_3$ equilibrium there are three phases: one solid phase consisting of pure calcium carbonate, another solid phase, which is pure calcium oxide, and a gas phase which is pure carbon dioxide. Because each phase is made up of a pure substance, there are no composition variables: the only variables necessary to define the equilibrium states are temperature and pressure. The equilibrium system is monovariant:

$$f = M[T,P] - N\left[\mu_{CaO} + \mu_{CO_2} = \mu_{CaCO_3}\right] = 2 - 1 = 1. \tag{21}$$

The experimentalist has the freedom to select the system's temperature. Thereafter the equilibrium pressure - the solution of the equilibrium equation in N, Equation (21) - is determined by the values of the chemical potentials of the three substances.

§ (007)

In his experiments, Debray selected the system's temperature by immersing the set-up - vessel provided with manometer - in the vapour of a liquid with a known boiling point. A survey of his findings, as far as the choice of materials and the measured pressures are concerned, is given in Table 1. A comparison of the temperatures in the second column (Debray) and the ones in the fourth column (modern values) clearly shows that, at Debray's time, high-temperature thermometry was scarcely out of the egg.

Table 1: Heterogeneous chemical equilibrium $CaCO_3 = CaO + CO_2$. First three columns are data reported by Debray (1867). The fourth column gives the normal boiling points found in modern tables

Vapour of	$t/°C$	$P/Torr$	$t/°C$
Hg	350	zero	357
S	444	hardly perc.	444
Cd	860	85	766
Zn	1040	510-520	907

From the results displayed in Table 1, and from a set of experiments in which he studied the effect of changing the relative amounts of the (solid) phases, Debray arrived at the following conclusions. i) The system's pressure is constant at a given temperature; ii) the pressure increases with increasing temperature; and iii) the pressure is independent of the fraction of $CaCO_3$ that has decomposed.

In the case of the salammoniac (NH_4Cl) equilibrium, studied for the first time by Horstmann (1869), there are two phases - solid and gas. The solid phase is pure NH_4Cl and the gas phase, in Horstmann's view, an equimolar mixture of NH_3 and HCl. As a result, the remaining variables to define the equilibrium states are temperature and pressure (obviously, an alternative approach to the equilibrium system would be a gas phase of varying composition, to be realized e.g. by preparing the system from unequal amounts of NH_3 and HCl). The formulation of the system as studied by Hortstmann is

$$f = M[T,P] - N\left[\mu_{NH_3} + \mu_{HCl} = \mu_{NH_4Cl}\right] = 2 - 1 = 1 \tag{22}$$

Again, the equilibrium system is monovariant: the equilibrium pressure is determined by the temperature chosen by the experimentalist.

The NH_4Cl equilibrium is much more accessible to experimentation than the $CaCO_3$ one: comparable equilibrium pressures are reached at temperatures 600 K lower. In the salammoniac case the temperatures are below the boiling point of mercury -temperatures that can be read from a mercury thermometer; see Table 2.

Horstmann's important finding was that the graphical representation of equilibrium pressure vs. temperature has the same form as the "pressure curve of liquids". And he concluded that it is permitted, therefore, to use the data to calculate the heat effect of the decomposition reaction: 'by applying the well-known formula which permits the calculation of the heat of vaporization from the pressure curve' (←004; →110).

Table 2:   Heterogeneous chemical equilibrium $NH_4Cl = NH_3 + HCl$. Temperature-pressure data pairs read from Horstmann's (1869) plot of pressure vs. temperature; more precisely, from the smooth curve through 20 data points

$t /°C$	$P /Torr$
240	25
260	58
280	122
300	235
320	435
340	736

idealized treatment of vapour pressures over solids

For the treatment of vapour pressures over solid phases in terms of chemical potentials, the change of the potentials of the solids with pressure has a minor influence. In most cases that influence is neglected, and it means that the measured pressure only enters in the expressions of the potentials of the gaseous species. As a result, the equilibrium equations for the $CaCO_3$ and NH_4Cl cases take the form

$$CaCO_3 : \ln P = -\frac{\Delta G^o}{RT} = -\frac{G^o_{CO_2} + G^*_{CaO} - G^*_{CaCO_3}}{RT}; \qquad (23)$$

$$NH_4Cl : \ln(0.5\ P)^2 = -\frac{\Delta G^o}{RT} = -\frac{G^o_{NH_3} + G^o_{HCl} - G^*_{NH_4Cl}}{RT}. \qquad (24)$$

In these equations G^*_B is the notation for the *molar Gibbs energy* of pure B (for pure substances molar Gibbs energy is identical with chemical potential (→108; Equation (108:13); →202; Equation (202:24)).

number of components

Up to this section chemical reactions were absent, and the correct variance of a given equilibrium situation could be obtained, not only by $f = M - N$, but also by means of the *Phase Rule*:

$$f = \text{number of components} - \text{number of phases} + 2, \qquad (25)$$

the number of components being equal to the number of substances in the definition of the system, like $\{(1 - X)$ mole of NaCl + X mole of KCl$\}$.

The reader certainly will have noticed that, for the equilibria discussed in this section, the correct variance is not obtained if, in the Phase Rule, Equation (25), the number of components is identified with the number of chemical species (e.g. three in the case of the ammonia equilibrium: NH_3, N_2, H_2). This is not at all surprising because in the derivation of the Phase Rule chemical reactions were not taken into account. In other terms, if one wants to maintain the rule - because of the beauty of its simplicity - and to apply it to chemical equilibria, one has to find a definition of number of components. '*Number of components in the sense of the Phase Rule*', so to say; and being the number that yields, when introduced in Equation (25), the correct value for f.

In our search for a definition of "c" - the symbol we use for number of components in the sense of the Phase Rule - we take the survey given in Table 3 as a starting point.

Table 3:   Selection of chemical equilibria. Cases (1) - (4), examples treated in the text, (2) being generalized to include T and P. Case (1a), section of case (1), having $X_{H_2} = 3 X_{N_2}$ and α being the degree of dissociation. Case (4a), generalization of case (4) in that the constraint of equimolar gas composition is removed

	chemicals	M variables	N	f	p	"c"	essential
system							
NH_3 (1)	$NH_3\ NH_2\ H_2$	$T P\ X_{N_2}\ X_{H_2}$	= 3	1	2		$N_2 + H_2$
HA (2)	HA H_2O	$T P\ [HA]\ [H^+]$	= 3	1	2		$HA + H_2O$
$CaCO_3$ (3)	$CaCO_3\ CaO\ CO_2$	$T P$	= 1	3	2		$CaCO_3$
NH_4Cl (4)	$NH_4Cl\ NH_3\ HCl$	$T P$	= 1	2	1		NH_4Cl
NH_3 (1a)	$NH_3\ N_2\ H_2$	$T P\ \alpha$	= 2	1	1		NH_3
NH_4Cl (4a)	$NH_4\ Cl\ NH_3\ HCl$	$T P\ X_{NH_3}$	= 2	2	2		$NH_3 + HCl$

In the table, the cases (1) - (4) are the examples given in the text. The example of the weak acid, case (2), is generalized to include T and P; to do justice to the number 2 in $f =$ "c" $- p + 2$. Case (1a) is a section of case (1): the equilibrium states that can be realized by filling the set-up with pure ammonia, and for which all of the mole fractions can be expressed in terms of the *degree of dissociation* α. The other way round, Horstmann's section of the salammoniac equilibrium, case (4), has been generalized by removing the constraint of equimolar gas composition.

In the last column of the table the *essential chemicals* are listed: the smallest section of the chemicals in the second column that allow the realization of all of the wanted or possible equilibrium states. With the exception of the $CaCO_3$ case, the number of the essential chemicals is equal to "c", the number of components in the sense of the Phase Rule. The five cases, excepting $CaCO_3$, have in common that each of the phases of the system can be prepared, from the essentials, in an arbitrary amount.

The very fact that the equilibrium states are independent of the (relative) amounts of the phases is one of the most important characteristics of equilibrium between phases. Therefore, we should turn this characteristic into a requirement. And state that the 'number of components in the sense of the phase rule' is the smallest section of the chemicals, taking part in the equilibrium, with the help of which each of the phases can be prepared such that its amount is arbitrary.

Equilibrium states related to a chemical reaction have to satisfy one equilibrium equation in terms of the chemical potentials of the reactants and products. The equilibrium constant is a means to rationalize chemical equilibria. In Phase Rule terms, chemical equilibria require an accentuation of the concept of 'component'.

EXERCISES

1.   *the exerted pressure*

Supposing that the experiment discussed above, and represented by Figure 1, is carried out at 450 °C, assess the pressure exerted on the system during the experiment.

Use the data in Exc 2, and give the calculated pressure as an integer in decabar.

2.   *the equilibrium constant of the ammonia equilibrium*

Values are given for the equilibrium constant of the ammonia equilibrium, when written as $0.5\ N_2 + 1.5\ H_2 = NH_3$, as a function of pressure, at 450°C (Larson and Dodge 1923; Larson 1924).

- By extrapolation, determine K's value at zero pressure - and therewith the value of the property ΔG° in Equation (10), referred to 1 bar.

$\dfrac{P}{\text{atm}}$	$\dfrac{10^3\,K}{\text{atm}^{-1}}$
10	6.595
30	6.764
50	6.906
100	7.249

$$K = \frac{X_{NH_3}}{X_{N_2}^{0.5} \cdot X_{H_2}^{1.5} \cdot P}$$

3.   Clausius-Clapeyron plot of Horstmann's data

Make a Clausius-Clayperon plot ($\ln P$ vs. $1/T$) of the data in Table 2, pertaining to the equilibrium $NH_4Cl = NH_3 + HCl$.

From the slope read from the plot, calculate ΔH°, the amount of heat needed to decompose one mole of NH_4Cl into NH_3 and HCl.

First, from Equation (24) derive the relationship between slope and ΔH°.

4.   the equipment to be used

Starting from pure salammoniac (NH_4Cl), an investigator wants to realize and study the equilibrium NH_4Cl (solid) = ($NH_3 + HCl$) (gas) under such circumstances that part (the fraction α) of the NH_3 will be dissociated into N_2 and H_2.

- What kind of equipment the investigator is going to use (a cylinder-with-piston, which can be controlled thermostatically and manostatically; or rather a vessel-with-manometer, which can be controlled thermostatically)?

Clue. First find out if all mole fractions can be expressed in α. Then write down the ($f = M - N$) system formulation.

5.   three pure substances taking part in a reaction

Is it possible for three substances A, B and C to be present, each in pure form, in a cylinder-with-piston at arbitrary temperature and arbitrary pressure, if they take part in the chemical equilibrium

$$A + B = C,$$

which establishes itself at high speed?

6. *Professor Denbigh's example from the zinc smelting industry*

An equilibrium system consists of three phases: solid I (zinc oxide); solid II (carbon); and gas (containing the species Zn, CO, and CO_2). The chemicals take part in two independent reactions

$$ZnO + C = Zn + CO, \text{ and}$$
$$ZnO + CO = Zn + CO_2.$$

- Determine the variance of the system, by writing down the $(f = M - N)$ system formulation.
- What is your selection of the "c" components in the sense of the phase rule?
- Repeat the exercise for the case that there is an additional phase of pure liquid zinc.

LEVEL 1

AN INTRODUCTION TO THERMODYNAMICS
AND PHASE THEORY

§ 101 DIFFERENTIAL EXPRESSIONS

Expressions of the kind MdX + NdY, here referred to as differential expressions, play an important part in thermodynamics; sometimes they represent the total differential of a function Z of X and Y; sometimes they do not.

the volume of the ideal gas as a function of T and P

The volume of one mole of ideal gas as a function of the two variables temperature and pressure is given by the *ideal-gas equation*

$$V = V(P,T) = \frac{RT}{P}, \tag{1}$$

where R is the so-called *gas constant*.

For any point in the PT plane there is a value for V and also for each of the *partial derivatives* of V.

The first derivative of V with respect to T, when P is constant, and the first derivative of V with respect to P, when T is constant, are

$$\left(\frac{\partial V}{\partial T}\right)_P = \frac{R}{P} \quad \text{and} \quad \left(\frac{\partial V}{\partial P}\right)_T = -\frac{RT}{P^2}. \tag{2}$$

Differentiating the latter with respect to P, when T is constant, we obtain

$$\left(\frac{\partial}{\partial P}\left(\frac{\partial V}{\partial P}\right)_T\right)_T = \frac{\partial^2 V}{\partial P^2} = \frac{2RT}{P^3}; \tag{3}$$

and in the same manner

$$\frac{\partial^2 V}{\partial T^2} = 0. \tag{4}$$

When we differentiate $(\partial V/\partial T)_P$ with respect to P and $(\partial V/\partial P)_T$ with respect to T, we obtain two identical results:

$$\frac{\partial^2 V}{\partial P \partial T} = \left(\frac{\partial}{\partial P}\left(\frac{\partial V}{\partial T}\right)_P\right)_T = -\frac{R}{P^2} = \frac{\partial^2 V}{\partial T \partial P}. \tag{5}$$

It is a general rule that in successive partial differentiation the result is independent of the order of differentiation. Following Guggenheim (1950), we shall refer to this property as the *cross-differentiation identity*. Generally, if Z is a function of X and Y, then

$$\frac{\partial^2 Z}{\partial X \partial Y} = \frac{\partial^2 Z}{\partial Y \partial X}. \tag{6}$$

Let us next consider, for the ideal gas example, changes in V caused by changes in T and P. Infinitesimal changes first and then finite changes.

The change in V caused by the infinitely small change dT in T when P is constant, i.e. the *partial differential* of V with respect to T, is given by

$(dV)_P = \left(\dfrac{\partial V}{\partial T} \right)_P dT$; the rate of change of V with T, multiplied by the change in T.

The *total differential* of V is the sum of two partial differentials

$$dV = \left(\frac{\partial V}{\partial T} \right)_P dT + \left(\frac{\partial V}{\partial P} \right)_T dP; \tag{7}$$

for the ideal gas

$$dV = \frac{R}{P} dT - \frac{R}{P^2} dP. \tag{8}$$

Changes in V caused by large changes in T and P are found by integration. For example, the finite change in V corresponding to the route marked 1 in Figure 1 is calculated as follows

$$V(B) - V(A) = \int_{T_1}^{T_2} \left(\frac{R}{P} \right) dT + \int_{P_1}^{P_2} \left(-\frac{RT}{P^2} \right) dP = \frac{R}{P_1} \int_{T_1}^{T_2} dT - RT_2 \int_{P_1}^{P_2} \left(\frac{1}{P^2} \right) dP$$

$$= \frac{R}{P_1}(T_2 - T_1) + RT_2 \left(\frac{1}{P_2} - \frac{1}{P_1} \right) = \frac{RT_2}{P_2} - \frac{RT_1}{P_1} \tag{9}$$

The same result is obtained for the route marked 2 and for any other route in the PT plane. By the way, this example of integration is somewhat silly as we knew the result beforehand: simply from $V(B)-V(A) = V(T_2,P_2)-V(T_1,P_1) = RT_2/P_2-RT_1/P_1$.

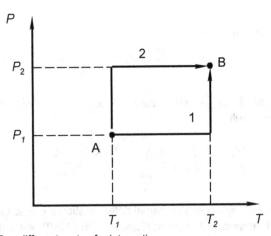

FIG.1.   Two different routes for integration

In thermodynamic practice, on the other hand, quantities often have to be found by integration, starting from *differential expressions* of the kind $M\,dT+N\,dP$, where M

and N are functions of T and P. The result of integration is only independent of the route in the PT plane if the differential expression belongs to a function, say V, of T and P. In that case M is the first partial derivative of V with respect to T and N the first partial derivative with respect to P. If, indeed, such a function of T and P exists, then the following condition must be satisfied

$$\left(\frac{\partial M}{\partial P}\right)_T = \left(\frac{\partial N}{\partial T}\right)_P. \tag{10}$$

This condition, often referred to as *Euler's criterion for integration*, is simply the above mentioned cross-differentiation identity.

An example of a differential expression for which the result of integration depends on the route through the PT plane is

$$R\,dT - \frac{RT}{P}\,dP. \tag{11}$$

Upon integration along the route marked 1 the change

$$R(T_2 - T_1) - RT_2 \ln\frac{P_2}{P_1} \tag{12}$$

is obtained; and along the route marked 2

$$-RT_1 \ln\frac{P_2}{P_1} + R(T_2 - T_1). \tag{13}$$

As an observation, the expression

$$R\,dT - \frac{RT}{P}\,dP$$

can be 'integrated', i.e. in the sense of yielding a function of T and P, when it is multiplied by $(1/P)$. In such case $(1/P)$ is a so-called *integrating factor*. After multiplication by $(1/P)$ the expression reads

$$\frac{R}{P}\,dT - \frac{RT}{P^2}\,dP, \tag{14}$$

being the total differential of the volume of the *ideal gas*.

$M\,dX + N\,dY$ is the total differential of a function $Z(X, Y)$ if $(\partial M / \partial Y)_X = (\partial N / \partial X)_Y$; this property is referred to as cross-differentiation identity; every $Z(X, Y)$ gives rise to $dZ = M\,dX + N\,dY$, in which case $M = (\partial Z / \partial X)_Y$ and $N = (\partial Z / \partial Y)_X$; not every $M\,dX + N\,dY$ gives rise to a $Z(X, Y)$.

EXERCISES

1. *the ideal gas*

For the ideal gas, taking V and T as the two independent variables, derive the first and second-order partial derivatives of P.
- Verify that the cross-differentiation identity is respected.
- Repeat the exercise for $T(P,V)$.

2. *the Van der Waals gas*

The *Van der Waals gas* is defined by the following so-called *equation of state*

$$\left(P + \frac{a}{V^2} \right) (V - b) = RT,$$

where a and b are substance-dependent constants.
- Take V and T as the two independent variables and formulate the total differential of P.

3. *integration of different expressions along different routes*

(A) and (B) are two differential expressions
- (A) $2XY^3\, dX + 3X^2Y^2\, dY$
- (B) $2Y^3\, dX + 3X\, Y^2\, dY$

(1) and (2) are two routes in the XY plane.
- Integrate (A) as well as (B) along each of the routes (1) and (2).
- What are your observations and conclusions?

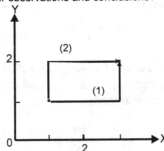

4. *a simplistic method*

A simplistic method of obtaining the total differential of the volume of the ideal gas in terms of T and P is the following.
- First, define dV by $\{V\,(T+dT,P+dP) - V\,(T,P)\}$; then enter the two pairs of coordinates $(T+dT,P+dP)$ and (T,P) in $V = RT/P$, and at a certain moment, replace $1\,/\,(1+dP/P)$ by $(1-dP/P)$.

By the way, with the help of a pocket calculator it can easily be verified that for $X \to 0$ $(1+X)^{-1} \to (1-X)$.

§ (101)

An ideal gas contained in a cylinder-with-piston is transferred from state $A(T_1,P_1)$ to state $B(T_2,P_2)$; are the amounts of heat and work, "exchanged" with its surroundings, independent of the route followed through the PT plane, going from A to B?

In the *PT* diagram, which is shown, A and B represent two states taken by one mole of helium gas. The positions of A and B are selected such that $V(B) = V(A)$.

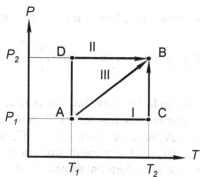

In different experiments the system is brought from A to B along different routes. These routes are marked by Roman numerals.

The first experiment is along route I, and the experiment is carried out as follows. From A to C: the cylinder with its contents at T_1 is immersed in a thermostat maintained at T_2 at constant external pressure P_1; the experimentalist waits until the piston has come to a stop. From C to B: the experimentalist gradually increases, at constant thermostat temperature T_2, the external pressure from P_1 to P_2.

work

In part AC the gas in the cylinder is expanding: it has to perform an amount of *work* on its surroundings. Or, in other words the surroundings extract an amount of work from the system i.e. the gas in the cylinder. In part CB the gas is being compressed: work is performed on the system; from the outside work is added to the system.

In thermodynamic usage *work added to a system* is taken positive: the work involved in part AC is negative and in part CB it is positive.

The amount of work added to the system is found by integration of $-P\,dV$:

$$\Delta W = -\int_A^B P\,dV . \tag{1}$$

The expression $-PdV$ is simply force x distance = pressure x area x distance = pressure x change in volume; the minus sign expresses the fact that when work is performed on (added to) the system there is a decrease in volume. ΔW can be calculated by substituting the total differential of the ideal gas ($\leftarrow$101):

$$\Delta W = -\int_A^B \left(R\,dT - \frac{RT}{P}\,dP \right). \tag{2}$$

For route I this is

$$\Delta W_I = -R\int_{T_1}^{T_2} dT + RT_2\int_{P_1}^{P_2}\frac{1}{P}\,dP = -R(T_2 - T_1) + RT_2\ln\frac{P_2}{P_1}. \tag{3}$$

In the second experiment route II is followed, and the amount of work along this route is

$$\Delta W_{II} = -R\int_{P_1}^{P_2}\frac{1}{P}\,dP - R\int_{T_1}^{T_2} dT = RT_1\ln\frac{P_2}{P_1} - R(T_2 - T_1). \tag{4}$$

We observe that, in spite of the fact that the two routes have the same initial and final coordinates, the amounts of work are not equal: $\Delta W_I \neq \Delta W_{II}$. Obviously the differential expression for $-PdV$, which is $-RdT + (RT/P)dP$, does not give rise to a function (W) of T and P! Expressed in a thermodynamic manner: work is not a *function of state*. Of course not: the differential expression does not obey the *cross-differentiation identity*.

In the case of route III the volume of the system does not change: there is no work at all; $\Delta W_{III} = 0$. The experiment can be carried out by immersing the system - in a closed vessel, and having temperature T_1 - into a thermostat set at T_2.

heat

From earlier findings it is clear that, in the case of route III, there is an intake of *heat*: heat is flowing from the thermostat to the system, until the temperature of the latter has reached T_2. We denote this amount of heat by ΔQ_{III}, and, heat being added to the system, it is positive.

If the experiments are carried out in such a manner that the amounts of heat are measured - which is not easy - one can observe that the amounts, just like the amounts of work, are different:

$$\Delta Q_I \neq \Delta Q_{II} \neq \Delta Q_{III}. \tag{5}$$

And just like work, heat is not a function of state.
However...

First Law of Thermodynamics, energy as a function of state

By careful experimentation it can be shown that the sum of heat and work added to the system is independent of the route followed from A to B!

§ (102)

$$\Delta Q_I + \Delta W_I = \Delta Q_{II} + \Delta W_{II} = \Delta Q_{III} + \Delta W_{III} \qquad (6)$$

And this experimental fact is the *First Law of Thermodynamics*. In other words, the sum of heat and work defines a function of state! This function of state is called *energy*.

Therefore, the sum $\Delta Q + \Delta W$ represents the difference in energy between the states *B* and *A*

$$\Delta Q + \Delta W \equiv \Delta U = U(B) - U(A) \ . \qquad (7)$$

For infinitesimal changes we can write

$$dU = q + w \ . \qquad (8)$$

In this expression, q (instead of dQ) and w (instead of dW) underline the fact that heat and work, in contrast to energy, are no functions of state.

The fact that energy is a function of state implies that it can be given as a function of T and P, its total differential being

$$dU = \left(\frac{\partial U}{\partial T}\right)_P dT + \left(\frac{\partial U}{\partial P}\right)_T dP \ . \qquad (9)$$

Later on, we will read more about the differential coefficients $(\partial U/\partial T)_P$ and $(\partial U/\partial P)_T$; the latter, for instance, is zero for the ideal gas.

The *SI unit* of work, heat and energy is the joule (J). The *thermochemical calorie*, which becomes more and more obsolete, is defined as

$$1 \ cal_{th} \equiv 4.184 \ J \qquad (10)$$

reversible changes

In the foregoing we used $-PdV$ (or $w = -PdV$) to calculate the amounts of work. In fact, this can only be done for experiments in which at any moment the external pressure is equal to the (internal) pressure of the system. In such a case the experiment can be reversed at any moment. One speaks of *reversible experiments*, *reversible processes*, and they are indicated by the subscript 'rev'. So, for reversible changes we have

$$dU = q_{rev} + w_{rev} = q_{rev} - PdV \ . \qquad (11)$$

The equality $dU = q + w$ is of general validity; no matter the degree of reversibility of the change!

When a system is transferred from $A(T_1, P_1)$ to $B(T_2, P_2)$ it exchanges heat and work with the surroundings; the amounts of heat and work depend on the route followed through the PT plane from A to B; the sum of heat and work, however, is independent of the route (First Law of Thermodynamics); the sum, therefore, represents a function of state: the energy.

EXERCISES

1.   q and the cross-differentiation identity

   Show, a posteriori, that $q = dU + PdV$ does not obey the criterion expressed by Equation (101:10).
   First substitute $dU = (\partial U/\partial T)_V \, dT + (\partial U/\partial V)_T \, dV$.

2.   units and conversion

   Show that 1J equals $1 \text{ kg·m}^2\text{·s}^{-2}$; show, in addition, that volume can be expressed in $J \cdot Pa^{-1}$. What is the conversion factor of $cm^3 \cdot atm$ to J?.

3.   reaction between zinc and sulphuric acid

   In a cylinder-with-piston, immersed in a thermostat at 25 °C, an amount of Zinc (Zn) is made to react with dilute sulphuric acid (H_2SO_4), the outside pressure being 1 atm.
   - Calculate the work exerted (added to) the contents of the cylinder (=the 'system') per mole of generated hydrogen (H_2, which may be taken as ideal gas). Express the answer in joule and in liter·atm.

The significance is examined of heat taken by a system at constant volume and heat taken at constant pressure.

heat at constant volume, and at constant pressure

Two experiments are examined, in which a given system, whose temperature is T_1, is immersed in a thermostat maintained at temperature T_2, such that $T_2 > T_1$. There will be an intake of *heat* until the system has reached the temperature T_2. The amount of heat is denoted by ΔQ and it equals the change of *energy* of the system minus the amount of *work* performed on the system ($\leftarrow$102)

$$\Delta Q = \Delta U - \Delta W = \Delta U + \int P \, dV. \tag{1}$$

In the first experiment the volume of the system is kept at a constant value ($dV=0$); there is no work and the amount of heat taken by the system is equal to the increase of its energy

$$(\Delta Q)_V = \Delta U. \tag{2}$$

In the second experiment the system is kept in a cylinder-with-piston under a constant external pressure P_o. In this case the heat taken by the system is given by

$$(\Delta Q)_{P=P_o} = \Delta U + P_o \int dV. \tag{3}$$

According to this expression, Equation (3), the *heat taken by the system at constant pressure* is needed to increase its energy and, at the same time, to perform an amount of work on the surroundings.

At this point we observe that at constant volume the heat taken by the system has the status of *increase of a function of state* - the energy.

It can be easily shown that, not only at constant volume, but also at constant pressure the heat taken by the system has the status of (the change of) a function of state. To do so, let us assume that in the second experiment the system is transferred from the state $A(P_o, T_1)$ to the state $C(P_o, T_2)$. Then

$$(\Delta Q)_{P=P_o} = \Delta U + P_o \int_A^C dV = U(C) - U(A) + P_o(V(C) - V(A))$$
$$= U(C) + P_o V(C) - (U(A) + P_o V(A)). \tag{4}$$

In words, the heat taken by the system is equal to the increase of $(U+P_oV)$, which clearly has the status of function of state: it combines two functions of state $U(T,P)$ and $V(T,P)$.

The combination of the two state functions is a new function of state, which is the *enthalpy*, symbol H:

$$H \equiv U + PV. \tag{5}$$

At constant pressure, therefore, the heat taken by the system is equal to the increase of its enthalpy:

$$(\Delta Q)_P = \Delta H. \tag{6}$$

heat capacity at constant volume, and at constant pressure

The quotient of the amount of heat taken by a system and its increase of temperature is referred to as the *heat capacity* of the system. A distinction is made between *heat capacity at constant volume*, symbol C_V, and *heat capacity at constant pressure*, C_P.
The heat capacity at constant volume is defined as

$$C_V = \lim_{(T_2 - T_1) \to 0} \frac{(\Delta Q)_V}{(T_2 - T_1)}. \tag{7}$$

The infinitesimal change which is meant in this expression can be indicated, as $(\Delta Q)_V = \Delta U$, as the change of energy due to an infinitesimal change in temperature at constant volume. Briefly formulated, therefore,

$$C_V = \left(\frac{\partial U}{\partial T}\right)_V. \tag{8}$$

Similarly, the heat capacity at constant pressure is identical with the partial differential coefficient of the enthalpy with respect to temperature under the condition of constant pressure:

$$C_P = \left(\frac{\partial H}{\partial T}\right)_P. \tag{9}$$

There is no special name for the (SI) *unit of heat capacity*. Heat capacities, therefore, are expressed in joule per kelvin, $J \cdot K^{-1}$. One can speak of the heat capacity of (a part of) an experimental set-up, a quantity expressed in $J \cdot K^{-1}$. When dealing with materials one can use either specific or molar heat capacities. *Specific heat capacity* is heat capacity divided by mass, a quantity expressed in $J \cdot K^{-1} \cdot kg^{-1}$ (or $J \cdot K^{-1} \cdot g^{-1}$ if one likes).
In this text we, invariably, will use *molar heat capacities*, which are expressed in $J \cdot K^{-1} \cdot mol^{-1}$.
The molar heat capacity has the same dimension as the *gas constant R*. For that reason it is occasionally preferred to display, e.g. in tables, C_P / R rather than C_P values.

The heat taken by a system at constant V is equal to its change in energy, which is a function of state; the heat taken by a system at constant P is also the change of a function of state; the latter is the enthalpy H, which is defined as $H = U + PV$; the heat capacities at constant volume and at constant pressure are $C_V = (\partial U / \partial T)_V$ and $C_P = (\partial H / \partial T)_P$.

EXERCISES

1. a classroom calorimeter

A 750 Watt ($=J \cdot s^{-1}$) immersion heater is part of a simple classroom calorimeter system and is used to determine the heat capacity of the set-up. In an experiment the heater was switched on for 120 seconds, as a result of which the temperature increased from 19.60 °C to 28.65 °C.
*   Calculate the heat capacity of the set-up.

2. a drop calorimeter

In a drop calorimeter a piece of 100g of copper (Cu; 63.5 $g \cdot mol^{-1}$) having a temperature of 800 K is made to drop in a thermally insulated container with water and ice having a temperature of 0 °C. Thereafter the temperature is still 0 °C and a measurement reveals that the mass of ice has decreased with 64.80 g.
*   Calculate for copper the value of $(H_T - H_{273})$; for $T = 800$ K, and expressed in $kJ \cdot mol^{-1}$. The heat of melting of ice is 333.5 $J \cdot g^{-1}$.
*   What is the mean specific heat capacity of copper in the interval $273.15 \le T/K \le 800$?

3. a cycle passed by a monatomic gas

One mole of a monatomic ideal gas, for which $C_P = C_V + R = 2.5 R$, passes the indicated cycle in a reversible manner.
*   For this cycle, complete the Table 1.

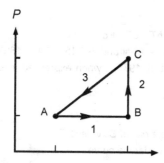

state	P /atm	V /dm³	T /K
A	1		298
B			596
C	2		

Table 1

- For each step (1, 2, 3) and the complete cycle (1+2+3) calculate q_{rev}, W_{rev}, ΔU, and ΔH.

4. *heat at constant pressure*

Ignoring the contents of this section, however knowing that $q_{rev} = dU + PdV$ (which implies that $(q_{rev})_V = dU$), show that when a function of state H is defined such that $H = U + PV$ it will have the property that its change at constant pressure is equal to (is caused by) the heat added to the system.

5. *choice of zero point*

Suppose you wish to assign the value of zero $J \cdot mol^{-1}$ to the enthalpy of gaseous oxygen for the conditions 1 Pa and 25 °C.
- If so, what is the energy of 1 mol of gaseous oxygen at 1 Pa and 25 °C.
- Repeat the exercise, changing 'gaseous oxygen' into diamond, for which $V = 3.417$ cm$^3 \cdot$ mol^{-1}.

6. *heat added to silver bromide*

For the substance silver bromide, AgBr, at P = 1 bar, the (molar) heat capacity in the range 298.15 $\leq T/K \leq$ melting point = 703 K is given by
C_P = {33.169 + 6.4434 x 10^{-2} (T /K)} J·K^{-1}·mol^{-1}. The (molar) heat of melting of the substance is 9.2 kJ·mol^{-1}.
- At 1 bar pressure, 1kJ of heat is added to a vessel containing 1 mol of silver bromide and originally having a temperature of 300 K. Calculate the temperature that will be reached by the system (vessel + contents). The heat capacity of the empty vessel is 20 J·K^{-1}.
- Calculate for liquid silver bromide and for T = 703 K the value of $H_T^o - H_{298}^o /T$, which is a quantity appearing in tables and in which the superscript o refers to P = 1 bar.

7. *an interpolation formula for C_P for diamond*

An *interpolation formula* for C_P data is $C_P(T) = a + bT + cT^2 + dT^{1/2} + eT^{-2}$.
When fitted to experimental data for diamond, for P = 1 bar and 298.15 $\leq T /K \leq$ 1800, the constants take the following values (Robie et al. 1978) - when expressed in SI units and per mole.
   $a = 98.445$; $b = -3.6554 \times 10^{-2}$; $c = 1.0977 \times 10^{-5}$;
   $d = -1.6590 \times 10^3$; $e = 1.2166 \times 10^6$
- For the validity range of the formula, make a plot of C_P versus T.
- Make an estimate, in kJ·mol^{-1}, of the area under the curve.
- Calculate for one mole of diamond, at 1 bar pressure, the value of $\{H(1800$ K$) - H(298.15$K$)\}$.

§ (103)

§ 104 THE IDEAL GAS,
EXPANSION AND COMPRESSION

For the ideal gas the dependence of energy on volume is examined.

Joule's experiment

When a gas in a cylinder-with-piston is made to expand, it has to push away the air above the piston (Figure 1, left). In other words, it has to perform an amount of *work*. If the gas were made to expand in an evacuated space, then there would be nothing to push away and, therefore, there would be no work to perform (Figure 1, right; when the lower clamps are removed, the piston will smash onto the upper ones!).

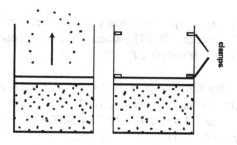

FIG.1.   Expansion of a gas. Left: against air. Right: 'against' vacuum

Similarly, in the case of the twin vessel (Figure 2) with an evacuated right-hand compartment: after opening the tap, the gas in the left-hand compartment will expand without having to perform work.

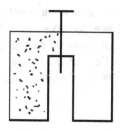

FIG.2.   Sketch of the twin vessel used by Joule in his experiment

In 1843 James Prescott Joule (1818-1889) studied the *expansion over vacuum* in a twin set-up immersed in a calorimeter bath. He measured the temperature of the bath before and after the opening of the tap. And, for the conditions of his experiment, he was not able to detect any change in temperature.

Today it is known that this result is valid for the *ideal gas*. A real gas, on the other hand, will, as a rule, produce a small effect.

For the ideal gas the outcome of *Joule's experiment* corresponds to a number of important properties. These properties come down to the principle that the energy of the ideal gas is independent of the extent of the space it occupies.

The discussion of the experiment is as follows. Because $\Delta W = 0$, ΔU equals ΔQ. The experiment reveals that the latter equals zero; if not, there would have been a change in temperature. As a result ΔU equals zero: the *energy of the ideal gas*, all other things being unchanged, does not depend on volume. One of the other things, which did not change during the experiment, is the temperature of the gas. For that reason one can also say that the experiment reveals that at constant temperature the energy of the ideal gas is independent of the volume it occupies – or, in other words, independent of its pressure:

$$\left(\frac{\partial U}{\partial V}\right)_T = 0 \; ; \; \left(\frac{\partial U}{\partial P}\right)_T = 0. \qquad (1a,b)$$

Otherwise expressed, the energy of the ideal gas is a function of temperature only: the partial differential coefficient $(\partial U/\partial T)_V$, which is the *heat* capacity C_V, reduces to the ordinary differential coefficient dU/dT.

The *enthalpy of the ideal gas* is $H \equiv (U + PV) = U + RT$; and clearly, like U, dependent on temperature only. Therefore, $(\partial H/\partial T)_P$, which is the *heat* capacity C_P, reduces to dH/dT. And, $dH/dT = d(U+RT) / dT = dU/dT + R = C_V + R$. For the ideal gas the *relation between C_V and C_P* is

$$C_P = C_V + R. \qquad (2)$$

isothermal expansion

When the ideal gas is made to expand at constant temperature (cylinder-with-piston in thermostat) it will have to take *heat* from the surroundings (the thermostat): the energy does not change, so that in the case of a *reversible* experiment $0 = q_{rev} + w_{rev} = q_{rev} - PdV$; or, $q_{rev} = PdV$.

adiabatic expansion

When the ideal gas is made to expand such that it cannot take heat from the surroundings (cylinder-with-piston provided with perfect *thermal insulation*), then the work to be performed is at the cost of its energy: the gas has to lower its temperature. Therefore, for $q = 0$, dU is equal to w_{rev}; or $C_V dT = -PdV = -(RT/V) dV$. After rearrangement, $(C_V/T) dT = -(R/V) dV$, or $C_V d\ln T = -R d\ln V$

a cycle

Let us now consider a cyclic experiment, see Figure 3, in which one mole of ideal gas first isothermally is expanded from A to B at $T = T_1$, then adiabatically from B to C (where the temperature has become T_2), next at $T = T_2$ isothermally compressed to D and finally adiabatically from D to A. Back in its original state A the energy of the gas is the same as it was before: the work performed by the gas during the cycle is equal to the net amount of heat taken by the gas.

The heat taken (from a reservoir) at T_1 follows from $q_{rev} = P\,dV = (RT_1/V)\,dV$ and is given by $RT_1 \ln(V_B/V_A)$. Similarly, the heat given off at T_2 corresponds to $RT_2 \ln(V_C/V_D)$. And because $V_B/V_A = V_C/V_D$, the net amount of heat is $R(T_1 - T_2)\ln(V_B/V_A)$. In terms of Figure 3, and from $w_{rev} = -P\,dV$, the amount of work performed curing the cycle (given by $R(T_1 - T_2)\ln(V_B/V_A)$) corresponds to the area enclosed by the four curves.

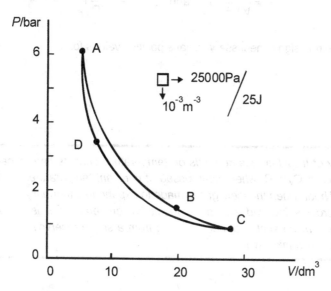

FIG.3.   The ideal gas is made to pass through the cycle ABCDA

When traversed repeatedly, the cycle - the famous *Carnot cycle* (Carnot 1824) - represents an *engine* in which heat is transformed into mechanical work. If we suppose that T_2 corresponds to ambient temperature, we may observe that the heat given off at T_2 is 'useless'. On the contrary, 'expensive' heat is needed to bring the other reservoir at T_1 and to maintain it there while the engine is taking the heat away. Therefore, the work $R(T_1 - T_2)\ln(V_B/V_A)$ performed by the engine is at the expense of the heat $RT_1 \ln(V_B/V_A)$. The *efficiency* η of the engine, as a result, is

$$\eta = \frac{T_1 - T_2}{T_1} = 1 - \frac{T_{low}}{T_{high}}. \tag{3}$$

The most important observation is that a difference in temperature is needed to operate an engine such that heat is transformed into mechanical work (one cannot run an engine just by withdrawing heat from a reservoir at ambient temperature!).

When the Carnot cycle is traversed in a counterclockwise sense mechanical work is used to take up heat from a low temperature reservoir and to give off heat at a higher temperature (*heat pump*).

expansion coefficient and compressibility

To end this section, the definitions are given of the *cubic expansion coefficient* α and the *isothermal compressibility* κ.

$$\alpha = \frac{1}{V}\left(\frac{\partial V}{\partial T}\right)_P ; \quad \text{and} \quad \kappa = -\frac{1}{V}\left(\frac{\partial V}{\partial P}\right)_T , \qquad \text{(4a, b)}$$

where the minus sign is necessary to have positive values for κ.

The energy of the ideal gas depends on temperature only; its heat capacities are related as $C_P = C_V + R$; when compressed at constant temperature, the ideal gas will give off heat; when the ideal gas is made to expand in a thermally insulated set-up, its energy is lowered and as a result its temperature; it is impossible to construct an engine such that heat is taken from a single reservoir and converted entirely into mechanical work.

EXERCISES

1. *a simulation of isothermal compression*

In an experiment the isothermal compression of the ideal gas can be 'simulated' as follows. A cylinder-with-piston containing argon is immersed in a vessel with about 7 kg of water. The external pressure is 1 bar. The system is thermally insulated and when thermal equilibrium is reached the temperature of the system is 25.000 °C and the volume of the gas in the cylinder 1 dm³. Next the gas is compressed reversibly until its pressure is 2 bar.

Calculate:
- the amount of gas;
- the work performed on the gas;
- the slight increase in temperature of the water in the vessel.

The heat capacity of the whole set-up is 34.65 kJ· K^{-1}.

2. *adiabatic compression*

In a thermally insulated cylinder-with-piston 1 mole of an ideal gas, with $C_v = 1.5R$, and originally having a temperature of 25 °C, is compressed to 20% of its original volume.

Calculate the temperature the gas will assume, considering the following two cases
- The heat capacity of cylinder and piston, i.e. the heat capacity of the set-up when empty, is neglected.
- The heat capacity of cylinder and piston is taken into account, its value being $C = 500$ J· K^{-1}.

3. *adiabatic compression of helium*

One mole of helium (He, with $C_P = 2.5\ R$) contained in a cylinder-with-piston is reversibly and adiabatically ($q_{rev} = 0$) compressed from the state A (298.15 K; 1 bar) to the state B (T; 10 bar).
- Calculate the temperature the gas will have in state B (all heat capacities except that of the gas itself may be ignored).
- Make a graphical representation of the trajectory - from A to B - of the gas in the PT plane.

4. *expansion of a Van der waals gas*

For a Van der Waals gas (←101:Exc 2) the differential coefficient $(\partial U / \partial V)_T$ is given by a/V^2

In a Joule-like experiment one mole of nitrogen is made to expand over vacuum, at 25 °C, from $V_1 = 1$ dm³ to $V_2 = 2$ dm³.
- Calculate the change in temperature. The heat capacity of the set-up is 14 kJ·K^{-1}; take $a = 0.14$ Pa· m⁶· mol⁻².

5. *volume quotients for a cycle*

Prove that for the cycle ABCD in the PV plane (with isotherms AB at T_1 and DC at T_2 and adiabatics AD and BC) for the ideal gas the following equality is valid
$$V_B/V_A = V_C/V_D.$$

6. *an equation for an adiabatic*

The equation for an isotherm in the *PV* plane for the ideal gas is, in its most elementary form, $P \cdot V = const.$
- Derive a similar equation, i.e. $P.V^{\alpha} = const.$, for an adiabatic in the *PV* plane for the ideal gas.
- Prove that two adiabatics cannot intersect one another.

7. *ideal gas, expansivity and compressibility*

Derive the equations for the cubic expansion coefficient α and the isothermal compressibility κ for the ideal gas.
Show that, generally,

$$\left(\frac{\partial \alpha}{\partial P} \right)_T = -\left(\frac{\partial \kappa}{\partial T} \right)_P$$

and, in addition, that for the ideal gas this identity comes down to 0 = 0.

8. *isothermal compression of Van der Waals gas*

The isothermal compressibility of the van der Waals gas κ_{vaw} can be given as $\kappa_{vaw} = \kappa_{ID} (1 + correction\ term)$, where κ_{ID} is the isothermal compressibility of the ideal gas.
- Derive an expression for the correction term.

In a closed, strong vessel - fully isolated for the transfer of heat and work - a mixture of hydrogen and oxygen is made to react, to change into water. There is an increase in temperature; the significance of this fact is assessed in the light of the law of conservation of energy.

a chemical reaction in an isolated system

If a system is completely isolated from the surroundings such that there is no transfer of heat to or from the system and there is no work performed on or by the system, then its energy will not change. This property is called *law of conservation of energy*.

That having remarked, we consider the case of a strong vessel, which is isolated from the surroundings, and in which a mixture of hydrogen and oxygen is made to react (by means of a built-in time mechanism).
After the reaction

$$2\,H_2 + O_2 \rightarrow 2\,H_2O$$

the temperature of the system has increased - an experimental fact which can easily be demonstrated. And, in spite of the increased temperature, the system still has the same energy. The explanation behind this apparent contradiction is in the existence of *chemical energy*.

chemical energy

The combination of the chemicals hydrogen and oxygen represents an amount of energy that can be released by their reaction to water. And the energy, which is released by the reaction, is stored in the experimental set-up (the vessel, and in it the water which is formed) as a result of which the temperature increases. It is like heat added from the outside to the system. And when the reaction is carried out in a set-up of known heat capacity (a calorimeter) the increase in temperature straightforwardly yields the heat effect (the heat effect of the reaction). At constant volume and referred to 25°C the heat effect of the reaction of two mole of hydrogen and one mole of oxygen to produce two mole of water is −564 kJ:

$$2 \text{ mol } H_2 + 1 \text{ mol } O_2 \rightarrow 2 \text{ mol } H_2O\,; \quad \text{heat effect } -564 \text{ kJ}$$
$$\text{or } 1 \text{ mol } H_2 + 1/2 \text{ mol } O_2 \rightarrow 1 \text{ mol } H_2O\,; \quad \text{heat effect } -282 \text{ kJ}$$

or, after division by 'mol',

$$2\,H_2 \quad + \quad O_2 \rightarrow 2\,H_2O\,; \quad \text{heat effect } -564 \text{ kJ·mol}^{-1}$$
$$\text{and} \quad H_2 + 1/2\,O_2 \rightarrow \quad H_2O\,; \quad \text{heat effect } -282 \text{ kJ·mol}^{-1}.$$

The heat added to a system at constant volume represents the change in energy;

$$H_2 + 1/2\ O_2 \rightarrow H_2O\ ;\ \Delta U = -282\ kJ\cdot mol^{-1}.$$

In this case therefore the heat effect of the reaction is the *reaction energy* $\Delta_R U$:

$$\Delta_R U \equiv U_{H_2O} - U_{H_2} - 1/2\ U_{O_2} = -282\ kJ\cdot mol^{-1}\ (at\ 25\ °C).$$

If the reaction is carried out at constant pressure - in a cylinder-with-piston situation - then the heat effect of the reaction is the *reaction enthalpy* $\Delta_R H$:

$$\Delta_R H \equiv H_{H_2O} - H_{H_2} - 1/2\ H_{O_2}.$$

Referred to 25 °C and 1 bar the enthalpy of reaction has the value of $-286\ kJ\cdot mol^{-1}$.

In most cases we will use reaction enthalpies rather than reaction energies. And whenever the therm heat of reaction is employed it will have the meaning of *reaction enthalpy*.

Hess's law

Let us next suppose that the water, which is formed, subsequently is decomposed into hydrogen and oxygen.

$$H_2O \rightarrow H_2 + 1/2\ O_2.$$

The energy effect of this reaction (R') referred to 25°C, must be $+282\ kJ\cdot mol^{-1}$, i.e. the reverse of the effect of the reaction in which H_2O is formed:

$$\Delta_{R'}\ U \equiv U_{H_2} + 1/2\ U_{O_2} - U_{H_2O} = -\Delta_R U = +282\ kJ\cdot mol^{-1}.$$

And it is clear that the same characteristic holds true for the enthalpy effect - the heat of reaction at constant pressure

$$\Delta_{R'}\ H = -\Delta_R H = +286\ kJ\cdot mol^{-1}.$$

From this line of argument is naturally follows that the heat effect of a reaction, carried out at constant volume or carried out at constant pressure, is independent of the fact whether the reaction is carried out directly or in a number of steps.

It goes without saying that this general fact, which was formulated in 1840 by Germain Henri Hess (1802-1850), is of vital importance in *thermochemistry*.

§ (105)

According to *Hess's law* heat effects of chemical reactions can be added and subtracted like the reactions themselves. As an example, we take the heat effect of the reaction of graphite (C) with oxygen to carbon monoxide. The effect which is relatively difficult to measure, because of the simultaneous formation of carbon dioxide, follows from the heat effects of the reactions (of graphite and carbon monoxide) with oxygen to carbon dioxide:

		ΔH(25 °C; 1 bar) $(kJ \cdot mol^{-1})$
$C + O_2 \rightarrow CO_2$		-393.51
$CO_2 \rightarrow CO + \frac{1}{2} O_2$		$+282.98$
$C + \frac{1}{2} O_2 \rightarrow CO$		-110.53

enthalpy of formation

The enthalpy effect of the formation of 1 mole of CO out of graphite and oxygen is -110.53 kJ. In thermochemical terminology this effect is named *enthalpy of formation (from the elements)*, $\Delta_f H$; in this case the enthalpy of formation of carbon monoxide $\Delta_f H_{CO}$.

Enthalpies of formation play an important part: the heat effect at constant pressure of any reaction is known if the enthalpies of formation of all of its reactants and products are known.

Chemicals represent amounts of energy - revealed by the heat effects of chemical reactions. Heat effects of chemical reactions sum up like the reactions themselves.

EXERCISES

1. *formation of liquid water*

In the text, while considering the reaction
$$H_2 + \frac{1}{2} O_2 \rightarrow H_2O,$$
the reaction energy was given as -282 kJ·mol^{-1} at 25 °C and the reaction enthalpy as -286 kJ·mol^{-1} at 25 °C and 1 bar.
- Are these two values mutually consistent, and wasn't it necessary to give a pressure indication for the reaction energy as well.

2. *formation of gaseous water*

The enthalpy of formation of liquid water at the normal boiling point of the substance, rounded to $kJ \cdot mol^{-1}$, is -284 $kJ \cdot mol^{-1}$. Under the same conditions the enthalpy of vaporization is 41 $kJ \cdot mol^{-1}$.
- What is the enthalpy of formation of gaseous water at the normal boiling point of the substance?
- What is for liquid H_2O the rounded value of $\Delta_f C_P$?

3. *reaction between graphite and carbon dioxide*

From   {C(graphite) + CO_2 $\rightarrow$ 2 CO} (25 °C; 1 atm); $\Delta H = +172.8$ $kJ \cdot mol^{-1}$ calculate, for the same reaction and under the given circumstances, the numerical values of
- the heat taken by the 'system';
- the work performed by the system;
- the energy change, ΔU.

The density of graphite is 2.26 $g \cdot cm^{-3}$.

4. *enthalpy of formation*

How is defined a substance's $\Delta_f H^\circ_{400}$?

Starting from the value for $\Delta_f H^\circ_{298}$, one wants to calculate for ethanol the value of $\Delta_f H^\circ_{400}$. What information is needed?

5. *reaction between hydrogen and chlorine*

In a thermally insulated vessel of constant volume an equimolar mixture of hydrogen and chlorine is made to react.
- Show that for this isolated system the enthalpy, in contrast to the energy, undergoes a change.

6. *formation of magnesite*

At 1 bar and 25 °C the heat of formation from the oxides ($\Delta_{fox} H$) of magnesite (magnesium carbonate, $MgCO_3$) is -118.28 $kJ \cdot mol^{-1}$ with an uncertainty of 1.30 $kJ \cdot mol^{-1}$ (Robie 1978).
- Calculate magnesite's heat-of-formation-from-the-elements, using the heats of formation of MgO and CO_2, which are for 1 bar and 25 °C.

$\Delta_f H_{MgO} = -601.49$ $kJ \cdot mol^{-1}$ and $\Delta_f H_{CO_2} = -393.51$ $kJ \cdot mol^{-1}$.

7. *standard formation energies*

Using the data given below, calculate the *standard* (298.15 K; 1 bar) *formation energies* of graphite (C, solid), silicon (Si, solid), oxygen (O_2, ideal gas), carbon monoxide (CO, ideal gas), carbon dioxide (CO_2, ideal gas) and quartz (SiO_2, solid).

substance	Volume $m^3 \cdot mol^{-1}$	standard enthalpy of formation $kJ \cdot mol^{-1}$
graphite	5.298×10^{-6}	0.000
silicon	12.056×10^{-6}	0.000
oxygen	RT/P	0.000
carbon monoxide	RT/P	-110.530 ± 0.170
carbon dioxide	RT/P	-393.510 ± 0.130
quartz	22.688×10^{-6}	-910.700 ± 1.000

8. *acetic acid from methanol and carbon monoxide*

With the help of a catalyst, gaseous methanol and carbon monoxide can be made to react to gaseous acetic acid. An appropriate temperature is 500 K.
Use the set of data given below to calculate, for standard pressure (1 atm), the value of the heat effect of the reaction at 500 K.

substance	(a) T_b (K)	(b) $\Delta H^o_{C,298}$ $(kJ \cdot mol^{-1})$	(c) $C^o_{P,298}$ $(J \cdot K^{-1} \cdot mol^{-1})$	(d) $\Delta H^o_{v,T_b}$ $(kJ \cdot mol^{-1})$	(e) $C^{o,vap}_{P,T_b}$ $(J \cdot K^{-1} \cdot mol^{-1})$
methanol	338	-725.7	81.6	35.3	44
carbon monoxide	81	-283.0	29.1		
acetic acid	391	-874.4	123.4	44.4	67

(a) normal boiling point;
(b) heat of combustion, to H_2O and CO_2, referred to 298.15 K;
(c) standard heat capacity;
(d) heat of vaporization at normal boiling point;
(e) heat capacity of vapour at normal boiling point.

9.   *combustion of benzoic acid*

The heat of combustion of benzoic acid (C_6H_5COOH; melting point 122.4 °C; heat of melting 17.3 kJ·mol^{-1}) to CO_2 and H_2O, referred to 25°C and 1 bar, is −3227 kJ·mol^{-1}.

- Calculate the enthalpy of formation of the substance, referred to 25°C and 1 bar and rounded to kJ·mol^{-1}.
- Estimate the heat of combustion of liquid benzoic acid at the melting point of the substance and 1 bar.

The necessary data, as far as they are not given here, can be found at other places in this section.

10.   *a reaction in a bomb calorimeter*

In a bomb calorimeter system, having a heat capacity of 15.25 kJ·K^{-1}, 1960 mg of a liquid organic substance with molecular formula $C_2H_4O_2$ was combusted to CO_2 and H_2O as a result of which the temperature of the system increased from 20.465 °C to 22.339 °C.

- Calculate the molar heat of combustion of the substance. Which value will be found in the thermodynamic tables for the substance's $\Delta_f H^\circ$?

An infinitesimal amount of reversible mechanical work, W_{rev}, is given by the product of the mechanical potential P and the change of a function of state (dV); (ana)logically, it is not foolish to ask the question "is q_{rev}, then, the product of the thermal potential T and the change of a certain function of state?".

a new function of state

Let us consider again a system which is one mole of *ideal gas* and let it be transferred in a reversible manner from some point $A(T_1, P_1)$ in the *PT* plane to another point $B(T_2, P_2)$.

In terms of temperature and pressure the *work* to be performed on the system follows from

$$W_{rev} = -P\,dV = -R\,dT + (RT/P)\,dP. \tag{1}$$

And the change of its *energy* simply from ($\leftarrow$104)

$$dU = C_v\,dT. \tag{2}$$

Now, for the *heat* taken by the system we have

$$q_{rev} = dU - W_{rev} = (C_V + R)\,dT - (RT/P)\,dP$$

$$= C_P\,dT - (RT/P)\,dP. \tag{3}$$

The amount of heat taken during the finite change from A to B depends on the route followed through the *PT* plane: the differential expression does not obey the *cross-differentiation identity.*

Suppose now that we were curious to know, for the transition from A to B, the integrated effect of q_{rev}/T: the sum of every small amount of heat taken divided by the temperature of the system at the moment of take up.

In the case of the ideal gas our curiosity can be gratified without delay: from the above expression for q_{rev} it follows

$$\frac{q_{rev}}{T} = \frac{C_P}{T}\,dT - \frac{R}{P}\,dP. \tag{4}$$

And we discover that this is a differential expression that obeys the cross-differentiation identify (be it in the form 0 = 0, but nonetheless). Or, in other words,

$$\int_A^B \frac{q_{rev}}{T} \quad \text{is independent of the route followed}$$

$$= C_P \ln\frac{T_2}{T_1} - R\ln\frac{P_2}{P_1} = \text{ the change of a function of state!} \qquad (5)$$

What we 'discover' easily for the ideal gas, is a truth of general validity. It is the *Second Law of Thermodynamics* and it expresses the general experimental fact that for *reversible changes* the heat divided by the thermodynamic temperature does - unlike the heat itself - correspond to the change of a *function of state*. The name of this function is *entropy* and its symbol S. The Second Law therefore states that when a system is transferred from state A to state B

$$\int_A^B \frac{q_{rev}}{T} = \int_A^B dS = S_B - S_A; \qquad (6)$$

in words: the integral over reversible heat divided by temperature is equal to the change of a state function of the system; that state function is the system's entropy; and the change, accordingly, is the difference in entropy between the states B and A.

The designation *entropy* was introduced by Rudolf Julius Emmanuel Clausius (1822-1888); Clausius (1854, 1865)

Entropy is a many-sided quantity which appeals to one's imagination. The remaining part of this chapter is meant to give just an idea of its fascinating characteristics.

absolute entropy

As an example of a system, whose entropy is raised by supplying heat to it in a reversible manner, we take one mole of mercury of which the temperature is raised from the *absolute zero* to 25 °C, at 1 bar pressure. The change in entropy consists of three parts, three contributions:

$$S(298.15K) - S(0K) = \int_0^{234.29K} \frac{C_P\,(\text{solid},T)}{T}\,dT + \frac{\Delta_s^l H}{234.29\text{ K}} \qquad (7)$$
$$+ \int_{234.29K}^{298.15K} \frac{C_P\,(\text{liquid},T)}{T}\,dT$$

$$= 75.90\text{J}\cdot\text{K}^{-1}\cdot\text{mol}^{-1}.$$

From the absolute zero to 234.29 K mercury is solid; the heat added to the substance is needed to raise its temperature - the relation between heat added and increase in temperature just is $q_{rev} = C_P\,dT$, and $dS = (C_P/T)\,dT$. The second contribution comes from the heat needed to change the substance from solid to liquid. The change from solid to liquid is an isothermal event: the change in entropy is equal to the *heat of melting* divided by the *melting point temperature*. The third contribution is related to the heat needed to increase the temperature of liquid mercury from its melting point to 298.15 K

It is a fortunate fact that for mercury and all other substances the entropy at zero kelvin can be put at zero! This important fact is rooted in the *Third Law of Thermodynamics* - also referred to as the *Heat Theorem* named after Walther Nernst (1864 - 1941). The Third Law expresses the general property of matter that entropy changes - such as the change involved in the transition from solid to liquid, or the change involved in a chemical reaction - become zero at the absolute zero of temperature

$$\lim_{T/K \to 0} \Delta S = 0. \tag{8}$$

According to this property it is not possible to detect - in the vicinity of zero kelvin - any entropy differences between substances. Therefore, one might just as well follow Max Planck (1858-1947) and say that the entropy of every substance is zero at zero kelvin. In other words, there is a *'natural zero point'* for the entropy, and the entropy value of a substance to appear in *tables* is given by

$$S\,(T = T') = \int_{T=0K}^{T=T'} \frac{q_{rev}}{T} \tag{9}$$

and referred to as the *absolute entropy* of the substance. The absolute entropy of mercury at 25 °C and for 1 bar pressure is 75.90 $J \cdot K^{-1} \cdot mol^{-1}$.

microscopic interpretation

An *isolated system* (U constant, V constant), from a macroscopic point of view, does not undergo any changes. Microscopically, however, it is constantly changing due to the changes of the positions, the velocities and the orientations of the molecules as well as their internal motions. The system, therefore, possesses an enormous number of microscopically different 'configurations'. The relation between the number W of *microscopic configurations* and the entropy of the system is

$$S = k \ln W, \quad \text{with} \quad k = R / N_{AV}, \tag{10}$$

which is called the *Boltzmann relation*, after Ludwing Boltzmann (1844-1906). The *Boltzmann constant k* is the quotient of the *gas constant* and *Avogadro's number*. The determination, in practice, of the number of microscopic configurations is not an easy job and in this text we will use the Boltzmann relation only in the form

$$\Delta S = \frac{R}{N_{AV}} \ln \frac{W_b}{W_a}, \tag{11}$$

where a and b represent two situations for which, all other things being equal, the W values easily can be compared.

As an example we examine the dependence-of-the-entropy-on-volume of the ideal gas at constant temperature (= constant energy). The difference in entropy between situation b with volume V_b and situation a with V_a follows from Equation (4):

$$\Delta S = -R \ln \frac{P_b}{P_a} = R \ln \frac{V_b}{V_a}. \tag{12}$$

The microscopic derivation of this relation is as follows and it is based on the positions of the *molecules* in space. The space in which the gas is contained can be divided in a number of small boxes, having the same, fixed volume. The number of the possible positions in space of a single molecule is equal to the number of boxes. That number is proportional to the volume of the system, say α times V. Then, for two molecules the number of configurations is $(\alpha V) \times (\alpha V)$: each position of the first molecule can be combined with all positions of the second. For N_{Av} molecules the number configurations is $(\alpha V)^{N_{Av}}$ and substitution of this number in Equation (11), for the two situations a and b, directly gives the result contained in Equation (12).

The exact formula for the entropy of the *monatomic ideal gas*, obtained by *statistical thermodynamics*, see e.g. Hill (1960), is the following

$$S = R \ln \left[\left(\frac{2\pi mkT}{h^2} \right)^{3/2} \frac{kT e^{5/2}}{P} \right]. \tag{13}$$

It is called the *Sackur-Tetrode equation* (Sackur 1913; Tetrode 1912). In this equation m represents the *atomic mass*, k the Boltzmann constant and h the *Planck constant*, while e is the base of the natural logarithm. After substitution of the numerical values of the fundamental physical constants and replacing atomic mass by *molar mass M*, the following formula is obtained

$$S = \frac{3}{2} R \ln \left(\frac{M}{\text{kg} \cdot \text{mol}^{-1}} \right) + \frac{5}{2} R \ln \left(\frac{T}{K} \right) - R \ln \left(\frac{P}{\text{Pa}} \right) + 20.73 \, R. \tag{14}$$

spontaneous changes in isolated systems

Suppose we have a space of given volume of which the lower part contains an amount of ideal gas; see Figure 1. It is clear that this is a *non-equilibrium situation* and it is also clear that the gas spontaneously will start to take all of the available space. After the *spontaneous change*, the manometers connected to the system will indicate a uniform pressure. And, moreover, if the system remains isolated (U, V constant), the pressure indicated by the manometers will not change anymore. From the foregoing, it is evident that during the spontaneous change the entropy of the gas increases. The gas reaches an equilibrium situation at which the entropy is maximal.

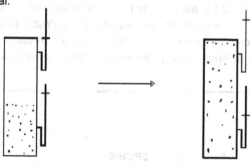

FIG.1.   In an isolated system a spontaneous change is accompanied by an increase in entropy

It is a general experimental fact that a spontaneous change in an isolated system goes together with an increase of entropy (there is no law of conservation of entropy!). Otherwise expressed, the equilibrium situation of an isolated system is the state at which the entropy is maximal. These two statements are contained in the expression

$$(dS)_{U,V} \geq 0. \tag{15}$$

Another example of a spontaneous change for which $(dS)_{U,V} > 0$, is the experiment considered in the very first part of this work ($\rightarrow$001). Two twin bodies with temperatures of $T + \Delta T$ and $T - \Delta T$ are put into contact, as a result of which they reach (U, V constant) the same temperature T. To calculate the change in entropy, one can 'simulate' the change by a *reversible experiment* (why not?: the entropy is a function of state and its change is determined by the 'coordinates' of initial and final state). Heat has to be withdrawn from the hotter body, the body with $T + \Delta T$, to bring it to the temperature T. The (negative amount of) heat added to the hotter body is

$$C \int_{T+\Delta}^{T} dT = C \cdot (-\Delta T) = -C \cdot \Delta T, \tag{16}$$

where C is the *heat capacity* of the body. Similarly, the heat taken by the colder body is $+C\,\Delta T$. The sum of the entropy changes of the two bodies is

$$C \int_{T+\Delta}^{T} \frac{dT}{T} + C \int_{T-\Delta T}^{T} \frac{dT}{T} = C \ln \frac{T}{T+\Delta T} + C \ln \frac{T}{T-\Delta T}$$

$$= C \ln \frac{T^2}{(T+\Delta T)(T-\Delta T)} = C \ln \frac{T^2}{T^2 - (\Delta T)^2}. \tag{17}$$

And because $T^2 > (T^2 - (\Delta T)^2)$, the entropy change, again, is positive.

§ (106)

Indeed, q_{rev} is the product of T and the change of a function of state - the Second Law; that function of state is the entropy (S); the entropy has a 'natural' zero point (substances have zero entropy at zero kelvin); through the Boltzmann relation, $S = k \ln W$, the entropy stands into relation to the number of microscopic configurations; spontaneous changes in isolated systems are characterized by an increase in entropy.

EXERCISES

1.   absolute entropy by graphical integration

Calculate by *graphical integration* the absolute entropy of diamond at 298.15 K from the heat-capacity values (De Sorbo 1953), which are given hereafter at intervals of 25 K from $T = 25$ K to $T = 300$ K. The unit is $cal_{th} \cdot K^{-1} \cdot mol^{-1}$ (Equation 102:10).

0.0012	0.0054	0.0203	0.0590	0.1295	0.2391
0.3829	0.5584	0.7653	0.9884	1.2319	1.4805

2.   absolute entropy of liquid sodium chloride

For the substance sodium chloride (NaCl, halite) the following information, valid for 1 bar pressure, is given (Robie 1978).
The absolute entropy at 25°C is 72.12 $J \cdot K^{-1} \cdot mol^{-1}$.
In the range from 25 °C to the melting point, which is 1073.8 K, the heat capacity is given by $C_P(T) = [45.151 + 1.7974 \times 10^{-2} (T/K)] \, J \cdot K^{-1} \cdot mol^{-1}$.
The heat of melting is 28158 $J \cdot mol^{-1}$.
  •   Calculate the absolute entropy of liquid sodium chloride at 1073.8 K.

3.   gaseous mercury and the Sackur-Tetrode equation

In the thermodynamic table (Robie 1978) for mercury (Hg, $M = 200.59$ $g \cdot mol^{-1}$) the following data are found for the entropy and heat capacity of the gaseous form at 1 bar pressure.

T (K)	S $(J \cdot K^{-1} \cdot mol^{-1})$	C_P $(J \cdot K^{-1} \cdot mol^{-1})$
629.0 (boiling point)	190.49	20.79
1000	200.12	20.79
1800	212.34	20.79

  •   Are these data in conformity with the Sackur-Tetrode equation?

4.   *orientations up and down*

Calculate $\Delta S = (R/N) \ln(W_b/W_a)$ for the following situations, (a) and (b), which differ in the orientations of N units on N lattice sites.
(a) all units have the *orientation* 'up';
(b) independent form the orientations of its neighbours, each unit has either the orientation 'up' or the orientation 'down'.

5.   *substitutional disorder of red and blue molecules*

Calculate $\Delta S = (R/N) \ln(W_b/W_a)$ for the following situations (a) and (b):
(a) $(1-X)$ mole of 'red' molecules contained in a lattice with $(1-X)$ N sites and X mol of 'blue' molecules contained in a lattice with $X \cdot N$ sites;
(b) $(1-X)$ mole of red molecules and X mol of blue molecules randomly distributed over a lattice with N sites.

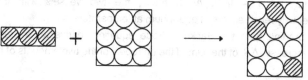

Make use of *Stirling's formula* $\ln(N!) = N \cdot \ln N - N$. By the way, the lattice characteristics invariably are the same.

6.   *cylinder with internal piston and entropy*

A cylinder of constant volume and containing an internal piston, which can be set at various positions, is immersed in a thermostat. The left-hand compartment, as well as the right-hand compartment, contains 1 mole of ideal gas.

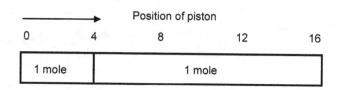

- For $4 \leq$ position ≤ 12, make a plot of entropy versus position of piston.

7. *uniform pressure at maximum entropy*

A two-compartment cylinder with movable piston is kept at constant temperature. The left-hand compartment contains n_1 mole of ideal gas and the right-hand compartment n_2 mole.

- Show that *maximal entropy* corresponds to $P_1 = P_2$, i.e. *equality of pressure*.

$n_1\, P_1\, V_1$	$n_2\, P_2\, V_2$

8. *two vessels with helium in thermal contact*

In a box, provided with material for thermal insulation, a vessel, of constant volume V, with a square cross-section and filled with 1 mole of helium having a temperature of $T + |\Delta T|$ is placed against an identical vessel with 1 mole of helium and having a temperature of $T - |\Delta T|$. As a result, the two vessels with their contents spontaneously reach a state of equilibrium at temperature T.

- Starting from the Sackur-Tetrode equation, derive an expression for the increase, ΔS, of the sum of the entropies of the two amounts of helium.

9. *supercooled water made to crystallize*

One mole of water having a temperature of $-5\,°C$ is, at atmospheric pressure, made to crystallize, as a result of which the temperature rises to the equilibrium value of $0\,°C$.

- Calculate the entropy effect of the spontaneous change.
- Which fraction of water will crystallize?

Rounded values for the heat capacity of liquid water and the heat of melting of water are $75\ J·K^{-1}·mol^{-1}$ and $6\ kJ·mol^{-1}$, respectively.

§ 107 CHARACTERISTIC FUNCTIONS

The differential expression dU = TdS – PdV, obtained by combining the First and the Second Law, corresponds to energy as a function of S and V; auxiliary thermodynamic quantities are introduced to replace S and V by variables that are more convenient from an experimental, and phase-equilibrium point of view.

fundamental equations

Following the *'natural'* route, which starts with the findings of the *First Law*

$$dU = q_{rev} + w_{rev} = q_{rev} - Pd\,V, \tag{1}$$

and then includes the *Second Law*

$$q_{rev} = TdS, \tag{2}$$

we obtain the following expression for the change of *energy* in *reversible changes*

$$dU = TdS - Pd\,V. \tag{3}$$

From a point of view of mathematics this is just a *total differential* of a function U whose variables are S and V. A fact which implies that $T = (\partial U/\partial S)_V$ and $P = -(\partial U/\partial V)_S$, and, by the *cross-differentiation identity*, $(\partial T/\partial V)_S = -(\partial P/\partial S)_V$. And, in order to find U by integration, one has to know both T and P as functions of S and V.

From a point of view of experimentation, however, one would like that this situation would be the other way round. That is to say, to know S and V as functions of T and P (we may think of the use of thermostats and manostats and thermometers and manometers).

To circumvent this inconvenience, thermodynamicists have extended the - what we like to call - natural route by the introduction of *auxiliary quantities*. These are the *enthalpy*, which we have met before, the *free energy* and the *free enthalpy*. The enthalpy (H) is defined as

$$H \equiv U + PV. \tag{4}$$

Its change, therefore, can be given as

$$dH = dU + Pd\,V + Vd\,P; \tag{5}$$

and upon substitution of Equation (1)

$$dH = TdS + Vd\,P. \tag{6}$$

The free energy (A) is defined as

$$A \equiv U - TS; \tag{7}$$

and it gives rise to

$$dA = -SdT - PdV. \tag{8}$$

Nowadays, the free energy is called *Helmholtz free energy*, or simply *Helmholtz energy*, after Hermann Ludwig Ferdinand von Helmholtz (1821-1894).
The free enthalpy (G) is now called *Gibbs free energy*, or simply *Gibbs energy*, after Josiah Willard Gibbs (1839-1903). The Gibbs energy is defined as

$$G \equiv H - TS \tag{9}$$

and it gives rise to

$$dG = -SdT + VdP. \tag{10}$$

The differential expressions, Equations (3), (6), (8) and (10) were referred to by Gibbs as the '*fundamental equations*'.

We now have the following four quantities with their *natural variables*

energy	$U(S,V)$
enthalpy	$H(S,P)$
Helmholtz energy	$A(T,V)$
Gibbs energy	$G(T,P)$

For our purposes, the most important auxiliary quantity is the Gibbs energy: its natural variables coincide with the experimentally convenient variables T and P (but not only for that matter, as we will observe soon!).
In terms of their natural variables each of the four quantities is said to be characteristic. The Gibbs energy is characteristic for T and P. It means that, if the Gibbs energy is known as a function of T and P, all thermodynamic quantities can be derived from it - as functions of T and P. The Helmholtz energy is characteristic for T and V; U for S and V and H for S and P.

the Maxwell relations

The two relations, Equations (11) and (12), are obtained by applying the cross-differentiation identity to Equations (8) and (10). They are known as the *Maxwell relations*, after James Clerk Maxwell (1831-1879):

$$\left(\frac{\partial S}{\partial V} \right)_T = \left(\frac{\partial P}{\partial T} \right)_V; \tag{11}$$

$$\left(\frac{\partial S}{\partial P} \right)_T = -\left(\frac{\partial V}{\partial T} \right)_P. \tag{12}$$

"These two relations are of great importance as they express the dependence of entropy on volume or pressure in terms of the more readily measurable quantities...", as one can read in Guggenheim (1950).

As an example of the use of one of the Maxwell relations, we examine the differential coefficient $(\partial U/\partial V)_T$ and prove that it is zero for the ideal gas (something we know from the Joule experiment). The example is also instructive in a mathematical sense, as it involves the 'translation' of a differential into a *differential coefficient*. From $dU = TdS - PdV$ it follows (division by dV at constant T, so to say) that

$$\left(\frac{\partial U}{\partial V}\right)_T = T\left(\frac{\partial S}{\partial V}\right)_T - P. \tag{13}$$

After substitution of Equation (11), and from $P = R \cdot T / V$

$$\left(\frac{\partial U}{\partial V}\right)_T = T\left(\frac{\partial P}{\partial T}\right)_V - P = T\frac{R}{V} - \frac{RT}{V} = 0. \tag{14}$$

After this observation, one can remark that no experiment is needed to demonstrate that, for a system defined by the *equation of state* $P \cdot V = RT$, the energy, at constant temperature, is independent of the dimensions of the space it occupies.

the Gibbs energy as characteristic function

From $dG = - SdT + VdP$ it follows that *entropy* and *volume* are obtained by partial differentiation with respect to T and P, respectively:

$$S = -\left(\frac{\partial G}{\partial T}\right)_P ; \tag{15}$$

$$V = \left(\frac{\partial G}{\partial P}\right)_T . \tag{16}$$

Next, the *Helmholtz energy*, the *enthalpy* and the *energy* are, in terms of Gibbs energy and its derivatives,

$$A = G - PV = G - P\left(\frac{\partial G}{\partial P}\right)_T ; \tag{17}$$

$$H = G + TS = G - T\left(\frac{\partial G}{\partial T}\right)_P ; \tag{18}$$

$$U = G + TS - PV = G - T\left(\frac{\partial G}{\partial T}\right)_P - P\left(\frac{\partial G}{\partial P}\right)_T. \tag{19}$$

The *heat capacity at constant pressure* is related to the second partial derivative with respect to temperature:

$$C_P = \left(\frac{\partial H}{\partial T}\right)_P = -T\frac{\partial^2 G}{\partial T^2} = \text{also } T\left(\frac{\partial S}{\partial T}\right)_P. \tag{20}$$

The *isothermal compressibility* is related to the first and second partial derivatives with respect to pressure:

$$\kappa = -\frac{1}{V}\left(\frac{\partial V}{\partial P}\right)_T = -\left(\frac{\partial G}{\partial P}\right)_T^{-1}\frac{\partial^2 G}{\partial P^2}. \tag{21}$$

And the *cubic expansion coefficient* is related as

$$\alpha = \frac{1}{V}\left(\frac{\partial V}{\partial T}\right)_P = \left(\frac{\partial G}{\partial P}\right)_T^{-1}\frac{\partial^2 G}{\partial T \partial P}. \tag{22}$$

The heat *capacity at constant volume* C_V, and the so-called *pressure coefficient* β, which is defined as

$$\beta = \left(\frac{\partial P}{\partial T}\right)_V, \tag{23}$$

both have the condition of constant V, rather than one of the Gibbs energy variables, T or P.
This inconvenience can be removed by using the following general expression between three variables of which two are independent ($\rightarrow$Exc 8). In terms of P, V and T the expression is

$$\left(\frac{\partial P}{\partial T}\right)_V \cdot \left(\frac{\partial V}{\partial P}\right)_T \cdot \left(\frac{\partial T}{\partial V}\right)_P = -1, \tag{24}$$

and it is clear that

$$\beta = \frac{\alpha}{\kappa} \tag{25}$$

a valuable equation

The partial differential coefficient expressing the change with temperature of the quotient of Gibbs energy and temperature is directly related to the enthalpy:

$$\{\partial(G/T)/\partial T\}_P = -H/T^2. \tag{26}$$

The equation derives its importance from the fact that the enthalpy is an experimentally accessible quantity.

$dU = TdS - PdV$; $dH = TdS + VdP$; $dA = -SdT - PdV$ and $dG = -SdT + VdP$ are the total differentials of a set of four thermodynamic functions in terms of their natural variables: energy as a function of entropy and volume, $U(S,V)$; enthalpy, $H(S,P)$; Helmholtz energy, $A(T,V)$ and Gibbs energy, $G(T,P)$; all thermodynamic quantities can be derived from each of these four functions, provided that the function is known in terms of its own natural variables.

EXERCISES

1. *the Maxwell relations differently*

For the derivation of the Maxwell relations it is not necessary to have the Helmholtz and Gibbs energies introduced first. The relations follow from $dU = TdS - PdV$ and $dH = TdS + VdP$ on replacing dS by the sum of two appropriate *partial differentials*.

2. *differential coefficients of energy*

Derive the following relations for the differential coefficients of the energy.

$$\left(\frac{\partial U}{\partial P}\right)_T = -\alpha TV + \kappa PV \ \ ; \ \text{ and } \ \ \left(\frac{\partial U}{\partial T}\right)_P = C_P - \alpha PV.$$

3. *total differential of enthalpy*

Starting form $dH = TdS + VdP$, show that the total differential of enthalpy in temperature and pressure is given by $dH = C_P dT + V(1 - \alpha T)dP$.

4. *model system with equation of state*

A model system satisfies the equation of state $P(V-b) = RT$, where b is a constant.
- Is the system's enthalpy at constant temperature dependent on pressure?
- And what about the energy?

5. *Van der Waals gas - change of enthalpy with volume*

Derive for a *Van der Waals gas*, which is defined by $(P + a/V^2)(V-b) = RT$, the expression for the differential coefficient $(\partial H/\partial V)_T$.

6. *change of heat capacity with pressure*

Show that $(\partial C_P/\partial P)_T$, the dependence of C_P on pressure at constant temperature, is given by $-T(\partial^2 V/\partial T^2)$.

7. *Gibbs energy of a hypothetical system*

The Gibbs energy, as a function of pressure and temperature, of a hypothetical gaseous system is given by $G(P,T) = \alpha + \beta T \ln T + \gamma T \ln P + \delta P + \varepsilon T + \varphi PT$ where α, β, γ, δ, ε and φ are constants.
- Derive the formulae for S, V, H, C_P, $(\partial P/\partial T)_V$, C_V, and the reversible heat involved in an experiment in which the system at constant temperature T_a is compressed from a pressure P_1 to a pressure P_2.

8. *the minus one identity*

X, Y and Z are a triplet of variables such that two of them can be set at arbitrary values as a result of which the value of the third is fixed (as an example, P, T and V for an amount of gas). Each of the three can be given as a function of the other two: $Z = Z(X,Y)$; $Y = Y(X,Z)$; $X = X(Y,Z)$.
- Prove that in this case.

$$\left(\frac{\partial X}{\partial Z}\right)_Y \cdot \left(\frac{\partial Y}{\partial X}\right)_Z \cdot \left(\frac{\partial Z}{\partial Y}\right)_X = -1,$$

Clue. First formulate the total differential of Z in terms of the variables X and Y and then imagine the meaning of $(\partial X/\partial Y)_Z$.

9. *Joule-Thomson coefficient*

Show that the *Joule-Thomson coefficient*, which is defined as the first partial derivative of the temperature with respect to pressure under the condition of constant enthalpy, obeys the following equality

$$\left(\frac{\partial T}{\partial P}\right)_H = -\frac{V(1-\alpha T)}{C_P}.$$

10. *difference between C_P and C_V*

Show that the *difference between C_P and C_V* is given

by $C_P - C_V = -T\left(\dfrac{\partial V}{\partial T}\right)_P^2 \left(\dfrac{\partial V}{\partial P}\right)_T^1$

For one mole of ideal gas this comes down to $C_P - C_V = R$.

11. *enthalpy as a characteristic function*

Find in terms of $H(S,P)$ and its derivatives with respect to S and P, i.e. from enthalpy as a characteristic function for the variables entropy and pressure, the expressions - the recipes - for G, U, C_P and (possibly) C_V.
The relation between the two heat capacity quantities is

$$C_V = C_P \left\{ 1 - \frac{\left(\dfrac{\partial^2 H}{\partial S\, \partial P}\right)^2}{\left(\dfrac{\partial^2 H}{\partial S\, \partial P}\right)^2 - \left(\dfrac{\partial^2 H}{\partial P^2}\right)\left(\dfrac{\partial^2 H}{\partial S^2}\right)} \right\}.$$

12. *energy expressed in T and P*

From which ideal-gas property it is directly clear that energy impossibly can be a function which is characteristic for the variables T and P?

13. *a cosmetic imperfection?*

In a fine thesis on the *ab initio* prediction of molecular crystal structures and the intriguing incidence of polymorphism, the following expression makes its appearance $\Delta G = \Delta U(T,P) + P\Delta V - T\Delta S$.

- Is this expression a logical one, in mathematical and thermodynamic respect?

§ 108 GIBBS ENERGY AND EQUILIBRIUM

A spontaneous change in an isolated system, i.e. a spontaneous change at constant U and V, is characterized by an increase in entropy. What we would like to know is: which thermodynamic quantity will act as the equilibrium indicator in the case of constant, or given T and P.

the equilibrium arbiter

In §106 we observed that whenever a spontaneous change is taking place in an isolated system it will be characterized by an increase in entropy;

$$(dS)_{U,V} > 0 \quad \text{(irreversible)} . \tag{1}$$

When, for given U and V, the system has reached the maximally possible entropy it is said to be in a state of (internal) equilibrium. This can be expressed by

$$\left(\frac{dS}{dX} \right)_{U,V} = 0 , \tag{2}$$

where X represents some kind of variable.

The system will stay in its state of equilibrium, unless it is freed from its isolation and, let us say by reversible experimentation, made to move to another state - another state of internal equilibrium. The change of its entropy is then related to the changes of U and V :

$$dS = \frac{q_{rev}}{T} = \frac{dU + P\,dV}{T} \quad \text{(reversible)} . \tag{3}$$

To arrive at a general statement about the change of entropy, we follow a system in its course, from time $= t_1$ to time $= t_4$, as depicted in Figure 1.
At t_1 the system is isolated from the rest of the world: neither heat nor work can be transferred to it; its energy and its volume are frozen. The state of the system at t_1 is represented by A: its energy and volume are U_A and V_A, its entropy S_A, and so on. In state A the system may or may not be in internal equilibrium. From t_1 on, the system is given enough time to change spontaneously into a state B of internal equilibrium (if a spontaneous change fails to take place: the system was in internal equilibrium already at t_1).
Now there are two possibilities: either state B = state A, or state B ≠ state A. In the first case we have, of course, $S_B = S_A$. In the other case there has been an increase in entropy; $S_B > S_A$. Generally, therefore,

$$(dS)_{A \to B} (= S_B - S_A) \geq 0. \tag{4}$$

Next, from t_2 on, the system is freed form its isolation and - by means of manostat and thermostat - the pressure exerted on the system is adjusted to P_B and the system's temperature to T_B. At t_3 the change from isolated system to closed system contained in cylinder-with-piston is complete. Finally, in the time span from t_3 to t_4, the system is transferred from (internal equilibrium) state B to (internal equilibrium) state C.

This time, Equation (3), the entropy change is given by

$$(dS)_{B \to C} (= S_C - S_B) = \frac{dU + PdV}{T}.$$ (5)

?

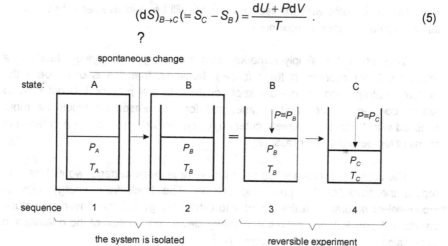

FIG.1.   The course of a system

The general statement, the general expression we are looking for is obtained by combining the two experimental truths:

state   state

A   C

arrow of time

$$(dS)_{A \to B} \geq 0$$

$$(dS)_{B \to C} - \frac{dU + PdV}{T} = 0$$

$$\overline{} \quad +$$

$$(dS)_{A \to C} - \frac{dU + PdV}{T} \geq 0$$

The general expression for the change-with-increasing-time of the entropy of a closed system is

$$dS - \frac{dU + PdV}{T} \geq 0.$$ (6)

§ (108)

This all-embracing expression for the entropy change of a closed system is the starting point for a number of statements.

First of all, if U and V are kept constant (if the system is isolated), dU and dV both are zero and the expression reduces to

$$(dS)_{U,V} \geq 0. \tag{7}$$

And the statement to be made is 'The entropy of an isolated system tends towards a maximum'. Or otherwise 'An isolated system will be in internal equilibrium when its entropy has reached a maximum'.

Quite often it is simply remarked that the energy of a system tends to a minimum. From expression (6) it follows, however, that this is only true if the system's entropy and volume are kept constant! In real practice it is not easy to realize constant entropy and volume, and for that reason we would be more satisfied with a more convenient criterion for equilibrium - preferably a criterion in terms of temperature and pressure.

The convenient criterion we are looking for is close at hand: we just have to repeat the track of the preceding section. The track from energy with its inconvenient *natural variables* S and V to Gibbs energy with its convenient natural variables T and P. If we make the following clever combination of the quantities in expression (6) - named Gibbs energy G -

$$G = U - TS + PV, \tag{8}$$

and substitute its change

$$dG = dU - TdS - SdT + PdV + VdP \tag{9}$$

in expression (6), we are going to observe that the change of the Gibbs energy of a closed system will be such that

$$dG + SdT - VdP \leq 0. \tag{10}$$

In other words if dT and dP are both equal to zero, i.e. if the temperature and the pressure of/on the system are kept constant, then

$$(dG)_{T,P} \leq 0. \tag{11}$$

Therefore, if a closed system is kept at constant T and P it will have reached internal equilibrium when its Gibbs energy is as low as possible. And any spontaneous change in a closed system at constant T and P will correspond to a lowering of its Gibbs energy. And also, when the temperature and the pressure of a closed system in internal equilibrium are changed by dT and dP, its Gibbs energy will change as

$$dG = -SdT + VdP. \tag{12}$$

For our purposes, having a criterion for equilibrium in terms of temperature and pressure is of the greatest importance. From the very beginning of the text we emphasized the very special status of the two quantities T and P. Uniformity of temperature and uniformity of pressure as the a priori equilibrium conditions. T and P as the ever-present variables of the M set, necessary to define the state of a system in equilibrium.

In the foregoing section we found that the 'natural' companion of the two variables is the Gibbs energy. And now we find that a system will be in state of equilibrium when at given, at fixed conditions of T and P its Gibbs energy has reached the lowest possible value.

the chemical potential

Up to this point, or more particularly in level 0, our approach to equilibrium relations was as follows. First the *variables of the set M* were defined and thereafter the *equilibrium conditions of the set N* in terms of *chemical potentials*. The N conditions represent a set of N equations between M variables of which $f = M - N$ are independent. It was observed that, in order to solve the set of equations, it is necessary to have the function recipes of the chemical potentials in terms of the variables of the M set.

In view of the findings above, we now know that the solution of the set of N equations has to correspond with the state of minimal Gibbs energy. More precisely, minimal Gibbs energy for every pair of selected T P coordinates. All this implies that there has to be a special relationship between Gibbs energy and chemical potentials. The nature of this relationship, as well as the proof that minimal Gibbs energy corresponds to the existence of the N conditions, will be the subject of sections in level 2.

For the time being, that is to say as long as we are in level 1, we limit ourselves to saying that for a pure substance B chemical potential is identical with *molar Gibbs energy*, Gibbs energy per mole: .

$$\mu_B \equiv G_B^* \quad \text{(pure substance B)}, \qquad (13)$$

where the asterisk refers to pure substance. In our terminology, Equation (13) is the most elementary *function recipe of chemical potential*.

The rest of this section, and in fact practically the whole remainder of this level, is devoted to equilibria that do need no more than pure substance chemical potentials.

Before considering two examples, we will precise the *molar Gibss energy of a pure substance in the ideal gas state.*

At constant temperature, $T = T_a$, the difference in Gibbs energy between the states of a system with $P = P_2$ and $P = P_1$ generally is given by

$$G(T_a, P_2) - G(T_a, P_1) = \int_{P_1}^{P_2} \left(\frac{\partial G}{\partial P} \right)_{Ta} dP = \int_{P_1}^{P_2} V(T_a, P) \, dP. \tag{14}$$

And, in the case of one mole of pure ideal gas B we have

$$\left(\frac{\partial G_B^*}{\partial P} \right)_{Ta} = V_B^*(T_a, P) = \frac{RT_a}{P}, \tag{15}$$

and with this relationship:

$$G_B^*(T_a, P_2) = G_B^*(T_a, P_1) + RT_a \ln \frac{P_2}{P_1}. \tag{16}$$

This result can be generalized as follows. First, we take for P_1 the unit of pressure, say $P_1 = 1$ p.u (*pressure unit*). And realizing that P_2 can be any pressure P and T_a any temperature T, we give P_2 and T_a the status of *running variable*; P and T, respectively,

$$G_B^*(T, P) = G_B^*(T, P = 1 \text{p.u.}) + RT \ln(P / \text{p.u.}). \tag{17}$$

This expression is usually written as

$$G_B^*(T, P) = G_B^o(T) + RT \ln P; \tag{18}$$

and in it G_B^o gives the value of the molar Gibbs energy at unit pressure (everyone is free to choose his/her own unit; it may, however, be wise to take 1 Pa, that is to say, a pressure at which the gas really will be virtually ideal).

examples

Just as an introduction to the remaining sections in this level, we will discuss two cases of equilibrium between phases. The first is the equilibrium between water and ice at atmospheric pressure. The second example involves the chemical change $CaO + CO_2 \rightarrow CaCO_3$. Instead of directly writing down the equations, we will start from the experimental side.

water and ice at atmospheric pressure

When a drop of (rain) water having a temperature of –5 °C falls on a ground equally having a temperature of –5 °C, the water will crystallize to ice, having a temperature of –5 °C. This event corresponds to a spontaneous change at constant T and P in a closed system (closed because of the fact that there is neither matter added to the system nor withdrawn from it; simply speaking, we are talking about a drop of water which changes completely into ice), and, consequently, to a lowering of the Gibbs energy. In other words, at –5 °C the Gibbs energy of an amount of ice (solid H_2O) is lower than the Gibbs energy of the same amount of water (liquid H_2O). At +5 °C the situation is reversed: water will not spontaneously change into ice. The other way round, a piece of ice put in an environment of +5 °C will melt spontaneously.

These facts are depicted in Figure 2, where, in a small temperature range, the molar Gibbs energies of solid and liquid H_2O are represented by straight lines. These lines have a negative slope, because of $(\partial G_B^* / \partial T)_P = -S_B^*$. And the difference in slope corresponds to $S_B^{*liq} > S_B^{*sol}$: heat is needed to melt a substance.

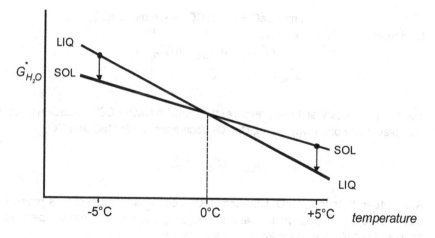

FIG.2.   The molar Gibbs energies of (liquid) water and ice in the vicinity of 0 °C

The two line intersect at 0 °C. Below this temperature $G_B^{*sol} < G_B^{*liq}$: water will change spontaneously into ice (the reverse is impossible). Above 0 °C ice will change spontaneously into water: $G_B^{*liq} < G_B^{*sol}$.

When a lump of ice having a temperature of 0 °C is immersed in water and the temperature of the water is 0 °C then nothing will happen. At 0 °C water and ice are in equilibrium: 0 °C is the solution of the equation

$$\mu_{H_2O}^{sol}(1\text{atm},T) = \mu_{H_2O}^{liq}(1\text{atm},T), \text{ or } G_{H_2O}^{*sol}(1\text{atm},T) = G_{H_2O}^{*liq}(1\text{atm},T). \quad (19)$$

Generally, the equilibrium between water and ice, with M = M [T,P], has to satisfy the condition

$$\mu_{H_2O}^{sol}(T,P) = \mu_{H_2O}^{liq}(T,P), \text{ or } G_{H_2O}^{*sol}(T,P) = G_{H_2O}^{*liq}(T,P). \tag{20}$$

The solution of this equation corresponds to a curve in the PT plane. The point 0 °C, 1 atm is just one point on that two-phase equilibrium curve.

calcium carbonate

In the case of calcium carbonate ($CaCO_3$, calcite) the following experimental observations can be made. Calcium oxide (CaO) and carbon dioxide (CO_2) are brought in an otherwise empty cylinder-with-piston under constant external pressure. The pressure, this time, is 1 bar. For $T < 1160$ K, CaO and CO_2 spontaneously react to $CaCO_3$:

$$n \text{ mol CaO} + n \text{ mol } CO_2 \rightarrow n \text{ mol } CaCO_3 .$$

Obviously

$$n \cdot G_{CaCO_3}^{*} < n \cdot G_{CaO}^{*} + n \cdot G_{CO_2}^{*}, \text{ or}$$
$$G_{CaCO_3}^{*} < G_{CaO}^{*} + G_{CO_2}^{*} \quad (T < 1160\text{K}).$$

For $T > 1160$ K (still at 1 bar pressure) the reaction CaO + $CO_2 \rightarrow CaCO_3$ does not take place. The other way round, $CaCO_3$ decomposes into CaO and CO_2:

$$G_{CaO}^{*} + G_{CO2}^{*} < G_{CaCO_3}^{*}.$$

At $P = 1$bar, therefore, CaO, CO_2 and $CaCO_3$ can exist in each other's presence - can coexist, be in equilibrium - at just one temperature; and that temperature is 1160 K, $T = 1160$ K is the solution of the equation

$$\mu_{CaO}(1\text{bar},T) + \mu_{CO_2}(1\text{bar},T) = \mu_{CaCO_3}(1\text{bar},T), \tag{21}$$

which in this case (each phase is a pure substance) is equivalent to

$$G_{CaO}^{*}(1\text{bar},T) + G_{CO_2}^{*}(1\text{bar},T) = G_{CaCO_3}^{*}(1\text{bar},T) \tag{22}$$

Generally, the chemical equilibrium

$$CaO + CO_2 \rightleftarrows CaCO_3$$

is represented by a curve in the PT plane, and is the solution of

$$G_{CaO}^{*}(P,T) + G_{CO_2}^{*}(P,T) = G_{CaCO_3}^{*}(P,T). \tag{23}$$

§ (108)

In this case it is not so difficult to say something more about the solution of the equation. It goes as follows. In the first place the molar volumes of the solid substances, CaO and $CaCO_3$, are negligibly small with respect to the molar volume of gaseous CO_2. That means that, in view of $(\partial G/\partial P)_T = V$, the dependence on pressure of G^*_{CaO} and $G^*_{CaCO_3}$ can be neglected. Secondly, for G_{CO_2} we can insert Equation (18), and, as a result

$$G^*_{CaO}(T) + G^0_{CO_2}(T) + RT\ln P = G^*_{CaCO_3}(T);\qquad(24)$$

we find, then, that the equilibrium pressure as a function of temperature is given by the equation

$$P(T) = e^{-\Delta G^0(T)/RT},\qquad(25)$$

where ΔG^0 is short for

$$\Delta G^0 = G^*_{CaO} + G^0_{CO_2} - G^*_{CaCO_3}.\qquad(26)$$

For $T < 1160$ K ΔG^0 is positive (the superscript ^{0} refers to 1 bar): P is less than 1 bar. At 1160 K $\Delta G^0 = 0$ and $P = 1$ bar, and for $T > 1160$ ΔG^0 is negative, and $P > 1$ bar.

The nature of the equilibrium curve can be read from Figure 3. The pure-substance data needed to do the calculations can be found in thermodynamic tables; the subject of the following section.

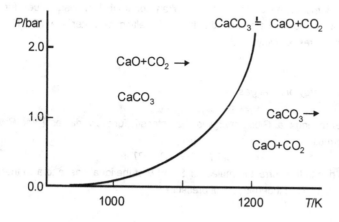

FIG.3.   PT diagram for the equilibrium between $CaCO_3$ and $(CaO+CO_2)$

A spontaneous change which proceeds at constant temperature and pressure is characterized by a decrease in Gibbs energy; at given T and P the equilibrium state of a thermodynamic system is the one with the lowest possible Gibbs energy; the principle of minimal Gibbs energy is discussed for the equilibrium between water and ice and the chemical equilibrium $CaO + CO_2 \rightleftarrows CaCO_3$.

EXERCISES

1.   *entropy versus energy diagram*

Make an S versus U diagram in which, for a series of equilibrium states of a system, the entropy as a function of energy is represented by a curve. Plot in the figure a few points that represent non-equilibrium states. Show that a spontaneous change at constant S is characterized by a decrease in energy.
NB. The volume of the system is considered to be constant.

2.   *a fancy device*

In an entropostat (a fancy device by means of which the entropy of a system can be kept constant) an amount of water having a temperature of $-5\ ^\circ C$ spontaneously changes into water and ice having a temperature of $0\ ^\circ C$.
   •   Show that ΔU is negative and calculate its numerical value.
The volume of the system may be taken constant. Rounded values for the heat capacity of liquid water and the heat of melting of water are $75\ J K^{-1} mol^{-1}$ and $6\ kJ mol^{-1}$, respectively.

3.   *Gibbs energy of ideal gas*

Remembering that Gibbs energy is characteristic for the variables T and P, verify that the formula
$$G(T,P) = G^\circ (T) + RT \ln P$$
gives rise to the correct formulae for S and V of the ideal gas, and also that it implies that H and U are independent of pressure.

4.   *ammonia's Gibbs energy of formation*

The Gibbs energy of formation of gaseous ammonia (NH_3), when treated as an ideal gas, referred to 25 °C and a standard pressure of 1 bar is $\Delta_f G^o_{NH_3} = 16.41\,kJ\cdot mol^{-1}$.

- What will be $\Delta_f G^o_{NH_3}$ for 1 Pa taken as standard pressure?

5.   *an equation of state for real gases*

Up to pressures of a few atm real gases can be described by (see Guggenheim 1950)

$$V(T,P) = \frac{RT}{P} + B(T)$$

- Use this *equation of state* to derive the formulae for $G(T,P)$ and $H(T,P)$.

Take 1 Pa as standard pressure, i.e. a pressure unit which is small with respect to the atm, as a result of which G^o will correspond to the ideal gas G^o.

6.   *the concept of fugacity*

In 1901 G. N. Lewis (see Prausnitz et al. 1986) introduced the concept of *fugacity, f*. In terms of fugacity, the Gibbs energy of a gas is given by

$$G_{T,P} = G^o(T) + RT\,\ln f.$$

For the ideal gas f is equal to P; for *real gases* it represents some kind of pseudopressure.

- Derive the expression for $f(T,P)$ which corresponds to the equation of state $V(T,P) = RT/P + B(T)$.
- Calculate the fugacity of nitrogen gas at 673 K and 1 atm. Nitrogen's value of B at 673 K is $24\times10^{-6}\,m^3$ (see Guggenheim 1950).

7.   *water + ice under a higher pressure*

How has Figure 2 (see above) to be changed and in which sense does the equilibrium temperature change, when the system is put under a higher pressure? Of course you know that ice has a lower density than water.

8.   *the calcium carbonate equilibrium under a higher pressure*

Knowing that for the heterogeneous chemical equilibrium $CaCO_3 = CaO + CO_2$ at 1 bar the equilibrium temperature is 1160 K, calculate the equilibrium temperature for $P = 1.25$ bar, using the following rounded values of the entropies (of the three substances involved) valid for 1160 K and 1 bar and expressed in $J\cdot K^-mol^{-1}$
CO_2 277; CaO 106; $CaCO_3$ 239.

9. *the calcite and aragonite forms of calcium carbonate*

Calcite and aragonite are two different solid forms (modifications) of the substance calcium carbonate. When heated from room temperature in a cylinder-with-piston under 1 bar pressure, calcite decomposes into CaO and CO_2 at 1160 K. Under the conditions of the experiment aragonite has a higher Gibbs energy than calcite (aragonite is said to be metastable: it has the capacity of changing spontaneously into calcite).

- If during the experiment aragonite does not change into calcite, will it then, like calcite, decompose at 1160 K?.

10. *liquid and gaseous water in equilibrium*

At 1 atm pressure liquid and gaseous water are in equilibrium at 373.15 K. Under these conditions, therefore, $\mu_{H_2O}^{liq} = \mu_{H_2O}^{vap}$, or, $G_{H_2O}^{*liq} = G_{H_2O}^{*vap}$. One can also say: for the transition $H_2O(liq) \rightarrow H_2O(vap)$ the value of ΔG (or $\Delta G_{H_2O}^{*}$, if you like) is zero for 373.15 K and 1 atm. At 373.15 K and 1 atm the value of ΔH, the heat of vaporization, is 40866 $J\,mol^{-1}$.

- For $T = 373.15$ K; $P = 1$ atm calculate the values of ΔS and ΔU.
- Calculate the values of ΔG for the following two states $T = 372.15$ K, $P = 1$ atm, and $T = 374.15$ K, $P = 1$ atm.

11. *lowest Helmholtz energy as a criterion for equilibrium*

Show that for a system, which is studied in a vessel-with-manometer, whose volume is fixed at a constant value, and whose temperature, by the experimentalist, is kept at a constant value, the equilibrium state is the state of lowest Helmholtz energy.

In the world of equilibria the availability of thermodynamic data is of great importance. Of course! This is a section on data and tables for pure substances. This time we will not use the asterisks, necessary on other occasions, to distinguish between pure substance property and property for mixture.

zero point matters

The molar *Gibbs energy* of a substance is equal to its molar *enthalpy* minus *T* times its molar *entropy*

$$G(T,P) = H(T,P) - TS(T,P). \tag{1}$$

This combination of enthalpy and entropy is such that its derivatives with respect to *T* and *P* simply are given by $-S$ and *V*.

$$\left(\frac{\partial G}{\partial T}\right)_P = -S \tag{2}$$

$$\left(\frac{\partial G}{\partial P}\right)_T = V \tag{3}$$

And, accordingly, the total differential in terms of *T* and *P*

$$dG = -S\,dT + V\,dP. \tag{4}$$

Enthalpy and entropy are two quantities which have much in common, and at the same time show a number of significant differences. A common feature is that their changes at constant pressure can be measured in one and the same experiment. Namely, when heat is added in a reversible manner to the system/substance. So, one can write

$$H(T) = H(\text{at } 0\,\text{K}) + \int_{0K}^{T} q_{rev}; \tag{5}$$

and similarly

$$S(T) = S(\text{at } 0\,\text{K}) + \int_{0K}^{T} \frac{q_{rev}}{T}. \tag{6}$$

One of the differences between the two is that entropy has a *natural zero point* - the entropy is zero at zero kelvin (←106) - whereas the enthalpy has not. Or, in other words, one can reduce Equation (6) to

$$S(T) = \int_{0K}^{T} \frac{q_{rev}}{T} \tag{7}$$

and speak of *absolute entropy*; there is, however, no absolute enthalpy. These facts are illustrated by Table 1, which displays data for diamond and graphite, two solid forms of carbon.

Table 1a: Molar thermodynamic properties of diamond and graphite at 1 bar, expressed in SI units (Robie 1978)

		diamond	graphite
entropy at zero K	$S(0K)$	0	0
entropy at 25 °C	$S(298.15K)$	2.38	5.74
enthalpy at zero K	$H(0K)$	?	?
enthalpy increase	$H(298.15K)-H(0K)$	523	1050
volume at 25 °C	$V(298.15K) \times 10^6$	3.417	5.298
heat capacity at 25 °C	$C_p(298.15K)$	6.13	8.53

Table 1b: Difference in molar thermodynamic properties for the two forms diamond and graphite of carbon. The numerical values are valid for 25 °C and 1bar and expressed in SI units; the difference is for diamond minus graphite

Volume	$\Delta V \times 10^6$	−1.881
heat capacity	ΔC_p	−2.40
Entropy	ΔS	−3.36
Enthalpy	ΔH	1895
Gibbs energy	ΔG	2900

On can safely speak of THE entropy of, e.g., diamond at 298.15 K and 1 bar, and it will be obvious that absolute entropy is meant. In the case of enthalpy, on the other hand, one has to proceed with caution. If one likes, one can say that THE enthalpy of diamond at zero K and 1 bar is −523 $J \cdot mol^{-1}$, but only so after having clearly specified that the zero point is at 298.15 K and 1 bar!
Furthermore, one is entitled to set, at the same time, the zero point for diamond and the zero point for graphite at 298.15 K and 1 bar. However, one is not entitled to do so if one wants to discuss the transition from diamond to graphite!

In that case one should use the data of Table 1b, which, by the way, are independent of zero points.

As a matter of fact, a *transition from diamond to graphite* is not a hypothetical event! At ambient conditions, or, more widely, at atmospheric pressure graphite has a lower Gibbs energy than diamond. From a thermodynamic point of view, therefore, diamond has the capacity of changing spontaneously into graphite (Although collectors of diamonds have no reason to fear that their precious stones will change into graphite, they should however know that it is not wise to heat these stones in a furnace. Because, at elevated temperature the *kinetic hindrance* for the transition may be less 'favourable'). At atmospheric pressure graphite is the *thermodynamically stable form*; diamond is the *metastable form*.

Next, let us consider a *chemical reaction* in which carbon is involved. As an example the reaction of graphite with oxygen to carbon dioxide, carried out at constant, given pressure and at a given temperature

$$C + O_2 \rightarrow CO_2 \qquad T,P \text{ const.}$$

The *enthalpy effect of* the *reaction* can be expressed in the enthalpies of the individual substances

$$\Delta H(T,P) = H_{CO_2}(T,P) - H_{O_2}(T,P) - H_C(T,P). \tag{8}$$

Similarly, for the entropy effect of the reaction

$$\Delta S(T,P) = S_{CO_2}(T,P) - S_{O_2}(T,P) - S_C(T,P). \tag{9}$$

In the case of Equation (9), the entropies of the individual substances have a well defined status and, as a result, equally so has ΔS. In the case of Equation (8), in spite of the zero point troubles of the individual substances, ΔH itself, like ΔS, has a well defined status. The heat of reaction, ΔH, is an experimental reality which, accordingly, has nothing to do with our choice of zero points. The other way round, the fact that *heats of reaction* are independent of zero points indicates the direction for a sensible *choice of zero points*. In, reality therefore, in the case of chemical reactions, the best thing to do is to assign zero values to the enthalpies of the elements - taking the form of the element which, at the selected T and P, has the lowest Gibbs energy. For example, in the case of the above reaction, at 298.15 K and 1 bar, the heat of reaction is −393.51 kJ per mole (CO_2). If we now say that H_{O2} and H_C (C for graphite) are zero for these T and P, then the enthalpy of CO_2, H_{CO2}, has the value of −393.51 kJ·mol^{-1}. This value is referred to as the *enthalpy of formation from the elements* of CO_2, for which we use the notation $\Delta_f H_{CO_2}$.

In the same terms, the entropy difference expressed by Equation (9) can be referred to as the *entropy of formation from the elements* of CO_2 :

$$\Delta_f S_{CO_2}(T,P) = S_{CO_2}(T,P) - S_{O_2}(T,P) - S_C(T,P).\tag{10}$$

And, for the *Gibbs energy of formation from the elements* of CO_2 we have

$$\Delta_f G_{CO_2}(T,P) = \Delta_f H_{CO_2}(T,P) - T \cdot \Delta_f S_{O_2}(T,P).\tag{11}$$

In the following we will use the expressions *enthalpy and Gibbs energy of formation* rather than *enthalpy and Gibbs energy of formation from the elements*.

The discussion, so far, on zero point matters is reflected in Table 2. It gives, for the substances considered, the numerical values of the properties C_p and S and the formation properties $\Delta_f H$ and $\Delta_f G$. It may be emphasized that the properties in Table 2 fall into two parts (A and B) with a different character. The A part consists of properties having an intrinsic nature, i.e. belonging to the substance (or the form) itself. The B part properties, on the other hand and as far as they are not zero, are related to other substances (or another form). The B part is needed for applications in which the appearance or disappearance of the substance (or form) is studied.

Table 2:   Molar thermodynamic properties and formation properties at 298.15 K and 1 bar, expressed in SI units (Robie 1978)

	C_P	S	$\Delta_f H$	$\Delta_f G$
C (graphite)	8.53	5.74	0	0
C (diamond)	6.13	2.38	185	2900
O_2 (as ideal gas)	29.37	205.15	0	0
CO (as ideal gas)	29.14	197.67	−110530	−137171
CO_2 (as ideal gas)	37.13	213.79	−393510	−394375
	A		B	

Gibbs energy at high temperatures and pressures

As next step, let us suppose that for a certain substance the Gibbs energy is known for the standard conditions of 25 °C and 1 bar, point A in Figure 1. Then, in order to obtain the Gibbs energy values at the points B, C and D we must carry out an integration on the basis of Equation (4)

$$dG = -S(T,P)\,dT + V(T,P)\,dP$$

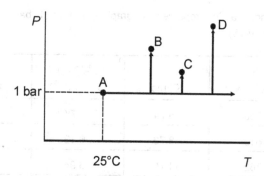

FIG.1.   Integration routes for the calculation of the Gibbs energies at B,
C, D

And, the most economical way to do this is to follow the indicated routes, i.e. to start isobarically and to finish isothermally. For the isobaric part of the routes the entropy of the substance at 1 bar pressure is needed as a function of temperature. For the various isothermal parts volume is needed at various temperatures as a function of pressure. To put it briefly, what is needed is S (1bar, T), or written as $S°(T)$, and $V(P,T)$.

influence of temperature at standard pressure

At this place we will discuss the obtainement of Gibbs energy values at 1 bar pressure as a function of temperature. Two items in particular: (i) the use of experimental data and (ii) the extrapolation of data to higher temperatures.
As remarked, the horizontal start corresponds to the determination of entropy at one bar pressure. The measurement therefore of the heat taken by the substance in equilibrium experiments (q_{rev}). And in practice this comes down to the measurement of *heat capacity C_p* and *heats of transition*.
As an example let us consider, again, the substance graphite. Its experimentally determined C_p, for $298.15 \leq (T/K) \leq 1800$ can be represented by the following formula (Robie et al.1978)

$$C_P^0 = \{63.160 - 1.1468 \times 10^{-2}(T/K) + 1.8079 \times 10^{-6}(T/K)^2 - 1.0323 \times 10^3 (K/T)^{1/2} \\ + 7.4807 \times 10^5 (K/T)^2 \} \ J \cdot K^{-1} \cdot mol^{-1} \quad (12)$$

In the interval from 298.15 K to 1800 K the substance does not undergo a phase transition, which means that the function values of the thermodynamic properties can be calculated by means of Equation (12). More precisely, from the data at 298.15 K and the use of the C_p-equation, for any temperature in the interval. Results are displayed in Table 3, which contains part of the numerical values from Robie's table.

Table 3:   Thermodynamic properties of graphite (C) at 1 bar pressure
             (Robie 1978)

T (K)	$(H^\circ - H_{298})/T$ $(J \cdot K^{-1} \cdot mol^{-1})$	S° $(J \cdot K^{-1} \cdot mol^{-1})$	$-(G^\circ - H^\circ_{298})/T$ $(J \cdot K^{-1} \cdot mol^{-1})$	C_P° $(J \cdot K^{-1} \cdot mol^{-1})$
298.15	0.000	5.74	5.74	8.53
400	2.610	8.73	6.12	11.92
500	4.762	11.70	6.94	14.70
1000	11.812	24.50	12.69	21.60
1500	15.488	33.73	18.24	23.70
1800	16.908	38.11	21.20	24.27

To go on, we consider a *compound* and as example silver oxide, Ag_2O. Thermodynamic data are available just to 500 K, see Table 4. The unavailability of data for higher temperatures is due to the fact that the substance decomposes into the element substances silver and oxygen.

To emphasize it again, the table consists of two parts having a different character. The data of the 'intrinsic' part A refer to the substance itself.

The data of the 'formation' part B are the enthalpy and Gibbs energy effects of the reaction

$$2 \, Ag + 1/2 \, O_2 \rightarrow Ag_2O \, .$$

The Gibbs energy effect, by the way, follows from the enthalpy effect $\Delta_f H$ and the entropies of the three individual substances Ag, O_2, and Ag_2O.

Table 4:   Thermodynamic table for silver oxide, Ag_2O. The data, taken from
             Barin (1989), are valid for the standard pressure (hence the
             superscript $^\circ$) of 1bar

T (K)	C_P° $(J \cdot K^{-1} \cdot mol^{-1})$	S° $(J \cdot K^{-1} \cdot mol^{-1})$	$H^\circ - H^\circ_{298}$ $(kJ \cdot mol^{-1})$	$\Delta_f H^\circ$	$\Delta_f G^\circ$	$\log K_f$ $(-)$
298.15	65.680	121.298	0.000	−31.049	−11.184	1.959
400.00	72.107	141.570	7.042	−30.729	−4.432	0.579
500.00	77.128	158.212	14.509	−30.002	2.065	−0.216
		A			B	

Let us suppose next that we would like to have for silver oxide the function values at 600 K for the properties in part A of Table 4 (needed e.g. for high-pressure work). Starting from the data at 500 K and taking C_P^o as a constant, we can write

$$H_{600}^o = H_{500}^o + C_{P\,500}^o (600 - 500)\,K\;;$$

$$S_{600}^o = S_{500}^o + C_{P\,500}^o \ln\frac{600}{500}\;.$$

And accordingly for $G_{600}^o = H_{600}^o - TS_{600}^o$, after rearranging

$$G_{600}^o = G_{500}^o - S_{500}^o(600 - 500)K - C_{P\,500}^o\left(600\ln\frac{600}{500} - 600 + 500\right)K\,.$$

Generally, for *extrapolation* from a *reference temperature* Θ to a temperature T

$$G(T) = G(\Theta) - S(\Theta)(T - \Theta) - C_P(\Theta)\left(T\ln\frac{T}{\Theta} - T + \Theta\right)\,. \tag{13}$$

This formula, which has the character of a *Taylor's series*, can easily be extended with terms that contain the derivatives of C_P with respect to temperature ($\rightarrow$Exc 6).

the influence of pressure

For the various vertical parts of the routes of integration, Figure 1, as was remarked before, the volume of the material (or $\Delta_f V$ in the case or formation properties) has to be known at various temperatures as a function of pressure. Or briefly, V has to be known as a function of T and P. Volume V itself, so to say, and its change with T and P

$$dV = \left(\frac{\partial V}{\partial T}\right)_P dT + \left(\frac{\partial V}{\partial P}\right)_T dP\;; \tag{14}$$

or,

$$dV = \alpha V\,dT - \kappa V\,dP\,, \tag{15}$$

where α is the *cubic expansion coefficient* and κ the *isothermal compressibility* ($\leftarrow$104). In terms of logarithm of volume

$$d\ln V = \alpha\,dT - \kappa\,dP\,. \tag{16}$$

In practice, therefore, V itself, α and κ have to be known - have to be measured - as a function of T and P.

From the point of view of knowing V as a function of T and P, the *ideal gas*, is the most cooperative kind of material: its volume as a function of T and P is fully known,

$$V = R\frac{T}{P}.$$ (17)

The corresponding change in Gibbs energy, at a given temperature T_a, as a result of a change in pressure form 1 bar to another pressure P, is simply given by ($\leftarrow$108)

$$G_{T_a}(P) = G^o_{T_a} + RT_a \ln P.$$ (18)

It is clear that the availability of recipes for V as a function of T and P, read *equations of state*, is of paramount importance. The famous *Van der Waals equation of state* for fluid materials has been the starting point of a whole family of *cubic equations*. This family of equations is generally represented by the following formula, which, implicitly, gives volume as a function of T and P; see Reid et al. (1987); Papon et al. (2002):

$$\left(P + \frac{a}{V^2 + ubV + wb^2}\right)(V - b) = RT.$$ (19)

For $u = w = 0$ the Van der Waals equation is obtained. The equation with $u=1$ and $w=0$ is the *Redlich-Kwong equation*. The *Peng-Robinson equation* has $u=2$, $w=-1$.

nota bene

In this work, numerical data - in examples and exercises - mainly are from the two data collections Barin (1989) and Robie et al. (1978), indicated as (Robie 1978).

Basic thermodynamic properties of pure substances are their entropies and heat capacities at standard pressure, volumes as a function of temperature and pressure, heats of formation and the heat effects involved in changes of state.

EXERCISES

1. *partial derivatives of H-TS*

    Ignoring that $H - TS$ equals G, show that for a function f which is defined by $f = H - TS$ the partial derivatives with respect to T and P are given by $-S$ and V, respectively.

2. *completion of a table*

Complete the table showing standard (25 °C, 1 bar) thermodynamic properties. $S°$ and $\Delta_r S°$ in $J \cdot K^{-1} \cdot mol^{-1}$ and $\Delta_f H°$ and $\Delta_f G°$ in $kJ \cdot mol^{-1}$.

substance	$S°$	$\Delta_f H°$	$\Delta_r S°$	$\Delta_f G°$
C (graphite)	5.74	0		
Cu	33.15			
O_2	205.15			
CO	197.67			-137.17
CO_2	213.79	-393.51		-394.38
CuO		-157.32		-129.56

3. *the essential information of a table*

Reduce the table to the one with the smallest possible number of independent numerical values. The data are for 25 °C and 1 bar. The subscript *fox* stands for *formation from the oxides*. $S°$ is in $J \cdot K^{-1} \cdot mol^{-1}$ and the other properties in $kJ \cdot mol^{-1}$.

	$S°$	$\Delta_f H°$	$\Delta_f G°$	$\Delta_{fox} H°$	$\Delta_{fox} G°$
Mg	32.68	0	0	X	X
O2	205.15	0	0	X	X
MgO	26.94	-601.49	-569.196	0	0
SiO_2,quartz	41.46	-910.70	-856.288	0	0
Mg_2SiO_4	95.19	-2170.37	-2051.325	-56.69	-56.645

4. *thermodynamic table for corundum*

At 25 °C and 1 bar the enthalpy effect of the reaction $4Al + 3O_2 \rightarrow 2Al_2O_3$ is -3351.4 $kJ \cdot mol^{-1}$. The entropies and heat capacities of the substances involved in the reaction are given below; valid for 25 °C and 1 bar and expressed in $J \cdot K^{-1} \cdot mol^{-1}$.

- Which values will be found in the thermodynamic table for corundum for $\Delta_f H°$ and $\Delta_f G°$ at 298.15 K and 1 bar, and at 400 K and 1 bar.

		S^o	C_P^o
aluminium	Al	28.35	24.31
oxygen	O_2	205.15	29.37
corundum	Al_2O_3	50.92	79.01

5. *thermodynamic table for helium from the Sackur-Tetrode equation*

Use the Sackur-Tetrode equation, Equation (106:14), to answer the following question.
- Which numerical values will be found in the thermodynamic table for helium (with M = 4.003 $g\,mol^{-1}$) for the quantities: $(H_T^o - H_{298}^o)/T$; S_T^o and $-(G_T^o - H_{298}^o)/T$ for T = 1000 K?

The superscript ^{o} refers to the standard pressure of 1 bar.

6. *extrapolation formula for Gibbs energy*

The *extrapolation formula for Gibbs energies*, Equation (3), can be extended with terms in $C'(\Theta)$, $C''(\Theta)$, and so on. C' is the first partial derivative of C_P with respect to temperature and C'' the second.

- Show that the term in C' is

$$C'(\Theta) = \left[\frac{1}{2}\Theta^2 - \frac{1}{2}T^2 + \Theta T \ln\left(\frac{T}{\Theta}\right) \right]$$

The term in C'' is

$$C''(\Theta) = \left[-\frac{1}{12}T^3 + \frac{1}{2}T^2\Theta - \frac{1}{4}T\Theta^2 - \frac{1}{6}\Theta^3 - \frac{1}{2}\Theta^2 T \ln\left(\frac{T}{\Theta}\right) \right]$$

See also Clarke and Glew (1966).

7. *silver oxide*

Fit the constants a and b in $C_P^o = a + bT$ to the C_P^o data for Ag_2O in Table 4. Use the result obtained, along with the data for 298.15 K, to calculate the values of $H_T^o - H_{298}^o$ and S_T^o for T = 400 K and T = 500 K. Take into consideration the number of digits, needed for a and b, in order to obtain the precision of Table 4, i.e. to three decimal places. Compare the values you calculate with the corresponding ones in Table 4.
See foregoing Exc.

8. *water versus steam*

In Robie's (1978) compilation there are two tables for the substance H_2O, which are headed *water* and *steam*, respectively. The headings of the tables are detailed as: 'water' (H_2O: liquid 298.15 to 372.8 K. Ideal gas 372.8 to 1800 K; 'steam' (H_2O: Ideal gas 298.15 to 1800 K).

- For which properties out of:

$$\{(H_T^\circ - H_{298}^\circ)/T;\ S_T^\circ - (G_T^\circ - H_{298}^\circ)/T;\ C_P^\circ;\ \Delta_f H^\circ;\ \Delta_f G^\circ\}$$

the same values will be found in the two tables; and for which properties different values will be found?

Answer this question for $T = 298.15$ K and for $T = 1800$ K.

9. *α-quartz - molar volume and compressibility*

The *crystallographic unit cell* of α-quartz contains three SiO_2 'molecules'. Its volume, at room temperature, is given as a function of pressure (D'Amour et al. 1979).

P (kbar)	V (10^{-30}m^3)
0.01	113.2
19	108.1
37	104.2
54	102.4
60	100.8
68	99.8
73	98.8

- Calculate the molar volume of α-quartz at room temperature.
- From $\ln V$ versus P calculate the isothermal compressibility of the mineral.

10. *α-quartz - Gibbs energy at high pressure*

For α-quartz at 25 °C calculate the change in Gibbs energy resulting from a change in pressure from 1 bar to 50 kbar,

- using only volume;
- using volume and isothermal compressibility.
- Mark the difference between the two results in terms of percentage.

Data: $V = 22.7 \times 10^{-6}$ m^3·mol^{-1}; $\kappa = 1.8 \times 10^{-11}$ Pa^{-1}.

11. *the change in Gibbs energy resulting from a change in pressure*

Derive an equation which allows, for isothermal conditions, the calculation of the Gibbs energy at pressure P from the Gibbs energy and its pressure derivatives at a reference pressure π, i.e. from $G(\pi)$, $V(\pi)$, $V'(\pi)$ and $V''(\pi)$, where $V' = (\partial V/\partial P)_T$ and $V'' = (\partial V/\partial P)_T$.

12. *bulk modulus*

In the Handbook of Physics and Chemistry one can find the following on *bulk modulus:* "The modulus of volume elasticity

$$M_B = \frac{P_2 - P_1}{\dfrac{V_1 - V_2}{V_1}}$$

where P_1, P_2; V_1, V_2 are the initial and final pressure and volume, respectively".

- Derive the relations between the set [κ(isothermal compressibility), $(\partial V/\partial P)_T$, and $(\partial^2 V/\partial P^2)_{TT}$] and the set [$M_B$, and $(\partial M_B/\partial P)_T$].

13. *forsterite at high pressure*

Y. Bottinga (1991) gives the following data for forsterite, Mg_2SiO_4, at its melting point (2163 K at 1 bar).
Volume = 47.13 cm^3·mol^{-1};
isothermal bulk modulus = 0.8372 Mbar;
pressure derivative of isothermal bulk modulus = 5.33

- Calculate {G(2163 K, 50 kbar) − G(2163 K, 1 bar)}.

See foregoing exercises.
Use the Taylor's series formula and compare the influences of the three terms.

§ 110 PURE SUBSTANCES

The equilibria considered are between different forms of a pure substance. There are just two variables: temperature and pressure. The thermodynamic potential in this case is identical with molar Gibbs energy.

system formulation - G^*PT surfaces

In the case of a pure substance, the variables of the *set M of variables* are just temperature and pressure:

$$M = M\,[P,\,T].$$

(1)

The equilibrium conditions, appearing in the *set N of equilibrium conditions*, invariably are of the type

$$\mu_B^\alpha = \mu_B^\beta,$$

(2)

in which B denotes the substance examined and α and β are symbols to denote different phases.

Difference in phase not necessarily comes down to difference in state of aggregation. For example, α and β can denote two different solid phases, such as graphite and diamond for B = carbon. As a matter of fact, the different forms in which a substance can manifest itself can be quite numerous. But no matter the numerousness of the forms, the maximum number of phases that can be in thermodynamic equilibrium is three, and no more. With three phases in equilibrium the *system formulation* is

$$f = M[T,P] - N[\,\mu_B^\alpha = \mu_B^\beta = \mu_B^\gamma\,] = 0;$$

(3)

and it means that the degrees of freedom are used up.

Let's first consider the coexistence of two phases, α and β. For two phases in equilibrium there is one degree of freedom:

$$f = M[T,P] - N[\,\mu_B^\alpha = \mu_B^\beta\,] = 2 - 1 = 1.$$

(4)

The set of equilibrium states is the solution of the equation

$$\mu_B^\alpha(T,P) = \mu_B^\beta(T,P),$$

(5)

which, in the case of a pure substance B, is equivalent to

$$G_B^{*\alpha}(T,P) = G_B^{*\beta}(T,P):$$

(6)

equality of (molar) Gibbs energies as condition for equilibrium. The molar Gibbs energies $G_B^{*\alpha}$ and $G_B^{*\beta}$, as functions of T and P, correspond to two surfaces in G^* TP space, see Figure 1. These surfaces have a positive slope in the P direction and a negative slope - increasing with T in an absolute sense - in the T direction, because of

$$(\partial G_B^* / \partial P)_T = V_B^* \text{ , and}$$ (7)

$$(\partial G_B^* / \partial T)_P = -S_B^* .$$ (8)

At the intersection of the two surfaces the equilibrium condition, Equation (6), is satisfied. Therefore, the projection of the intersection on the PT plane is the representation of the set of equilibrium states. We will refer to it as the *two-phase equilibrium curve*.

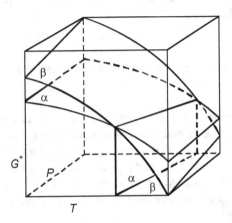

FIG.1. G^*PT surfaces for two forms, α and β, of a pure substance

It may be stressed, once more, that in a PT phase diagram the two-phase equilibrium curve has a double function. In the first place it is the representation of the solution of Equation (6); the representation of the set of equilibrium states. In the second place it has the property of dividing the PT plane into two parts: an α-field and a β-field. For α-field conditions the form α will NEVER change spontaneously into the form β, as $G^{*\alpha} < G^{*\beta}$. The β form, on the other hand, when exposed to α-field conditions, can change spontaneously into α.

The addition of a third form γ gives rise to a third surface in G^*TP space and, as a consequence, to two additional two-phase equilibrium curves in the PT projection, see Figure 2. The point of intersection of the tree two-phase equilibrium curves is the so-called *triple point*. The triple point represents the invariant equilibrium, Equation (3), between the three phases α, β and γ.

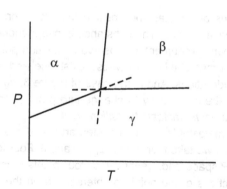

FIG.2. Elementary PT phase diagram of a pure substance

stable and metastable

The two diagrams in Figure 3 are isobaric sections of G^*TP space: G^*T *diagrams* in which for convenience the Gibbs energies are represented by straight lines. G^*T (and G^*P) diagrams are very instructive when discussing the relative stabilities of the forms; especially when their number is increasing.

In the case of the diagram at the left-hand side, α and β are two solid forms and the third form is liquid: α is the stable form below T_1; β is the stable form between T_1 and T_3; and liquid is the stable form above T_3. For the situation of the diagram at the right-hand side, α is stable below T_3 and liquid is stable above T_3, whereas β for no temperature has the lowest Gibbs energy. Notwithstanding that, β may have a real existence and change into liquid at T_2.

In what follows, we will use the qualification *metastable* for everything above the solid lines in the G^*TP space. In the left-hand case the form β is metastable below T_1 and above T_3; and T_2 is a *metastable melting point*. The temperature T_1 in the case of the right-hand diagram, is a *metastable transition point*.

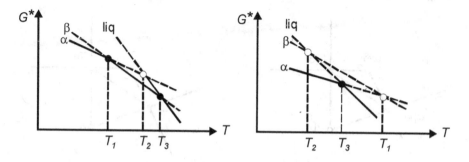

FIG. 3. G^*T diagrams showing stable (filled circles) and metastable
          (open circles) transition points

Metastable forms and metastable melting and transition points can have a real as well as a virtual existence. The substance benzophenone has a metastable form, which has its own, experimentally observable melting point. Naphthalene on the other hand has a metastable form whose existence and melting point follow from theoretical considerations. And in the case of Figure 3, right-hand side, α has melted before it gets the opportunity to change into β at T_1. And if one likes, one can refer to T_1 as a *virtual metastable transition point*.

The adjectives metastable and its counterpart stable are also used for triple points. To discuss this, we take four forms, α, β, γ and δ. Four forms correspond to four surfaces in G^*TP space and, generally, to four points of intersection (of three surfaces). The projections of the points of intersection on the PT plane are triple points. The four points of intersection in G^*TP space are the vertexes of a tetrahedron. With respect to the 'lower side' (the lowest possible G^* as a function of T and P) of G^*TP space, that tetrahedron can have three different positions: it can share with that lower side (i) only one vertex, (ii) two vertexes and (iii) three vertexes. The projections of the three possibilities are shown in Figure 4; a filled circle represents a stable triple point and an open circle a metastable one.

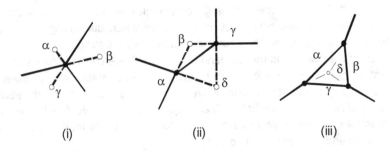

(i)   (ii)   (iii)

FIG.4.   The three possible configurations of stable (filled circles) and
           metastable (open circles) triple points shared by four forms

The case (ii), with two stable and two metastable triple points, is shown two times more.

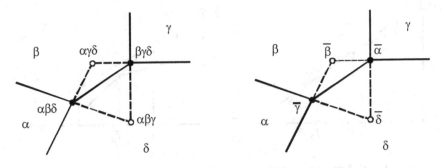

FIG.5.   Two notations for triple points: left, 'normal' notation; right, absent
           notation

In each of the two figures the 'names' of the triple points are inserted. At the left-hand side an obvious notation is used: α β δ represents the triple point at which α, β and δ coexist. The fourth form γ is absent at the α β δ triple point: its absence can be indicated by γ̄. In the right-hand figure the *absent notation* is used to name the triple points. And herewith we observe and important property: the metastable triple point φ̄ (φ is for form) is in the field where φ is the stable form. Similarly, the *metastable extension* of the two-phase equilibrium curve has to run into the field at which φ is the stable form.

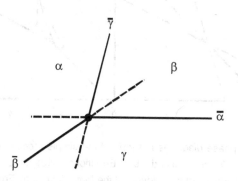

FIG.6.   Pertaining to the rule for metastable extensions of two-phase
          equilibrium curves

a simple model

The Gibbs energy of one mole of pure substance, in a given form α, is defined by

$$G^{*\alpha} = U^{*\alpha} - TS^{*\alpha} + PV^{*\alpha}, \tag{9}$$

in which energy U, entropy S and volume V are functions of T and P. If we treat U^*, S^* and V^* as constants, we obtain a simple model with the help of which a number of phenomena can be surveyed in a handy manner. And to accentuate the simplification, G^* is replaced by Z and T and P by X and Y.
To start with, we take three forms, α, β and γ, defined by

$$Z^{\alpha} = -X + Y + 1 \tag{10}$$
$$Z^{\beta} = -4X + 2Y + 6 \tag{11}$$
$$Z^{\gamma} = -6X + 6Y + 2. \tag{12}$$

The phase diagram for this set of properties is Figure 7, left-hand side. Its (α + β) equilibrium line is the solution of the equation $Z^{\alpha} = Z^{\beta}$. The two equilibrium lines for (α + γ) and (β + γ) follow from $Z^{\alpha} = Z^{\gamma}$, and $Z^{\beta} = Z^{\gamma}$, respectively. The triple point is

§ (110)

the solution of the two equations implied in $Z^\alpha = Z^\beta = Z^\gamma$. The three equilibrium lines divide the XY plane into three fields. In each of these fields the Z values of one of the forms are lower than the Z values of the other two forms.

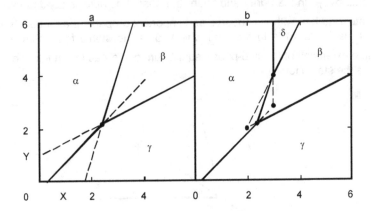

FIG.7.   Two phase diagrams in terms of the simple model. Left, for the three forms defined by Equations (10)-(12); right, the consequence of the addition of the form, defined by Equation (13)

The addition of a fourth form δ will give rise to three more triple points and possibly to a fourth single-phase field which is stable. The phase diagram resulting from the addition of the form δ defined by

$$Z^\delta = -2X + Y + 4 \tag{13}$$

is shown in Figure 7, right-hand side.

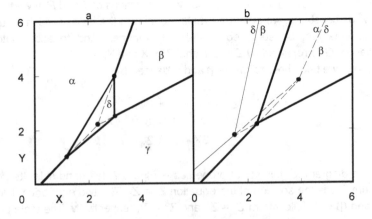

FIG.8.   Phase diagrams representative of enantiomorphism (left) and monotropism (right)

Two more phase diagrams are shown in Figure 8. The diagram at the left-hand side is obtained for the fourth form defined by

$$Z^\delta = -2\frac{2}{3}X + 2Y + 2. \tag{14}$$

And the diagram at the right-hand side by

$$Z^\delta = -2.35X + 1.59Y + 4.25. \tag{15}$$

It is evident that the forms α, β and γ, defined above and when taken together, are representative, of solid, liquid and vapour in that order.

To go on, we take δ as a second solid form. The phenomenon that a substance can manifest itself in more than one solid form is referred to as *polymorphism*. In the case of Figures 7, right-hand side, and 8, left-hand side, which show stability fields for α as well as for δ, it is possible to transform the forms into one another in a reversible manner. In the case of Figure 8, right-hand side, on the other hand, it is not possible to transform α into δ in a direct manner, and the transition form δ to α is irreversible, whenever it takes place. These two kinds of different behaviour are referred to as *enantiomorphism* when a reversible transition is possible, and *monotropism*, when a reversible transition is impossible.

Clapeyron's equation

For any point A on the $(\alpha + \beta)$ equilibrium curve, see Figure 9, the (molar) Gibbs energy of the form α is equal, has to be equal to the (molar) Gibbs energy of the form β

$$G^{*\alpha} = G^{*\beta}. \tag{16}$$

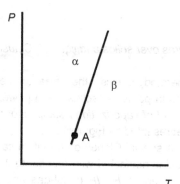

FIG.9.   Two-phase equilibrium

The fact that the equilibrium condition must be fulfilled for A and for all other points on the equilibrium curve means that, along the curve, the change in $G^{*\alpha}$ must be equal to the change in $G^{*\beta}$:

$$dG^{*\alpha} = dG^{*\beta};$$ (17)

or, after inserting the total differentials in terms of T and P,

$$-S^{*\alpha}dT + V^{*\alpha}dP = -S^{*\beta}dT + V^{*\beta}dP;$$ (18)

or

$$(S^{*\beta} - S^{*\alpha})\,dT = (V^{*\beta} - V^{*\alpha})\,dP.$$ (19)

As a consequence, the slope of the equilibrium curve is given by

$$\frac{dP}{dT} = \frac{(S^{*\beta} - S^{*\alpha})}{(V^{*\beta} - V^{*\alpha})} = \frac{\Delta S^*}{\Delta V^*}.$$ (20)

This relation is called Clapeyron's equation.
In the Δ notation the equilibrium condition, Equation (16), is

$$\Delta G^* = G^{*\beta} - G^{*\alpha} = 0.$$ (21)

And because $\Delta G^* = \Delta H^* - T\Delta S^*$, we have for conditions on the equilibrium curve $\Delta S^* = \Delta H^*/T$ (of course: on the equilibrium curve α can be transformed into β in a reversible manner, and, change in entropy is reversible heat divided by temperature) so that the relation can be given as

$$\frac{dP}{dT} = \frac{\Delta H^*}{T\Delta V^*}.$$ (22)

As an example, for the melting curve in the PT phase diagram (β = liquid; α = solid), ΔH^* is the heat of melting and ΔV^* the difference in volume of the liquid and the solid.

equilibrium vapour pressures over solids and liquids - Clausius-Clapeyron plot

In experimental thermodynamics, the measurement of the equilibrium pressure over liquids - and in particular solids - is a powerful, indirect method for the determination for the *heat of vaporization / sublimation*. An example of a set of vapour pressure data is represented by Figure 10.
For this class of equilibria a special Clapeyron equation can be derived ($\leftarrow$004). If the volume of solid or liquid is neglected with respect to the volume of the vapour and if, next, the latter is taken as RT/P (ideal-gas approximation), the general equation, Equation (22) will change into

$$\frac{dP}{dT} = \frac{P\Delta H^*}{RT^2} \quad \text{or} \quad \frac{d\ln P}{dT} = \frac{\Delta H^*}{RT^2}.$$ (23)

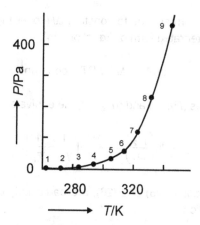

FIG.10. Showing that vapour pressures increase in an exponential
manner with temperature. The nine data points pertain to the
vapour pressure over crystalline naphthalene (Ambrose et al.
1975)

A further simplification is obtained with the help of $d(1/T) = -T^{-2} dT$:

$$\frac{d\ln P}{d(1/T)} = -\frac{\Delta H^{\bullet}}{R} \tag{24}$$

From the last expression, Equation (24), it follows that it is advantageous to plot
vapour-pressure data in an $\ln P$ versus $1/T$ diagram, i.e. a so-called *Clausius-Clapeyron plot*. In that case the data points, in first approximation, are on a straight
line; and the more so the smaller the temperature range of the data and the lower
the pressure; see Figure 11.

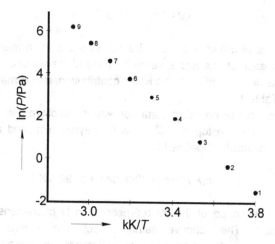

FIG.11.   Clausius-Clapeyron plot of the data points in Figure 10

The straight line, which corresponds to constant ΔH^* over the (small) temperature range, is given by the integrated form of Equation (24):

$$\ln P = -\Delta H^* / RT + \text{constant} . \qquad (25)$$

For two equilibrium states (P_1, T_1) and $((P_2, T_2)$, the equivalent equation is

$$\ln\left(\frac{P_2}{P_1}\right) = -\frac{\Delta H^*}{R}\left(\frac{1}{T_2} - \frac{1}{T_1}\right) . \qquad (26)$$

These equations, Equations, (25) and (26), are particularly useful for interpolation and (modest) extrapolation.

The Clausius-Clapeyron plot is a powerful representation of the dependence on temperature of quantities - like pressure, *equilibrium constant*, and *distribution coefficient* - of which the logarithm is nearly linear in reciprocal temperature. The representation is rooted in thermodynamics, and owing to its straight-line appearance it has become very popular. Its attractiveness, in a sense, is also its weakness - because of the fact that the representation does not reveal the fine structure in a set of experimental data.

the arc representation

A fact is that, due to the rapid change of (e.g.) pressure with temperature, several units on the vertical axis are needed to represent, in a Clausius-Clapeyron plot, the outcome of an average investigation. The 'uneconomical' use of the vertical axis can be repaired by the addition of a *linear contribution* in $1/T$, that is to say, by replacing $\ln P$ by

$$\ln f = \ln P - \alpha + \beta / T , \qquad (27)$$

of which the constants α and β have to be adjusted in such a manner that $\ln f$ is (close to) zero for each of the two extreme TP pairs of the set of data (Oonk et al. 1998). To demonstrate the effect of the linear contribution, use is made of the data set displayed in Table 1.
In Figure 12 the vapour-pressure data for water, displayed in Table 1, are represented two times: on top the Clausius-Clapeyron plot, and at the bottom side in the *$\ln f$ representation*, defined by

$$\ln f = \ln(P / \text{Torr}) - 20.93641 + 5299.5 \text{ K} / T . \qquad (28)$$

The rainbow/arc like shape of the $\ln f$ representation is characteristic of a high-quality set of data. The concave nature reflects the general fact that for liquid+vapour equilibria the heat-capacity difference is negative: heats of vaporization decrease with increasing temperature

Table 1:   Vapour pressures over liquid water in the range from 10 to 40 °C

T /K	P /Torr
283.15	9.209
288.15	12.788
293.15	17.535
298.15	23.756
303.15	31.824
308.15	42.175
313.15	55.324

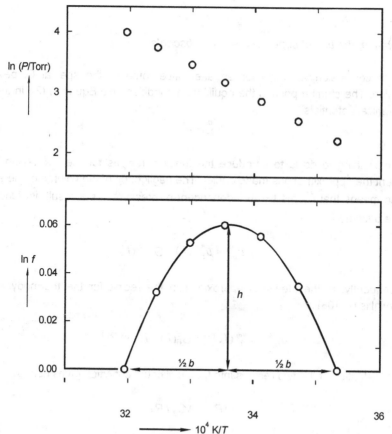

FIG.12.   Two graphical representations of the vapour-pressure data in
           Table 1. Top: the traditional Clausius-Clapeyron plot. Bottom: the
           arc representation

After differentiation of Equation (27), and substitution of Equation (24), the following equation is obtained for the slope of the ln$f$ function

$$d\,(\ln f)/d(1/T) = -\Delta H_B^o / R + \beta.$$ (29)

For the temperature at the top of the *arc* the heat of vaporization is equal to the product of the gas constant R and the parameter β. At the left-hand side, which is the high-temperature side, the slope is positive and the heat of vaporization is smaller than at the top. At the right-hand side these things are the other way round. As a matter of fact, the difference in heat capacity between vapour and liquid can be calculated from the characteristics of the arc (see Figure 12): its height h, base length b, and the reciprocal temperature at the top $1/T_{max}$:

$$\Delta C_{P_B}^o = -8R \{T_{max} \cdot b\}^{-2} \cdot h$$ (30)

decoration, the use of superscripts and subscripts

In conclusion we consider an alternative route to the special Clapeyron equation. The starting point is the equilibrium condition, like Equation (2), in terms of chemical potentials

$$\mu_B^{sol} = \mu_B^{vap}$$ (31)

The next thing to do is to introduce the function recipes for the μ's, taking into account the approximations made earlier. The negligence of the volume of the solid phase means that μ_B^{sol} is taken independent of pressure. As a result, its function recipe becomes

$$\mu_B^{sol} = \mu_B^{sol}(T) = G_B^{*sol}(T).$$ (32)

Subsequently, in the ideal-gas approximation the recipe for the thermodynamic potential is ($\leftarrow$108)

$$\mu_B^{vap} = \mu_B^{vap}(T,P) = G_B^o(T) + RT \ln P.$$ (33)

The substitution of the two recipes into the equilibrium condition gives rise to

$$\ln P = -\Delta G_B^o / RT,$$ (34)

and with $\Delta G^o = \Delta H^o - T\Delta S^o$ we have

$$\ln P = -\Delta H_B^o / RT + \Delta S_B^o / R.$$ (35)

Differentiation of ln$P$, Equation (34), with respect to T - remembering that the derivative of G/T is given by $-H/T^2$ (Equation 107:26) - yields

$$\frac{\mathrm{d}\ln P}{\mathrm{d}T} = \frac{\Delta H_B^o}{RT^2}.$$

(36)

The analogue of this equation is Equation (23) and the analogue of Equation (35) is Equation (25). We observe that the two equations and their analogues differ as to the *decoration* of the symbols; and, in addition, the constant in Equation (25) is equal to $\Delta S_B^o / R$.

As regards decoration: in Equations (23) and (36) the superscripts of H are different; and in the former of the two equations the subscript for substance B is missing. It is obvious that in a section on one single pure substance it is not of eminent importance to take along the subscript-for-substance all the time. In the case of Equation (36), B was introduced in Equation (31) as a subscript of μ. It is a good thing to realize that μ, i.e. the chemical potential, is a property that belongs to a given substance, or chemical, or chemical species (either in pure form or as a component of a mixture). It is appropriate to stress this fact when working with the symbol μ; in particular when formulating the equilibrium conditions of the N set. In a subsequent step, that is to say when function recipes are going to be introduced, it is early enough to decide if it is necessary or not to take along the subscript(s). The ΔH^* in Equations (23-26) and ΔH^o in Equations (35-36), and defined as $\Delta H^* = H^{*\,vap} - H^{*\,sol}$ and $\Delta H^o = H^{o\,vap} - H^{*\,sol}$, respectively, differ in $H^{*\,vap}$ and $H^{o\,vap}$. Again, the appearance of each of the two superscripts is related to the route followed during the derivation. In fact, the two quantities $H^{*\,vap}$ and $H^{o\,vap}$ are identical for the case considered: the enthalpy of the ideal gas is independent of its pressure, i.e. $(\partial H/\partial P)_T = 0$. As a result the superscript in H^o, which refers to a chosen unit of pressure, is meaningless. Of course, this observation does *not* hold true for the superscripts in S^o and G^o.

conservation of decoration

As a final remark, once a certain decoration has been selected, it should be carried along from the beginning to the end of the operation. The other way round, all thermodynamic operations, like the derivation of thermodynamic properties from $G(T,P)$, remain valid no matter the decoration of G.

Forms, in which a pure substance manifests itself, may have their own stability field in the PT phase diagram; outside its field the form is metastable. There are forms that are metastable under all circumstances. Two stability fields, when neighbours, share a two-phase equilibrium curve; the slope of the curve is related to the differences in entropy and volume of the two forms (Clapeyron's equation). The maximum number of forms that can coexist is three; the three-phase equilibrium corresponds to a triple point in the phase diagram.

---

§ (110)

EXERCISES

1. **$G*T$ diagrams around a triple point**

For the three forms α, β and γ of a pure substance sketch $\overset{\cdot}{G}T$ diagrams - for a temperature interval around the α β γ triple point temperature - for three different pressures; that is to say, above, at, and below the pressure of the triple point.

2. **benzophenone, a monotropic substance**

Benzophenone (diphenylketone, $C_6H_5COC_6H_5$) has two solid forms, denoted by α and β.
- For $0 < t\,/^{\circ}C < 70$ make a sketch of the $\overset{\cdot}{G}T$ diagram of the substance, using the data given.
- Where do you expect the metastable (α to β) transition point?

	α	β
melting point (°C)	48.0	26.1
heat of melting (kJ·mol^{-1})	16.7	13.8

3. **a phase diagram analogue**

For five forms, in terms of the simple model defined by
$Z^{\alpha} = -X + Y + 1$
$Z^{\beta} = -4X + 2Y + 6$
$Z^{\prime} = -6X + 6Y + 2$
$Z^{\delta} = -2X + Y + 4$
$Z^{\varepsilon} = -X + 3Y - 1$
- Construct the (stable) phase diagram. Take $0 \le (X,Y) \le 6$.

Clue: first calculate the coordinates of the triple points and find out which of the triple points are stable.

4. **a negative degree of freedom?**

In the realm of thoughts it is quite easy to let the $\overset{\cdot}{G}TP$ surface of a form δ pass through the point of intersection of the three $\overset{\cdot}{G}TP$ surfaces of the forms α, β and γ. In that case - of which no real counterpart exists - the four triple points (α β γ), (α β δ), (α γ δ), and (β γ δ) coincide to one quadruple point, corresponding to a variance $f = 1-4+2$ of minus one?!

- To get an idea, calculate the 'phase diagram' for the four forms, which, in terms of the simple model, are defined by

$$Z^\alpha = -X + Y + 1$$
$$Z^\beta = -4X + 2Y + 6$$
$$Z^\gamma = -6X + 6Y + 2$$
$$Z^\delta = -2X + 5Y - 5.4$$

5. Antoine equation for 1-aminopropane

The vapour pressure of liquid 1-aminopropane as a function of temperature, for $20 < t/°C < 80$, can be represented by the Antoine equation

$$\ln(P/\text{Torr}) = A - B\{(T/K) - C\}^{-1};$$

with A = 15.9576, B = 2408.66, C = 62.060 ($\leftarrow$Exc 004:9).

- Calculate the heat of vaporization of the substance at its normal boiling point.

6 heat of melting along the melting line

Derive an expression for the change of the heat of melting with pressure along a substance's melting line. *Clue*: first formulate the total differential of enthalpy in terms of temperature and pressure.

7. vapour pressure over 1,4-dibromobenzene

In the temperature range $0 \le t/°C \le 25$ the equilibrium vapor pressure over solid p-dibromobenzene (B) can be represented by

$$\ln(P/\text{Pa}) = -8900 \text{ kelvin}/T + 31.928$$

The entropy of solid B at 25 °C is 193.10 J·K^{-1}·mol^{-1}.

- Supposing that you wish to assign the value zero to the enthalpy of solid B at 25 °C, what are the values of the following properties of ideal gaseous B: H°, S° and G° valid for 25 °C and 1 Pa.
- How do these three quantities change when the standard pressure is changed from 1 Pa to 1 atm?.

8. monoclinic and orthorhombic sulphur

Referred to 1 bar and 25 °C the entropy difference between monoclinic sulphur (m) and orthorombic sulphur (o) is $\Delta S = S^m - S^\circ = 1.09$ J·K^{-1}·mol^{-1}. Under the same circumstances the heats of combustion (to SO_2) are −297.21 kJ·mol^{-1} for m and −296.81 kJ·mol^{-1} for o.

- Calculate ΔG for 1 bar and 25 °C.
- Neglecting C_P, calculate the temperature at which the two forms will be in equilibrium under 1 bar pressure.
- Which value will be found in the thermodynamic table for SO_2 for its enthalpy of formation at 1 bar and 25 °C?.

9. *caesium chloride*

The substance caesium chloride (CsCl) has two different solid forms, which are denoted here by α and β. The difference in Gibbs energy between liquid and α and between liquid and β as a function of temperature and at 1 bar pressure are

$$\Delta_\alpha^{liq} G \equiv G^{liq} - G^\alpha = \{\ 20116 - 22.03\,(T/K)\ \}\ \text{J·mol}^{-1}$$

$$\Delta_\beta^{liq} G \equiv G^{liq} - G^\beta = \{\ 23016 - 25.94\,(T/K)\ \}\ \text{J·mol}^{-1}$$

- Calculate the temperatures and the heat effects of the three transitions
  $\alpha \rightarrow$ liq,
  $\beta \rightarrow$ liq and $\alpha \rightarrow \beta$.

10. *diamond out of graphite*

Let's suppose that you think of using high pressure to transform at 1700 K graphite into diamond. What order of magnitude of pressure do you need? The information which is put at your disposal is minimal: just the Gibbs energy of formation of diamond from graphite at 1700 K, 1 bar, which is 9034 J·mol⁻¹, and the molar volumes of graphite and diamond at 298 K, 1 bar, which are 5.298 and 3.417 cm³·mol⁻¹, respectively.

11. *heat capacity change from the shape of the arc*

The relation between the heat-capacity difference and the characteristics of the arc, Equation (30), can be found by Taylor's series expansion of ln f from the maximum, and truncating after the second-derivative's term.

- Write down the derivation of the equation.

12. *the water arc*

Apply Equation (30) to the arc, Figure 12, for the vapour pressures over water in the range from 10 to 40 °C. NB at 25 °C the heat capacities of gaseous and liquid water are (33.590 and 75.288) J·K⁻¹·mol⁻¹, respectively.

13. *a different arc*

In this exercise the trick of the linear contribution - to construct the arc representation - has to be applied to the logarithm of vapour pressure as a function temperature (not its reciprocal):

$$\ln f = \ln P - \alpha + \beta\,T.$$

- Show that in this case the relation between the arc characteristics and the difference in heat capacity is given by

$$\Delta C_P = -2\, R\, \Theta \,\{(4h\Theta/b^2) + \beta\},$$

where Θ stands for the temperature at the top of the arc.

- Construct the arc for the data in Table 1, and from its characteristics evaluate the value of the difference in heat capacity ΔC_P.

14. naphthalene: the assessment of a data set

The nine data points in Figure 10, representing the vapour pressures over crystalline naphthalene, are given by the following pairs (T /K; P /Pa) of numbers: (263.61; 0.23); (273.16; 0.74); (283.14; 2.41); (293.24; 6.93); (303.29; 18.45); (313.24; 44.73); (323.14; 104.14); (333.34; 238.73); (343.06; 488.58).

- To start with, make an ln$f$ plot of the data points, such that ln f = ln (P /Pa) $-$ 31.610 + 8720 K /T; and, in a freehand manner, construct a representative arc which has its maximum at (1/298.15) K^{-1}.
- Next, use the arc to calculate, for the change solid-to-vapour at T = 298.15 K, the values of $\Delta G°$; $\Delta H°$; and ΔC_P°, referred to P = 1 Pa; and, to end with, use the transition properties to calculate the temperature at which the equilibrium pressure will be equal to 800 Pa.

15. second-order transition according to Ehrenfest

In Ehrenfest's (1933) classification of phase transitions, a first-order transition is one for which G^{*} is continuous on crossing the equilibrium curve, whereas its derivatives V^{*}, and minus S^{*} are not. In the case of second-order transitions, G^{*}, V^{*}, and S^{*} are continuous, whereas the second derivatives of G^{*} are not.

- Derive two "Clapeyron equations" for second-order transitions - one from d(ΔV^{*}) = 0, and the other from d(ΔS^{*}) = 0.

NB. Combination of the two equations yields a relationship between the second derivatives, that is to say between ΔC_P^{*}, $\partial(\Delta V^{*})/\partial P$, and $\partial(\Delta V^{*})/\partial T$.

Ignoring kinetic matters, substances when brought together at specified conditions of temperature and pressure will give rise to a chemical reaction if that reaction offers the opportunity to lower the Gibbs energy. Under certain circumstances the reactants and the products of the reaction can coexist - be in equilibrium with one another. It is not impossible that the substances of a given system are involved in more than one chemical reaction.

considerations

At ordinary conditions the Gibbs energy of 1 mole of water is lower than the sum of the Gibbs energies of 1 mole of *hydrogen* and 0.5 mole of *oxygen*. This observation implies that H_2 and O_2 *can* change spontaneously into H_2O. Why not simply "this observation implies that H_2 and O_2 change spontaneously into H_2O"? The problem is that thermodynamics has no control of *kinetics*. Thermodynamics can say that a certain reaction can proceed spontaneously, but it cannot predict whether the reaction will really proceed.

In the case of *oxyhydrogen* a spark will do the trick. On the other hand, graphite+hydrogen+oxygen, even in the right proportions, never will change spontaneously into sugar, whatever tricks one may invent; and in spite of the fact that the change would correspond to a lowering of the Gibbs energy.

In the opposite direction thermodynamics has luck on its side. It can firmly pronounce that *water* at ambient conditions *never* will change spontaneously into a mixture of hydrogen and oxygen. By means of thermodynamics one can reveal/ calculate the boundary in the *M* [*variables*] *space*, which separates the region where a *spontaneous change* is possible from the region where a spontaneous change is impossible. No more and no less.

Incidentally, the fact that a chemical reaction does not proceed spontaneously under certain conditions, does not mean that under those conditions the reaction cannot be carried out. To appreciate this, we start again with the spontaneous reaction of hydrogen with oxygen. The chemical energy which is present in that combination, is released by the reaction

$$2\,H_2 + O_2 \;\rightarrow\; 2\,H_2O\;;$$

and can be stored in materials, e.g. for heating purposes. In a more sophisticated way H_2 and O_2 are made to react in an *electrochemical cell* (without being in direct contact with one another). In that case the chemical energy is converted (in part) in electric energy ($\rightarrow$202:Exc 2). The electrochemical reaction between H_2 and O_2 can be detailed as follows, e^- denoting the electron.

anode reaction $H_2 \rightarrow 2\,H^+ + 2\,e^-$
cathode reaction $1/2\,O_2 + H_2O + 2\,e^- \rightarrow 2\,OH^-$
neutralization reaction $2\,H^+ + 2\,OH^- \rightarrow 2\,H_2O$

---

sum reaction $H_2 + 1/2\,O_2 \rightarrow H_2O$

When such a cell, the *oxyhydrogen cell*, operates at 25 °C and with H_2 and O_2 at 1bar pressure, its *electromotive force* will be 1.23 Volt. When the oxyhydrogen cell is connected - in the proper way - to another electric cell with a higher electromotive force, water is forced to change into hydrogen and oxygen.

anode reaction $H_2O \rightarrow 2H^+ + 2\,e^- + 1/2\,O_2$
cathode reaction $2H_2O + 2\,e^- \rightarrow 2OH^- + H_2$
neutralization reaction $2H^+ + 2OH^- \rightarrow 2H_2O$

---

sum reaction $H_2O \rightarrow H_2 + 1/2\,O_2$

The forced reaction in this case is called *electrolysis*; one can speak of the electrolysis of water.

The important conclusion is that there is a region in the M [variables] space where a chemical reaction can proceed spontaneously and another region where the reaction has to be forced to proceed. Thermodynamics has the power to allocate these regions and to pinpoint their interface.

a reaction between pure substances and its equilibrium curve

In the simplest case the M [variables] space is simply the *TP* plane, namely when the substances involved in the chemical reaction do not dissolve in one another.
We consider the reaction

$$A + B \rightarrow 2C. \tag{1}$$

At equilibrium, the equality

$$\mu_A + \mu_B = 2\,\mu_C \tag{2}$$

is respected and there is one degree of freedom:

$$f = M[T,P] - N[\mu_A + \mu_B = 2\,\mu_C] = 2 - 1 = 1. \tag{3}$$

As a result, the solution of this system is represented by a curve in the *TP* plane.

Dealing with pure substances, we can replace Equation (2) by

$$G_A^* + G_B^* = 2G_C^*;\qquad(4)$$

and to reduce notation we will write

$$\Delta G = 2G_C^* - G_B^* - G_A^*.\qquad(5)$$

The equilibrium condition now takes the simple form

$$\Delta G = 0.\qquad(6)$$

The curve in the *TP* plane which is the solution of Equation (6), or the solution of Equation (2) if you like, has either a positive or a negative slope; Figure 1.

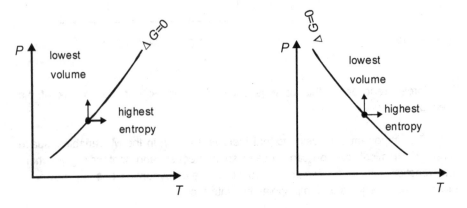

FIG. 1.   The locus of the equilibrium states divides the *PT* plane into two fields;   left: highest/lowest   entropy   goes   together   with highest/lowest   volume;   right: highest/lowest   entropy   goes together with lowest/highest volume

For a point on the *equilibrium curve* ΔG is equal to zero, and from $G = U + PV - TS$ it follows

$$\Delta G = \Delta U + P\,\Delta V - T\,\Delta S = 0.\qquad(7)$$

If, starting from the point on the curve, the system at constant pressure is brought at a higher temperature, ΔG (which was zero) will become negative only if ΔS is positive. So, if for the reaction considered, indeed $2S_C^* > (S_A^* + S_B^*)$, i. e. ΔS >0, then the spontaneous reaction at the right-hand side is A + B → 2C. In that case A and B react to C until one of the two, or both at the same time, will be exhausted. If, on the other hand, $(S_A^* + S_B^*) > 2S_C^*$ then, at the right-hand side, C will decompose spontaneously and entirely into A and B. In the latter case ΔG is negative for the reaction 2C → A + B. Putting these things together, we can observe that at the high-temperature side of the equilibrium curve the spontaneous reaction is the one that corresponds to an increase in entropy.

On the same lines of reasoning, the stability field at the high-pressure side of the equilibrium curve is the field where the system takes the lowest volume.

The equilibrium curve, Figure 1, has a positive slope when ΔS and ΔV have the same sign and a negative slope when the signs of ΔS and ΔV are different.

In a somewhat different approach, one can observe that along the equilibrium curve ΔG is equal to zero and remains equal to zero. Its total differential, therefore, is equal to zero:

$$0 = d(\Delta G) = \left(\frac{\partial \Delta G}{\partial T}\right)_P dT + \left(\frac{\partial \Delta G}{\partial P}\right)_T dP = -\Delta S\, dT + \Delta V\, dP. \tag{8}$$

As a result, for the slope of the equilibrium curve,

$$\frac{dP}{dT} = \frac{\Delta S}{\Delta V}. \tag{9}$$

There are many reactions between pure substances in which one of them is gaseous, the others being solid. In these cases the signs of ΔS and ΔV, generally, are determined by the gaseous substance: ΔS and ΔV generally have the same sign and are positive when the gaseous substance is at the right-hand side of the reaction equation. An example is found in the calcium carbonate equilibrium, discussed already in § 007, and §108.

$$CaCO_3(solid) \rightleftarrows CaO(solid) + CO_2(gas) \tag{10}$$

In the case of homogeneous equilibria, there is one phase which contains all of the substances involved in the equilibrium reaction. A well-known example is the gaseous *ammonia equilibrium*,

$$N_2 + 3\,H_2 \rightleftarrows 2\,NH_3 \tag{11}$$

discussed already in § 007, and elaborated for the ideal-gas mixture.

reactions and dependent reactions

Between the four substances carbon (graphite), oxygen, carbon monoxide, and carbon dioxide one can formulate four reactions such that every time one of the substances is not involved:

C + 0.5 O_2 → CO   (CO_2 absent)
CO + 0.5 O_2 → CO_2 (C absent)
C + O_2 → CO_2 (CO absent)
C + CO_2 → 2 CO   (O_2 absent)

In terms of thermodynamics and thermodynamic properties only two of the four reactions are independent. The third reaction can be seen as the sum of the first two reactions; and the fourth as their difference.

To discuss a situation like this, we take the abstract case of four substances A, B, C, and D which are not soluble in one another, and between which there are two independent reactions:

$$(\bar{D}) \quad A + B \rightarrow C$$
$$(\bar{A}) \quad B + C \rightarrow D$$

The *absent notations* $(\bar{D})$ and $(\bar{A})$ are used to denote that the substances D and A do not take part in the first and second reaction, respectively.
For $(\bar{D})$ the equilibrium curve in the PT plane follows from

$$\Delta_{\bar{D}}G = G_C^* - G_A^* - G_B^* = 0. \tag{12}$$

At the equilibrium curve, the substances A, B and C coexist: when brought into a vessel and put under the conditions corresponding to a point on that curve the substances will not react, their amounts remaining the same (however.....).
Similarly, there is an equilibrium curve for the second reaction $(\bar{A})$. The point of intersection of the two curves is an *invariant point*, as follows from the *system formulation*

$$f = M\,[T,P] - N\,[\Delta_{\bar{D}}G = 0;\ \Delta_{\bar{A}}G = 0] = 0. \tag{13}$$

At the invariant point all four A, B, C, and can coexist.
Out of the two (independent) reactions $(\bar{D})$ and $(\bar{A})$ we can derive, by eliminating C and B, the dependent reactions $(\bar{C})$ and $(\bar{B})$:

$$(\bar{C}) \quad A + 2B \rightarrow D$$
$$(\bar{B}) \quad A + D \rightarrow 2C$$

The two dependent reactions give rise to another pair of equilibrium curves in the TP plane. These two curves, obviously, intersect at the invariant point defined by $(\bar{D})$ and $(\bar{A})$. Taken together, the four curves divide the TP into eight fields.

The stability relations between these fields, as always, follow from Gibbs energy considerations

To understand these matters we will use the *simple model* again ($\leftarrow$110). In the simple model Z stands for molar Gibbs energy, X replaces T and Y replaces P.
In this model let's define the four substances by

A	$Z_A = -X + Y + 1$	(14)
B	$Z_B = -4X + 2Y + 6$	(15)

$$C \qquad Z_C = -6X + 6Y + 2 \tag{16}$$
$$D \qquad Z_D = -2X + Y + 4 \tag{17}$$

For the equilibrium corresponding to the reaction $(\bar{A})$ the equilibrium line in the XY plane follows from $Z_D = Z_B + Z_C$:

$$(\bar{A}) = -8X + 7Y + 4 = 0. \tag{18}$$

On the same lines for the other equilibria:

$$(\bar{B}) = -9X + 10Y - 1 = 0 \tag{19}$$
$$(\bar{C}) = -7X + 4Y + 9 = 0 \tag{20}$$
$$(\bar{D}) = -X + 3Y - 5 = 0. \tag{21}$$

The four equilibrium lines are shown in Figure 2, the coordinates of the invariant point being $X = 2.765$; $Y = 2.588$.

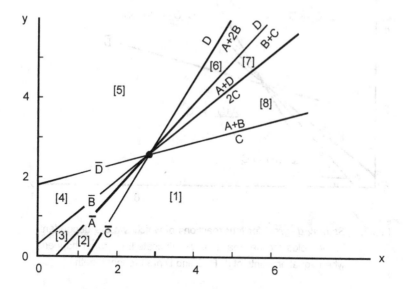

FIG.2.   Equilibrium lines corresponding to four chemical reactions, only two of them being independent

Next, for the reaction $(\bar{C})$ it follows - from the principle of minimal Z - that D is stable at the left-hand side of the line; and (A + 2B) at the right-hand side. Similarly for the other lines, see Figure 2.

As a following step we have to find out where the stable fields are situated. To that end the diagram shown in Figure 2 is split up in eight regions. The easiest thing to do is to answer the question: what will happen when the four substances are brought together under the conditions of a given region. Let's take one mole of each substance and put the combination under the conditions of the region marked [1]. Region [1] is at the lower side of the line for ($\bar{D}$), which means that (A + B) will react to C and as a result the system will have 2 mole of C and 1 mole of D. In region [1] the substance D, however, is not stable and it follows from the line for ($\bar{A}$) that it will react to (B + C). In region [1], therefore, A + B + C + D change into B + 3C. On the same lines of reasoning it follows that in region [2] the result will be the same as in region [1] : A + B + C + D change into B + 3C. In fact, therefore [1] and [2] together constitute a stable field. Region [3] is a stable field on its own; in that field A + B + C + D change into 2C + D. The complete result of this exercise is shown in Figure 3.

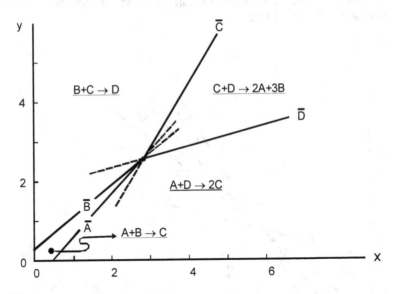

FIG.3.   *Stability diagram for four reactions of which two are dependent. In the fields the net reactions are indicated, i.e. what happens when equal amounts of A, B, C and D are brought together*

Again, we may appreciate the usefulness of the absent rotation. We observe, Figure 3, that the *metastable extension* of the equilibrium curve corresponding to the absence of a certain substance runs into a region where that substance is produced. This general fact is referred to as *Schreinemaker's Rule* (after F.A.H. Schreinemakers 1864-1945). A consequence of Schreinemaker's Rule: at the invariant point, the angle between the stable parts of two successive equilibrium curves has to less than 180º.

Another observation which we can make is that, for instance, the substances C and D react with one another in the field enclosed by the stable parts of ($\bar{C}$) and ($\bar{D}$). In other words, there is a field in the XY (read *TP*) plane that corresponds to circumstances at which C and D are not found in each other's presence. The other way round, a (geological) observation - e.g. that in a given setting C and D are not found together - can be a clue to the *TP* circumstances of that setting. In geology, to conclude with, *Schreinemakers analysis* is an important tool in *barothermometry*; see E-an Zen (1966); Powell (1978).

For most of the relatively simple cases considered above, the equilibrium between the reactants and the products of a chemical reaction corresponds to a curve in the PT plane. For conditions outside that curve either the reactants (or part of them if they are not present in stoichiometric amounts) disappear spontaneously or the products. At the high-temperature side of the equilibrium curve the spontaneous change corresponds to an increase in entropy; at the high-pressure side the change is to lower volume. When substances are involved in more than one reaction, the single PT plane, with a single curve, separating two fields, changes into a complex stability diagram with one or more invariant points and several stability fields.

EXERCISES

1.   *the strontium oxides*

The formation properties of strontium oxide (SrO) and strontium peroxide (SrO_2) at 25 °C and 1 bar, expressed in $J \cdot mol^{-1}$, are (Barin 1989)

	$\Delta_f H^\circ$	$\Delta_f G^\circ$
SrO	−592036	−561404
SrO_2	−633458	−573279

An amount of strontium is brought into a cylinder-with-piston along with an equal amount of oxygen. Next the external pressure is set to 1 bar and the temperature to 25 °C.

- Which substance(s) will be present when equilibrium has been reached?
- Like previous question, however, with the double amount of strontium.
- Like first question, the temperature, however, being set at 500 K (first calculate $\Delta_r S^\circ$ and next assume that $\Delta_r S^\circ$ and $\Delta_f H^\circ$ are independent of temperature).

2. *Alexander von Humboldt's discovery*

In 1799 Alexander von Humboldt (1769-1859) reported that barium oxide, when heated in air, absorbs oxygen from the air: formation of barium peroxide (BaO_2). On heating to higher temperatures the 'absorbed' oxygen is released and can be collected. The process was used in the nineteenth century to isolate *oxygen* from air on a technical scale.

Hildebrand (1912) measured the equilibrium pressure of the system $2BaO_2(s) \rightarrow 2BaO(s)+O_2(g)$ as a function of temperature

T /K	800	900	1000	1111	1250
P /atm	1.48×10^{-3}	1.95×10^{-2}	1.54×10^{-1}	9.85×10^{-1}	6.32

- From the experimental *PT* data calculate for each temperature ΔG°, the Gibbs energy change of the reaction, referred to 1 bar; next by linear least squares of $\Delta G^\circ(T) = \Delta H^\circ - T\Delta S^\circ$, calculate ΔH° and ΔS°.
- Up to what temperature BaO will 'absorb' oxygen from the air with 20 mole % of that gas?

3. *magnesium carbonate*

The table gives for magnesite (magnesium carbonate, $MgCO_3$) values of the Gibbs energy of formation from the oxides, for $P = 1$ bar (Robie 1978).

T K	$\Delta_{f,ox}G^\circ$ kJ·mol^{-1}
400	−48.077
500	−30.736
600	−13.592
700	3.343
800	20.050
900	36.529

- Calculate, for the temperatures of the table, the values of the equilibrium pressure P and make a plot of $\ln P$ versus $1/T$.

- In an experiment 1 mol $MgCO_3$ and 1 mol CO_2 are brought in a cylinder-with-piston. Next, at constant external pressure of 1 bar, the cylinder with its contents is slowly heated from room temperature to 100 K. Make a plot of the height of the piston as a function of temperature; the height of the piston at 400 K is 20 cm.

4. *air as a CO$_2$ buffer*

From the data on $MgCO_3$ given in Exc 3 calculate the values of $\Delta_{f,ox}H^\circ$ and $\Delta_{f,ox}S^\circ$, assuming that the dependence on temperature of the two quantities may be ignored. Next, calculate the temperature at which $MgCO_3$ will decompose in the *open air*. The air in this case may be considered as a CO_2 buffer, containing 0.03 mole percent of that substance.

5. *the ammonia equilibrium - the role of pressure*

Four values are given of the equilibrium constant

$$K = \frac{(X_{NH_3} P)}{(X_{N_2} P)^{1/2}(X_{H_2} P)^{3/2}}$$

of $1/2 N_2 + 3/2 H_2 \rightarrow NH_3$ for $t = 450\ °C$ and as a function of pressure (Larson and Dodge 1923; Larson 1924).

- By extrapolation determine the value of K at zero pressure; and from that the Gibbs energy of formation of NH_3 - in the ideal-gas approximation and referred to 1 bar.

P /atm	$10^3 K$ / atm^{-1}
10	6.595
30	6.764
50	6.906
100	7.249

- Compare the result with the data given by Robie et al. (1978) for the formation properties of NH_3 (ideal gas; 1 bar). For $T = 700K$ these data are, expressed in kJ· mol^{-1}, $\Delta_f H^° = -52.684$; $\Delta_f G^° = 27.155$.

6. *ammonia's degree of dissociation*

The homogeneous gas equilibrium $NH_3 \rightarrow 1/2\ N_2 + 3/2\ H_2$ can be realized starting from pure ammonia. In that case the equilibrium state can be fully described by the set of variables M $[T,P,\alpha]$, where α is the *degree of dissociation*.

- In the ideal gas approximation, derive the equilibrium equation, from which α, for given T and P, can be solved.
- Calculate the α values for the twelve circumstances indicated below. The Gibbs-energy-of-formation values are for 1 bar (Robie et al. 1978).

$\Delta_f G_{NH3}$ / ($J·mol^{-1}$)	T /K
−16410	298.15
−5984	400
4760	500
15841	600

P /bar →	1	10	100
	0.336		

7.   *vapour in equilibrium with solid salammoniac*

In a vessel-with-manometer, immersed in a thermostat at 600 K, the equilibrium is
realized between solid salammoniac (NH_4Cl) and vapour formed from NH_4Cl,
containing HCl, NH_3, N_2 and H_2.
- Calculate the mole fractions of the species in the vapour, and also the
  pressure indicated by the manometer.

The available information consists just of the Gibbs energies of formation, valid for
600 K and 1 bar and expressed in $kJ \cdot mol^{-1}$ (Robie 1978), of solid NH_4Cl
(–92.837) gaseous HCl (–97.963) and gaseous NH_3 (+15.841). First see Exc 007:4.

8.   *dissociation of water at 1800 K*

An amount of steam (H_2O) is kept in a cylinder-with-piston at 0.5 bar external
pressure and at 1800 K. At 1800 K and 1 bar the Gibbs energy of formation of
gaseous H_2O is –147.032 $kJ \cdot mol^{-1}$ (Robie 1978).
- Calculate the degree of dissociation α, i.e. the fraction of H_2O which has
  dissociated in H_2 and O_2.

9.   *the simple model – interdependent reactions*

In terms of the simple model, four substances are defined by
$$Z_A = -X + Y + 1 \qquad\qquad Z_B = -4X + 2Y + 6$$
$$Z_C = -6X + 6Y + 2 \qquad\qquad Z_D = -2X + Y + 4$$
And they give rise to the following independent reactions
$$(\bar{C}) \quad A + D \rightarrow B$$
$$(\bar{A}) \quad B + D \rightarrow C$$
- Formulate the dependent reactions ($\bar{B}$) and ($\bar{D}$) and calculate the complete
  stability diagram.

10.   *virtual experiments related to Figure 3*

This exercise pertains to the system of four substances A, B, C and D, of which
Figure 3 is the stability diagram. Each position in the scheme below represents a
virtual experiment in which equal amounts of two of the four substances are brought
at conditions that correspond to a point in one of the stability fields.
- Give the composition - in terms of substances and their amounts - the
  system will have, after equilibrium has been reached.

combination	FIELD			
	$\bar{C}\,\bar{D}$	$\bar{D}\,\bar{A}$	$\bar{A}\,\bar{B}$	$\bar{B}\,\bar{C}$
A + B →			C	
A + C				
A + D		2C		
B + C				D
B + D				
C + D	2A + 3B			

11. *one of Professor Schuiling's favourites*

In this exercise five substances are considered which are involved in five reactions such that in each reaction one of them is missing. The substances are water (W), and the four minerals kaolinite (K = $Al_2Si_2O_5(OH)_4$), andalusite (A = Al_2SiO_5), pyrophyllite (P = $Al_2Si_4O_{10}(OH)_2$) and quartz (Q = SiO_2). For the two independent reactions we take

$(\bar{K})$ $Al_2Si_4O_{10}(OH)_2 \rightarrow Al_2SiO_5 + 3SiO_2 + H_2O$
$(\bar{P})$ $Al_2Si_2O_5(OH)_4 \rightarrow Al_2SiO_5 + SiO_2 + H_2O$

The equilibrium curves of the five reactions run more or less as indicated. The substances Q and W are formed at the right-hand side of the equilibrium curves, except for reaction ($\bar{A}$) where Q is formed at the left-hand side.

- Formulate the dependent reactions and find the five stability fields.

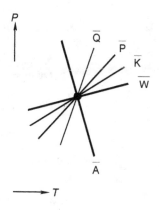

PHASE THEORY: THE THERMODYNAMICS
OF EQUILIBRIUM BETWEEN PHASES

§ 201 MIXTURES AND PARTIAL QUANTITIES

When, at a certain T and P, a mixture of 2 mole of substance A and 3 mole of substance B has a volume of 200 cm³ it is sure that, at unchanged T and P, a mixture of 4 mole of A and 6 mole of B will have a volume of 400 cm³. The significance of this "simple" but important property is going to be examined.

partial volumes

To keep things simple, let's take two miscible liquid substances A and B and talk about mass and volume. By m_A^* and m_B^* we denote the *molar masses* of pure A and pure B, respectively. The *molar volumes* of the pure substances are V_A^* and V_B^*.

For mass $m(n_A, n_B)$ and volume $V(n_A, n_B)$ of a mixture of n_A mole of A and n_B mole of B, we can make the following general observations, the properties 1, 2 and 3.

$$m(n_A, n_B) = n_A \cdot m_A^* + n_B \cdot m_B^* \tag{1}$$

$$V(n_A, n_B) \neq n_A \cdot V_A^* + n_B \cdot V_B^* \quad (P, T \text{ constant}) \tag{2}$$

$$V(t \cdot n_A, t \cdot n_B) = t \cdot V(n_A, n_B) \quad (P, T \text{ constant}) \tag{3}$$

Property 1 is a consequence of the law of *conservation of mass*. There is no law of conservation of volume: generally the volume of a mixture is not equal to the sum of the volumes before mixing - property 2. Property 3 expresses the general fact that, when the amounts of A and B are multiplied by the same factor, the volume is also multiplied by that factor, provided that P and T are kept constant. Everyone who buys liquors is familiar with this general fact.

The change the volume of the mixture will undergo on the addition of infinitesimal amounts of A and B is - obviously, and at constant P and T - given by

$$(dV)_{P,T} = \left(\frac{\partial V}{\partial n_A}\right)_{P,T,n_B} dn_A + \left(\frac{\partial V}{\partial n_B}\right)_{P,T,n_A} dn_B. \tag{4}$$

The partial differential coefficients are named *partial (molar) volumes*, symbols V_A and V_B. The quantity V_A is the partial volume of A in the mixture of A and B and V_B is the partial volume of B in that mixture. Note that, as a consequence of property 2, V_A as rule is not equal to V_A^* and V_B not equal to V_B^*.

Generally, in a *multicomponent mixture* and for a given *extensive quantity Z*, the partial Z of the i-th component is defined as

$$Z_i = \left(\frac{\partial Z}{\partial n_i} \right)_{P,T,n_i'} . \tag{5}$$

In this definition the accent rotation n_i' is used to indicate that, with the exception of n_i, the amounts of all components have to be kept constant. It should be stressed that for the definition of partial quantities the condition of constant P and T is imperative. As an example, the differential coefficient $(\partial S/\partial n_B)_{V,T}$ is not the partial entropy of B!

relation between integral volume and partial volumes

For lack of a relation between the volume of a mixture and the molar volumes of the *pure* components, we are now going to find out if we can find a relation between the volume of the mixture and the *partial* volumes of its components - and so - by integration of Equation (4), rewritten as

$$dV = V_A \, dn_A + V_B \, dn_B , \quad (P, T \text{ constant}). \tag{6}$$

It is clear that the volume of a mixture of n_A' mol A and n_B' mol B is given by

$$V = \int_0^{n_A'} V_A \cdot dn_A + \int_0^{n_B'} V_B \cdot dn_B , \tag{7}$$

no matter how the integration is carried out. No matter how A and B are mixed together; no matter the route followed in the $n_A n_B$ plane form point (0;0) to point $(n_A' ; n_B')$; see Figure 1.

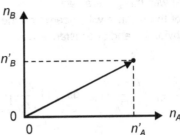

FIG. 1.   Preparation of a mixture of A and B

The problem, however, is that, to be able to carry out the integration, we must know how, along the route followed, V_A and V_B depend on n_A and n_B - something not so obvious.

In order to find a solution for this problem, we will make use of property 3 and compare the following two situations, indicated by (s) and (d)

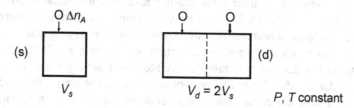

(s) V_s

(d) $V_d = 2V_s$ P, T constant

In situation (s) a drop of pure A is added to a mixture of A and B, having a certain composition, i. e. a given *mole fraction* of B ($X_B = n_B / (n_A + n_B)$). In situation (d) the composition of the mixture is the same as in (s); the difference with (s) is that the amounts are doubled. After the addition of the drops of A, (s) and (d) have the same composition and the volume of (d) is two times the volume of (s) and it is obvious that the increase in volume in (d) is two times the increase in (s). Now for very small Δn_A, the partial volume of A follows from

in the case of (s) $(V_A)_s = \dfrac{(\Delta V)_s}{(\Delta n_A)_s}$;

and in the case of (d) $(V_A)_d = \dfrac{(\Delta V)_d}{(\Delta n_A)_d} = \dfrac{2(\Delta V)_s}{2(\Delta n_A)_s} = (V_A)_s.$

We observe, therefore, and as a result of property 3, that, for a given mole fraction of B, the partial volume of A is independent of the total amount of mixture. Of course this observation is also valid for V_B.

The easiest way to carry out the integration, as may be clear, now, is to follow the straight line between $(0;0)$ and $(n_A'; n_B')$, i.e. the route indicated by the arrow in Figure 1. Along that route A and B are mixed at constant mole fraction: along the route V_A is constant and the same holds true for V_B. Consequently, Equation (7) changes into

$$V = V_A \int_0^{n_A'} dn_A + V_B \int_0^{n_B'} dn_B, \qquad (8)$$

from which it follows that the *relation between volume and partial volumes* simply is

$$V = n_A' \cdot V_A + n_B' \cdot V_B. \qquad (9)$$

The *integral volume* of a mixture is the sum of the products of amount of component and partial volume of that component:

$$V = \sum_B n_B \cdot V_B. \qquad (10)$$

Besides, the adjective *integral* is sometimes used to avoid confusion with partial volumes (in that sense it is superfluous in the foregoing sentence).

homogeneous functions and Euler's Theorem

An alternative, a mathematical manner to obtain the result expressed by Equation (9) is to use *Euler's Theorem on homogeneous functions*. A function U of x and y is said to be a homogeneous function of the k-th degree when it has the following property: when *all* variables are multiplied by the same factor t it has for result that the function value is multiplied by t-to-the-power-k (t^k). Euler's Theorem is (see e.g. Mellor 1955): "In any homogeneous function, the sum of the products of each variable with the partial differential coefficient of the original function with respect to that variable is equal to the product of the original function with its degree".

$$x\left(\frac{\partial U}{\partial x}\right) + y\left(\frac{\partial U}{\partial y}\right) = k \cdot U. \tag{11}$$

Therefore, from property 3 it follows that volume is a homogeneous function of the first degree of the amounts of substance. And Equation (9) is its Euler's Theorem expression - as simple as that.

Gibbs-Duhem equation

Going on with a little bit of mathematics, we notice that from Equation (9) it follows that the *differential of V* can be given as

$$dV = n_A \cdot dV_A + V_A \cdot dn_A + n_B \cdot dV_B + V_B \cdot dn_B. \tag{12}$$

On the other hand, we know already that at constant *T* and *P* the differential of *V* is given by Equation (6)

$$dV = V_A \cdot dn_A + V_B \cdot dn_B.$$

Therefore, as a result of these two truths, the following relation has to exist between the changes of the partial quantities.

$$n_A \cdot dV_A + n_B \cdot dV_B = 0 \quad (P, T \text{ constant}). \tag{13}$$

From this equation - the volume exponent of the important *Gibbs-Duhem equation* which we will meet soon - it follows that at constant *P* and *T* any changes in V_A and V_B are related to one another and, in addition, have opposite signs, i.e. when V_A increases, V_B has to decrease.

two observations

To conclude this section, two observations are made. The first is that the other extensive thermodynamic quantities U, H, S, A, G... like V are homogeneous functions of the first degree of the amounts of substance. Taking entropy for example, it is obvious that in $\int_{T_1}^{T_2} q_{rev} / T$ the heat needed to bring ten mole of a given mixture from T to $T+\Delta T$ will be ten times the heat needed to bring one mole - 10% of the same mixture - from T to $T+\Delta T$. It implies, among other things, that we can speak of the partial entropy of substance B in a mixture with other substances. The other observation is on pure substances. When applied to a pure substance, the definition of partial quantity tells us that the partial quantity - the change of the quantity per mole of the substance itself - obviously is equal to the molar quantity. For example, for the partial volume of a pure substance B

$$V_B = \left(\frac{\partial V}{\partial n_B} \right)_{P, T} = V_B^*. \tag{14}$$

The extensive thermodynamic quantities Z, like U, S and V, are first-degree homogeneous functions of the amounts of substance. The partial quantity Z_B of a substance B in a mixture with other substances is the partial differential coefficient, taken at constant T and P and constant amounts of the other substances, of Z with respect to n_B, i.e. the amount of B. The function value of Z is equal to the sum over substances of $n_B \times Z_B$.

EXERCISES

1.   *partial volumes from densities*

Calculate the partial volume of alcohol ($m^* = 46$ g·mol^{-1}) in a mixture of water ($m^* = 18$ g·mol^{-1}) and alcohol containing 72 wt% of the latter (about equimolar mixture). The available data are the densities for two compositions:

composition (wt%)	density (g·cm^{-3})
72	0.86710
73	0.86470

2.   *from integral volume to partial volumes*

The integral volume of a mixture of n_A mole of A and n_B mole of B can be approximated by the formula

$$V = n_A \cdot V_A^* + n_B \cdot V_B^* + C \frac{n_A \cdot n_B}{(n_A + n_B)},$$

where V_A^* and V_B^* are the molar volumes of pure A and pure B, respectively; and C is a system-dependent constant.

- Find the corresponding formulae for the partial volumes of A and B.

3. zeroth degree homogeneous functions

It can be proved quite easily that partial quantities are zeroth-degree homogeneous functions of the amounts of substance. As an example, substance A's partial volume V_A in a mixture of A and B is a zeroth-degree homogeneous function of n_A and n_B. In other terms, you are invited to show that

$$n_A \left(\frac{\partial V_A}{\partial n_A} \right) + n_B \left(\frac{\partial V_A}{\partial n_B} \right) = 0 \cdot V_A = 0.$$

4. Van Laar and Euler

In a section on homogeneous functions U of x and y, van Laar (1935) gives the following examples of a homogeneous function of the first degree (A) and a function of the zeroth degree (B):

$$(A) \quad U = (x + y) \ln \frac{x}{y} - y \, e^{- \sin (y/x)}$$

$$(B) \quad U = \frac{x}{y} - \ln \frac{x^2 + y^2}{2xy}.$$

- Show that the functions (A) and (B) obey Euler's Theorem, i.e.

$$x \frac{\partial U}{\partial x} + y \frac{\partial U}{\partial y} = U \text{ for (A); and } = 0 \text{ for (B).}$$

5. molality (m) makes it easy

The volume of the homogeneous liquid system {m mole of substance B + 1kg of water} is given by the expression $V = V^\circ + a \, m + b \, m^2 + c \, m^3$.

- Derive the expressions for the partial molar volumes of the two components: B and water; the molar mass of the latter is 18.015 $g \cdot mol^{-1}$.
- Check your result with the help of the Gibbs-Duhem equation, unless you used the equation already for the step from V_B to V_{H_2O}

6. *partial volumes of sodium chloride and water in their liquid mixture*

The data in the table below are densities, at $t = 20\ °C$, of aqueous solutions of sodium chloride (NaCl, molar mass 58.443 $g \cdot mol^{-1}$) as a function of wt % NaCl, and up to saturation.

The purpose of this exercise is to use the data to calculate the partial molar volumes of NaCl and water in the system {m mole of NaCl + 1 kg of water}; at

  i)   infinite dilution ($m \to 0$);
  ii)   saturation ($m = m^{sat}$);
  iii)   half saturation ($m = 0.5\ m^{sat}$).

- To that end, first transform the table to one in which the volume of the system is given as a function of m; then make a plot of volume as a function of m and, by drawing tangent lines, determine, for each of the three situations, the numerical value of (dV/dm); see foregoing Exc.

NB In a somewhat more subtle manner the plot of volume against m is replaced by a plot in which the deviation from linear behaviour is plotted against m. Let V^{sat} represent the volume at saturation and a' the quotient ($V^{sat} - V^o)/m^{sat}$, then the deviation can be visualized by plotting $V' = V - V^o - a'm$ against m.

wt %	$\rho /(g \cdot cm^{-3})$
0	0.99823
2	1.0125
4	1.0268
6	1.0413
8	1.0559
10	1.0707
12	1.0857
14	1.1008
16	1.1162
18	1.1319
20	1.1478
22	1.1640
24	1.1804
26	1.1772

The systems are opened up for the transfer of matter. Next to the potentials temperature and pressure for the transfer of heat and space, the chemical potentials of the substances make their appearance.

chemical potentials

The system considered in this section is a *homogeneous mixture* of the substances A and B. The change of the energy of the system in a reversible experiment, as we know, can be expressed as

$$dU = q_{rev} + w_{rev} = TdS - PdV. \tag{1}$$

Energy in this way - the *natural route* as we did call it - appears as a function of the two variables S and V.

If we open up the system, that is to say by making it accessible to the addition or withdrawal of A and/or B, by which the amounts of A and B are changed, we will find ourselves in a new situation. And in that new situation - the *open system* - the amounts of A and B are going to act as additional variables. The function $U(S, V)$ changes into $U(S, V, n_A, n_B)$; and Equation (1) can be extended as

$$dU = TdS - PdV + \mu_A dn_A + \mu_B dn_B. \tag{2}$$

The quantities μ_A and μ_B are the *chemical potentials* of A and B, respectively; and quite analogous to *thermal potential* for T and *mechanical potential* for P (←001).

Purely mathematically, from Equation (2), the chemical potentials are the partial derivatives of the energy with respect to the amount of substance:

$$\mu_B = \left(\frac{\partial U}{\partial n_B}\right)_{S,V,n_A} \quad \text{and} \quad \mu_A = \left(\frac{\partial U}{\partial n_A}\right)_{S,V,n_B}. \tag{3}$$

Observe that S and V have to be kept constant and not T and P: the chemical potentials of A and B are *not* identical with the *partial energies* of A and B.

If, again, by the introduction of the 'auxiliary quantity' Gibbs energy

$$G = U + PV - TS, \tag{4}$$

we are going to replace the 'inconvenient' variables S and V by the desirable variables T and P, we get

$$dG = -SdT + VdP + \mu_A dn_A + \mu_B dn_B. \tag{5}$$

This time we can make the important observation that the chemical potentials are identical with the *partial Gibbs energies* ($\leftarrow$ 201 for partial quantities):

$$\mu_A = \left(\frac{\partial G}{\partial n_A}\right)_{T,P,n_B} \equiv G_A \quad \text{and} \quad \mu_B = \left(\frac{\partial G}{\partial n_B}\right)_{T,P,n_A} \equiv G_B. \tag{6}$$

The partial derivatives of the chemical potentials with respect to temperature are equal to the *partial entropies*, with opposite sign; and those with respect to pressure are equal to the *partial volumes*. This is the result of the cross-differentiation identity; applied to μ_A it gives.

$$\left(\frac{\partial \mu_A}{\partial T}\right)_{P,n_A,n_B} = -\left(\frac{\partial S}{\partial n_A}\right)_{T,P,n_B} \equiv -S_A \tag{7}$$

$$\left(\frac{\partial \mu_A}{\partial P}\right)_{T,n_A,n_B} = \left(\frac{\partial V}{\partial n_A}\right)_{T,P,n_B} \equiv V_A. \tag{8}$$

And again, as a result of the cross-differentiation identity the following identity is obtained

$$\left(\frac{\partial \mu_A}{\partial n_B}\right)_{T,P,n_A} = \left(\frac{\partial \mu_B}{\partial n_A}\right)_{T,P,n_B}. \tag{9}$$

Gibbs energy, like volume, is a first-degree homogeneous function of the amounts of substance. From Euler's Theorem, therefore

$$G = n_A G_A + n_B G_B = n_A \mu_A + n_B \mu_B. \tag{10}$$

Repeating the lines of the foregoing section, we observe that from Equation (10) it follows that change of G can be given as

$$dG = n_A d\mu_A + \mu_A dn_A + n_B d\mu_B + \mu_B dn_B. \tag{11}$$

The 'confrontation' of this expression with Equation (5) yields the Gibbs-Duhem equation

$$-SdT + VdP - n_A d\mu_A - n_B d\mu_B = 0. \tag{12}$$

All in all we now have a function G, the Gibbs energy, which combines the properties of G (T,P) - G with its *natural variables* T and P - of the closed system and the properties of G (n_A, n_B) which is a first degree homogeneous function of the variables n_A and n_B.

It means, among other things, that the combined function G (T, P, n_A, n_B) is characteristic for T and P: it contains all information necessary to determine the other thermodynamic quantities as a function of T, P, n_A and n_B. V (T, P, n_A, n_B), for example, is the first partial derivative with respect to P; minus S (T, P, n_A, n_B) the first partial derivative with respect to T; and so on, and so forth.

The circumstance that G is a first-degree homogeneous function of the amounts is of great experimental importance. It means, for example, that, when G has been measured for a mixture of 1/2 mole of A + 1/2 mole of B, G will be known for all equimolar mixtures of A and B, i.e. whatever their total amount. Expressed in other terms, the function G (T, P, n_A, n_B) can be replaced by $(n_A + n_B)$ $G_m(T, P, X)$. The new function G_m is the *molar Gibbs energy*: the Gibbs energy of a mixture of one mole of (A + B), consisting of $(1 - X)$ mole of A + X mole of B. The new variable X is the *mole fraction of B* ($\leftarrow$002) in the mixture; and obviously $0 \le X \le 1$. In the next section the transcription is made from G (T, P, n_A, n_B) to G_m (T, P, X).

virtual changes in closed systems reveal equilibrium conditions

One of the most exciting properties of G (T, P) for the closed system is its function, its capacity as *equilibrium arbiter*. Irrespective of its contents, the closed system will be in equilibrium for given T and P if its Gibbs energy has reached the lowest possible value ($\leftarrow$108). We will now see how the vehicle of the *open system* can be used to translate the minimum principle into the equilibrium conditions we are familiar with - the equilibrium conditions of the N set. We will consider two cases; the first without, and the second involving a chemical reaction.

The first case is the equilibrium between liquid and vapour in the system alcohol (A) + water (W). The starting point is one of the possible equilibrium situations: P and T chosen by the investigator; the compositions of the two phases according to the choice of P and T. Now, in a virtual experiment we let δn_A mole of alcohol pass from the liquid to the vapour state. The δ is the operator for a small *virtual change* applied to a system already in equilibrium. 'Already in equilibrium' has to say that the corresponding change in Gibbs energy has to be zero:

$$(\delta G)_{P,T} = 0. \tag{13}$$

The liquid as well as the vapour state are now considered as two systems of which the amount of alcohol is changed. In other words, the *closed system in equilibrium* is split up into *two open subsystems*. At constant T and P the Gibbs energy change of the liquid subsystem is given by (Equation (5) with $dT = 0$; $dP = 0$; $dn_B = 0$; $dn_A \neq 0$)

$$(\delta G)_{T,P}^{liq} = \mu_A^{liq} \, \delta n_A^{liq}. \tag{14}$$

The change of the vapour subsystem is

$$(\delta G)_{T,P}^{vap} = \mu_A^{vap} \, \delta n_A^{vap}. \tag{15}$$

Obviously we have

$$-\delta n_A^{liq} = \delta n_A^{vap} = \delta n_A \tag{16}$$

so that

$$(\delta G)_{T,P} = (\delta G)_{T,P}^{liq} + (\delta G)_{T,P}^{vap} = (-\mu_A^{liq} + \mu_A^{vap})\delta n_A. \tag{17}$$

And because the total change has to be zero, $(-\mu_A^{liq} + \mu_A^{vap})$ has to zero. In other words, we obtain the *equilibrium condition*

$$\mu_A^{liq} = \mu_A^{vap}. \tag{18}$$

At equilibrium, the chemical potential of A in the liquid phase is equal to the chemical potential of A in the vapour phase. And, of course, the same can be said of the chemical potentials of W (water still) in the two phases.

Generally, if a substance B is present in more than one phase of a system in equilibrium its chemical potential will have (must have) the same value in all of these phases.

As an example of equilibrium involving a chemical reaction we take the homogeneous (gaseous) ammonia equilibrium

$$NH_3 \rightleftarrows \tfrac{1}{2} N_2 + \tfrac{3}{2} H_2.$$

Let's assume that in the equilibrium situation the amounts n_{NH3}, n_{N2} and n_{H2} are present in a cylinder-with-piston and that the system is kept at a constant temperature and under a constant pressure. The Gibbs energy, which has reached its lowest possible value, is given by the three-substances equivalent of Equation (10):

$$G = n_{NH_3} \cdot \mu_{NH_3} + n_{N_2} \cdot \mu_{N_2} + n_{H_2} \cdot \mu_{H_2}. \tag{19}$$

In a *virtual experiment*, at constant T and P and without adding matter to or withdrawing from the system, we let an amount of δn of NH_3 dissociate into N_2 and H_2. The amounts of NH_3, N_2 and H_2 change as $\delta n_{NH3} = -\delta n$; $\delta n_{N2} = 1/2 \delta n$; and $\delta n_{H2} = 3/2 \delta n$. The change in Gibbs energy, which has to be zero, is given as

$$(\delta G)_{T,P} = \mu_{NH_3} \delta n_{NH3} + \mu_{N_2} \delta n_{N2} + \mu_{H_2} \delta n_{H2}$$
$$= (- \mu_{NH_3} + \frac{1}{2} \mu_{N_2} + \frac{3}{2} \mu_{H_2}) \delta n = 0 \qquad (20)$$

The resulting equilibrium condition is

$$\frac{1}{2} \mu_{N_2} + \frac{3}{2} \mu_{H_2} - \mu_{NH_3} = 0. \qquad (21)$$

Generally, for the *chemical equilibrium* (either homogeneous or heterogeneous)

$$\sum_B \nu_B \cdot B = 0, \qquad (22)$$

in which B stands for substance and ν_B for *stoichiometric coefficient* (being positive for *products* and negative for *reactants*), the equilibrium condition is

$$\sum_B \nu_B \cdot \mu_B = 0. \qquad (23)$$

As a final observation, for a pure substance B the change of Gibbs energy with amount of substance expressed in moles is nothing more or less than the molar Gibbs energy of the substance ($\leftarrow$108, Equation 108:13; $\leftarrow$ 201, Equation 201:14):

$$\text{(pure substance B)} \quad \mu_B \equiv G_B^*. \qquad (24)$$

The chemical potentials of the substances are going to play a principal part, especially as regards the formulation of equilibrium conditions.

EXERCISES

1.   *the ammonia equilibrium from different angles*

   $(1-\alpha)$ mole of ammonia (NH_3), $1/2\alpha$ mole of nitrogen (N_2) and $3/2\alpha$ mole of hydrogen (H_2) are present in a cylinder-with-piston under 1 bar external pressure and kept at a temperature of 400 K. Under these conditions the Gibbs energy of formation of NH_3 is equal to $- 5984$ J·mol^{-1} (Robie 1978).

- First, express in α the mole fractions of NH_3 (X_{NH_3}), N_2 (X_{N_2}) and $H_2(X_{H_2})$ and formulate the expression for the Gibbs energy of the system as a function of α.
- For $0.1 \leq \alpha \leq 0.8$ calculate the Gibbs energy of the system, in α-steps of 0.1, and make a plot of Gibbs energy versus α. What will be the equilibrium value of α?
- Next, for the reaction $NH_3 \rightarrow \frac{1}{2}N_2 + \frac{3}{2}H_2$, formulate in terms of α the expression for the function $\sum_B \nu_B \cdot \mu_B$. For $0.1 \leq \alpha \leq 0.8$ calculate the numerical values of the function in steps of 0.1 and plot these values against α (for which value of α the function goes through zero?).
- To end with, show algebraically that the equilibrium condition $\sum_B \nu_B \cdot \mu_B = 0$ follows from the criterion of minimal Gibbs energy, which is to say from $(\partial G / \partial \alpha)_{T,P} = 0$.

2. the electrochemical cell

A beautiful example to demonstrate the power of the open-system concept is provided by the electrochemical cell.

When an electrochemical cell is operated in a reversible manner, an amount of electrical work is performed - equal to the electromotive force (emf) of the cell E multiplied by the transported charge de. To deal with electrical work, the fundamental equation, Equation (1), has to be extended with Ede:

$$dU = TdS - PdV + Ede.$$

And, after the introduction of the Gibbs energy

$$dG = -SdT + VdP + Ede :$$

at constant T and P the amount of work is equal to the change in Gibbs energy.

The working of the cell is coupled to a chemical reaction, and from a given initial state to a given final state, the amounts of the reactants have decreased and the amounts of products have increased. Realizing that the change from initial to final state can be brought about in a non-electrical manner by removing from the system the reactants and adding to the system the products, it becomes clear that the amount of electrical work is equal to the Gibbs energy effect of the reaction.

In the case of the oxyhydrogen cell ($\leftarrow$111), the formation of 1mol water, from 1 mol hydrogen and 0.5 mol oxygen, goes together with the transfer of 2 mol electrons. This corresponds to an amount of electricity of 2 Faraday (F, named after Michael Faraday, 1791-1867), which is 2 x 96485.309 coulomb, and volt x coulomb = joule.

- For the oxyhydrogen cell, operating reversibly at 25 °C with hydrogen and oxygen at 1 bar pressure, calculate - passing over the delicate problem of signs - the emf E, the temperature coefficient dE/dT, the sign of the pressure coefficient dE/dP, and the amount of heat exchanged with the surroundings.

The standard formation properties of water are (Barin 1989):
$\Delta_f H° = -285.830$ kJ·mol^{-1}; $\Delta_f G° = -237.141$ kJ·mol^{-1}

3. *the influence of gravity*

Taking into account the influence of gravity, the Gibbs energy change of an open system composed of the substance B, has to be extended with a term for gravitational work, given by $Mg\,dh$ where M is the system's mass, g the acceleration of free fall and h the symbol for altitude:

$$dG = -S\,dT + V\,dP + \mu_B\,dn_B + M\cdot g\cdot dh.$$

- How does B's chemical potential, all other things remaining unchanged, change with altitude? Give the function recipe, including the gravity term, for the chemical potential of ideal gaseous, pure B.
- Realizing that in two adjacent layers of ideal gas, the one at altitude h and the other at $h + dh$, B's chemical potential has to have the same value, it is clear that its change with altitude has to be compensated by a change in pressure. Demonstrate that the barometric formula ($\leftarrow$002) is a direct consequence of this statement.

§ 203 CHANGE TO MOLAR QUANTITIES, MOLAR GIBBS ENERGY

A simplification of the thermodynamic description of systems is obtained by switching to molar quantities; by replacing k amount-of- substance variables by (k-1) mole fractions. This is shown for the Gibbs energy: the principle of minimal Gibbs energy corresponds to a lucid geometric criterion.

molar Gibbs energy

The fact that the Gibbs energy of a mixture of two substances A and B is a first-degree homogeneous function of the amounts of A and B, makes, that to know $G\,(T, P, n_A\,n_B)$, it is sufficient to know $G_m\,(T, P, X)$. That is to say the molar Gibbs energy as a function of temperature, pressure and the mole fraction of B, which is defined as

$$X = \frac{n_B}{n_A + n_B} \ . \tag{1}$$

Obviously, the mole fraction of A is equal to (1–X):

$$X_A = \frac{n_A}{n_A + n_B} = (1 - X) \ . \tag{2}$$

For constant $n_A + n_B$ (= 1, when one mole of mixture is taken), the differentials of the mole fractions are

$$dX = \frac{dn_B}{n_A + n_B}; \ \ d(1 - X) = \frac{dn_A}{n_A + n_B} = -dX \ . \tag{3}$$

The three general expressions (←202; Equations 202:5;10;12)

$$dG = -SdT + VdP + \mu_A dn_A + \mu_B dn_B \tag{4}$$

$$G = n_A \mu_A + n_B \mu_B \tag{5}$$

$$-SdT + VdP - n_A d\mu_A - n_B d\mu_B = 0 \tag{6}$$

can be made, upon division by $(n_A + n_B)$, to refer to one mole of mixture:

$$\frac{dG}{n_A + n_B} = -\frac{S}{n_A + n_B}\,dT + \frac{V}{n_A + n_B}\,dP + \mu_A \frac{dn_A}{n_A + n_B} + \mu_B \frac{dn_B}{n_A + n_B} \tag{7}$$

$$\frac{G}{n_A + n_B} = \frac{n_A}{n_A + n_B}\,\mu_A + \frac{n_B}{n_A + n_B}\,\mu_B \tag{8}$$

$$-\frac{S}{n_A+n_B}dT+\frac{V}{n_A+n_B}dP-\frac{n_A}{n_A+n_B}d\mu_A-\frac{n_B}{n_A+n_B}d\mu_B=0.\tag{9}$$

Obviously, the quantities G, S and V divided by ($n_A + n_B$) yield the molar quantities: the *integral molar quantities* G_m, S_m and V_m. Herewith, and with Equations (1), (2) and (3) we obtain the molar equivalents of Equations (4), (5) and (6): the differential expression (the *fundamental equation*; ←107) for the change of molar Gibbs energy

$$dG_m=-S_m dT+V_m dP+(\mu_B-\mu_A)dX;\tag{10}$$

the relation between the integral molar Gibbs energy and the chemical potentials

$$G_m=(1-X)\mu_A+X\mu_B;\tag{11}$$

and the Gibbs-Duhem equation

$$-S_m dT+V_m dP-(1-X)d\mu_A-Xd\mu_B=0.\tag{12}$$

Purely mathematically it follows from Equation (10) that ($\mu_B - \mu_A$) is the first partial differential coefficient of molar Gibbs energy with respect to mole fraction

$$\left(\frac{\partial G_m}{\partial X}\right)_{T,P}=\mu_B-\mu_A.\tag{13}$$

Equations (11) and (13) can be looked upon as two equations with two unknowns (μ_A and μ_B); the solution is

$$\mu_A=G_m-X\left(\frac{\partial G_m}{\partial X}\right)_{T,P}\tag{14}$$

$$\mu_B=G_m+(1-X)\left(\frac{\partial G_m}{\partial X}\right)_{T,P}.\tag{15}$$

molar quantities in general – geometric representation

The Equations (11), (13), (14) and (15) are of general validity for any molar, integral molar quantity Z_m and its corresponding partial quantities Z_A and Z_B. The first of the general equations

$$Z_m=(1-X)Z_A+XZ_B\tag{16}$$

gives Z_m when Z_A and Z_B are known. The two Equations (14) and (15) are the recipes for the determination of the two partial quantities from a known integral molar quantity; rewritten in the Z notation

$$Z_A = Z_m - X\left(\frac{\partial Z_m}{\partial X}\right)_{T,P} \tag{17}$$

$$Z_B = Z_m + (1-X)\left(\frac{\partial Z_m}{\partial X}\right)_{T,P}. \tag{18}$$

From an *algebraic* point of view, Equation (16) shows a great resemblance with its general counterpart

$$Z = n_A Z_A + n_B Z_B. \tag{19}$$

Equations (17) and (18), on the other hand, look more complex than

$$Z_A = \left(\frac{\partial Z}{\partial n_A}\right)_{T,P,n_B}; \tag{20}$$

and

$$Z_B = \left(\frac{\partial Z}{\partial n_B}\right)_{T,P,n_A}. \tag{21}$$

From a *geometric* point of view, the change-from-general-to-molar is exceptionally fruitful. First, there is the reduction of variables - as a result of which the Gibbs energy, at T, P constant, can be represented by a curve in the flat G_m-X plane. And second, in that flat G_m-X plane the partial properties Z_A and Z_B have a very clear appearance: Z_A and Z_B for a given $X = X'$ are the *intercepts of the tangent line*, at $X = X'$, on the axis $X = 0$ and $X = 1$, respectively. This is shown in Figure 1.

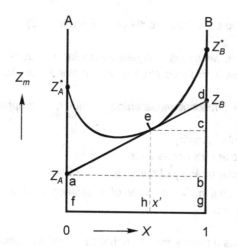

FIG. 1.   Relation between the integral molar quantity Z_m and the two partial quantities Z_A and Z_B

In Figure 1, $(\partial Z_m/\partial X) = bd / ab$, which equals bd, because $ab = 1$;
$fh = X$ and $hg = (1-X)$; $cd = (1-X)bd = (1-X)(\partial Z_m/\partial X)$;
$gd = he + cd = Z_m + (1-X)(\partial Z_m/\partial X) = Z_B$, Equation (18);
$fa = he - bc = Z_m - X(\partial Z_m/\partial X) = Z_A$, Equation (17).

One can read from Figure 1 that a change in X produces changes of opposite sign in Z_A and Z_B ; and also that the quotient of the differential coefficients is given by

$$\frac{\left(\dfrac{\partial Z_A}{\partial X}\right)}{\left(\dfrac{\partial Z_B}{\partial X}\right)} = -\frac{X}{(1-X)} \,. \tag{22}$$

This variant of the Gibbs-Duhem relation also follows from the general relation

$$\left(\frac{\partial Z_m}{\partial T}\right)_{P,X} dT + \left(\frac{\partial Z_m}{\partial P}\right)_{T,X} dP - (1-X)dZ_A - X\,dZ_B = 0: \tag{23}$$

'divide' this equation by dX and keep T and P constant. A third way to obtain Equation (22) starts by differentiation of Z_A and Z_B, Equations (17) and (18), with respect to X:

$$\left(\frac{\partial Z_A}{\partial X}\right) = -X\left(\frac{\partial^2 Z_m}{\partial X^2}\right); \tag{24}$$

$$\left(\frac{\partial Z_B}{\partial X}\right) = (1-X)\left(\frac{\partial^2 Z_m}{\partial X^2}\right). \tag{25}$$

The division of the two equalities directly gives Equation (22).

From now on we will almost uniquely use molar quantities - integral as well as partial (the latter were introduced already on a molar $\neq$ specific = related to mass base).
For that reason the subscript m, referring to molar, is dropped. From now on, therefore, we have

Z for molar quantity;
$Z(X)$ for molar quantity as a function of X;
Z_B^{*} for molar quantity of pure substance B;
Z_B for partial molar quantity of substance B in a mixture with other substances.

The Gibbs energy has a unique status, in that its partial quantities are identical with the chemical potentials:

$$\mu_B \equiv G_B \,. \tag{26}$$

the principle of minimal Gibbs energy

Having introduced the molar Gibbs energy function, we are in a new position (←202) to show that the *principle of minimal Gibbs energy* gives rise to *equilibrium conditions* in terms of chemical potentials. As an example we take, again, the equilibrium between liquid and vapour in the system $\{(1-X)\, A + X\, B\}$. We assume that A and B are completely miscible in the liquid as well as in the vapour state. Complete miscibility implies that, no matter the proportions in which A and B are added together, the substances mix spontaneously. Therefore, when the mixing is carried out at constant T and P, it will be accompanied by a lowering of the Gibbs energy. This is shown in Figure 2, where the Gibbs energy of $(1-X)$ mol A + X mol B before mixing, which is

$$(1-X)\, G_A^* + X\, G_B^* = G \text{ (unmixed, } X) \,, \tag{27}$$

is represented by the straight line between G_A^* and G_B^*. The fact that the mixing of A and B is accompanied by a lowering of the Gibbs energy implies that the molar Gibbs energy of the mixture has to be a *convex curve*, a curve of which the second derivative with respect to X is positive.

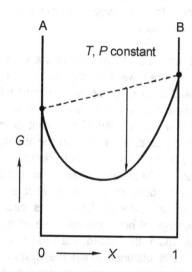

FIG. 2.   Molar Gibbs energy as a function of mole fraction X in the system $\{(1-X)\, A + X\, B\}$. The entirely convex curve is evidence of the fact that A and B mix in all proportions

In Figure 3 two such Gibbs energy curves are shown: one for liquid mixtures and one for vapour mixtures. The conditions of T and P are such that the two curves intersect. Now let us ask the question "what will happen when, for the selected T and P, $(1-X)$ mole of A and X mole of B are added together?"

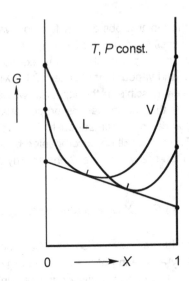

FIG. 3. The G-curve for the liquid state and the one for the vapour state intersect. In a cylinder-with-piston, the system is either homogeneous (liquid or vapour) or heterogeneous (liquid in equilibrium with vapour)

The answer is a function of X as we will see. In the vicinity of $X = 0$ the Gibbs energies of vapour mixtures are lower than the Gibbs energies of liquid mixtures: when added together A and B will give rise to a gaseous mixture. Similarly, in the vicinity of $X = 1$ the substances A and B will form a liquid mixture. A different situation arises when X is in the vicinity of the point of intersection of the two curves. More precisely expressed, when X is in the vicinity of the mole fraction value of the point of intersection. This situation is sketched in Figure 4, which is a blow-up of a part of Figure 3. To avoid confusion with the things that follow, let's say that $(1–X_o)$ mole of A is added to X_o mole of B. It is obvious that the *whole* amount of matter will not be gaseous (point a. is above point b., the latter representing the Gibbs energy of one mole of liquid mixture). Neither the whole amount of matter is liquid (point b). Indeed, the lowest possible Gibbs energy is represented by point c.: it is obtained when the mixture of overall (or global) composition X_o splits up, separates into an amount of vapour mixture of composition X_e^{vap} and an amount of liquid mixture of composition X_e^{liq}.

Let the amount vapour, the amount of the vapour phase be n^{vap} and the amount of the liquid phase n^{liq} $(= 1 – n^{vap})$, then the Gibbs energy of the equilibrium system is $n^{vap} \cdot G_e + n^{liq} \cdot G_d = G_c$. Next, the amounts of liquid and vapour follow from the *lever rule* ($\leftarrow$003):

$$\frac{n^{liq}}{n^{vap}} = \frac{X_o - X_e^{vap}}{X_e^{liq} - X_o} . \qquad (28)$$

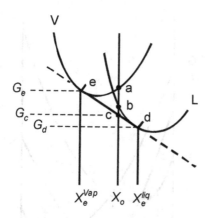

FIG. 4. Blow-up of the central part of Figure 3. For overall composition X_o the
system is heterogeneous: liquid in equilibrium with vapour

The most important observation to make is, that the equilibrium compositions of the
vapour and liquid phases, X_e^{vap} and X_e^{liq}, are given by the *points of contact of the
common tangent line* to the two Gibbs energy curves. And because a tangent line
is related to partial quantities, see Figure 1, this observation implies that, for the
compositions represented by the points of contact, the partial quantities will be
equal (see Figure 3):

$$G_A^{liq}(X_e^{liq}) = G_A^{vap}(X_e^{vap})\qquad\qquad(29a)$$

$$G_B^{liq}(X_e^{liq}) = G_B^{vap}(X_e^{vap}).\qquad\qquad(29b)$$

In other words, and with $G_A \equiv \mu_A$ and $G_B \equiv \mu_B$, the equilibrium condition implied in
the common tangent line is the double condition

$$\mu_A^{liq} = \mu_A^{vap}\qquad\qquad(30a)$$

$$\mu_B^{liq} = \mu_B^{vap}.\qquad\qquad(30b)$$

In summarizing, the *equilibrium conditions* in terms of chemical potentials - and
figuring in the *set N of conditions* - naturally follow from the *principle of minimal
Gibbs energy*.

GX diagrams, i.e. Gibbs energy versus mole fraction at constant T and P,
play a key role in the treatment of binary heterogeneous equilibria. The situations
that correspond to the lowest possible Gibbs energy are readily found: by stretching
in one's mind a cord along the underside of the G-curves, see Figure 5.

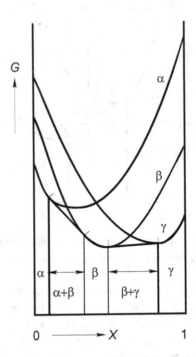

FIG. 5.   The *phase composition* of a system is readily found by stretching
a cord along the underside of the G-curves. Along the cord the
Gibbs energy has its lowest possible value

---

*The definition of the chemical potentials of the substances, in terms of molar Gibbs
energies and mole fraction variables, especially in the case of binary systems,
corresponds to a simple geometric interpretation. For that reason, graphical
representations of Gibbs energies are the most lucid opening to understanding
phase-equilibrium matters - the consequences of the principle of minimal Gibbs
energy.*

EXERCISES

1.   *integral molar quantities are zeroth degree homogeneous functions*

The integral molar quantities Z_m are *homogeneous functions* of the zeroth degree of
the variables n_A and n_B, the amounts of A and B.

- Or, in other words, prove that the following equality is valid

$$n_A \left(\frac{\partial Z_m}{\partial n_A} \right) + n_B \left(\frac{\partial Z_m}{\partial n_B} \right) = 0.$$

2.   *recipes for partial quantities in a ternary system*

For mixtures composed of $(1-X-Y)$ mole of substance A, X mole of substance B and Y mole of substance C, the equations relating partial quantities to integral molar quantities are

$$Z_A = Z_m - X\left(\frac{\partial Z_m}{\partial X}\right) - Y\left(\frac{\partial Z_m}{\partial Y}\right)$$

$$Z_B = Z_m + (1-X)\left(\frac{\partial Z_m}{\partial X}\right) - Y\left(\frac{\partial Z_m}{\partial Y}\right)$$

$$Z_C = Z_m - X\left(\frac{\partial Z_m}{\partial X}\right) + (1-Y)\left(\frac{\partial Z_m}{\partial Y}\right).$$

- Write down the derivation of these relations.

3.   *a Gibbs-Duhem exercise*

For a molar quantity Z_m at constant T and P it is given: $Z_A (X) = 8 - 10X^2$; and $Z_B (X = 0) = 0$.
- Derive $Z_B (X)$ by *Gibbs-Duhem integration* (dZ_B follows from dZ_A) and give the expression for $Z_m(X)$.

4.   *water + methanol: volumes, integral and partial*

Molar volumes, expressed in $cm^3 mol^{-1}$, are given for liquid mixtures of $(1-X)$ mole of water (A) + X mole of methanol (CH_3OH; B), valid for $t = 20$ °C.
- Calculate the volume change on mixing $\Delta_m V_m(X) = V_m(X) - (1-X)\,V_A^* - X\cdot V_B^*$ and make a plot of $\Delta_m V_m$ versus X. Use that plot to calculate the partial volumes of A and B for the mixture having $X = 0.591$. Check the result by means of $V_m = (1-X)V_A + X\cdot V_B$.

Data:

X	0	0.059	0.123	0.240	0.342	0.458	0.591	0.692	0.866	1
V_m	18.047	19.20	20.42	22.70	24.80	27.32	30.34	32.73	37.01	40.46

5.   *the ideal mixture*

The molar Gibbs energy of a so-called *ideal mixture* of $(1-X)$ mole of substance A and X mole of substance B is given by the expression ($\rightarrow$204)
$$G_m(T,P,X) = (1-X)G_A^*(T,P) + XG_B^*(T,P) + RT\{(1-X)\ln(1-X) + X\ln X\},$$
where R is the gas constant.
- Derive the expressions for the chemical potentials of A and B.
- Calculate the *total Gibbs energy* of a mixture of (10 mol A + 30 mol B) having a temperature of 400 K; the molar Gibbs energies of pure A and B, for this temperature and the selected pressure, are 1000 and 2000 $Jmol^{-1}$, respectively.

A formula is derived for the Gibbs energy change of mixing (1–X) mole of substance A with X mole of substance B, assuming that the only thermodynamic effect of mixing is the entropy effect of mixing $(1-X)N_{Av}$ A molecules with $X N_{Av}$ B molecules in a random manner.

the ideal gas mixture

   With the help of Figure 1 two situations are compared for given constant temperature and pressure. In Figure 1a the left-hand cylinder contains (1–X) mole of the ideal-gaseous substance A; and the right-hand cylinder X mole of the ideal-gaseous substance B. The cylinder in Figure 1b contains a mixture, having ideal-gas properties, of (1–X) mole of A and X mole of B.

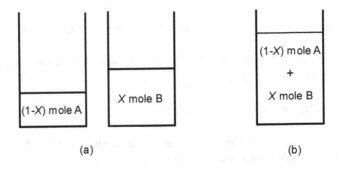

(a)   (b)

FIG. 1.   Amounts of A and B; before (a), and after mixing (b)

In the case of a. the contents of the vessels represent an 'amount' of Gibbs energy given by (←203; ←108)

$$G = (1- X)G_A^* + XG_B^*$$

$$= (1- X)\{G_A^o + RT \ln P\} + X\{G_B^o + RT \ln P\} \tag{1}$$

$$.. = (1- X)G_A^o + XG_B^o + RT \ln P.$$

This Gibbs energy corresponds to the dashed line in Figure 2 (left-hand side).

The mixing of the two gases is a spontaneous event: after mixing, the Gibbs energy, as a function of X, must have the convex appearance of Figure 2 (right-hand side (←203).

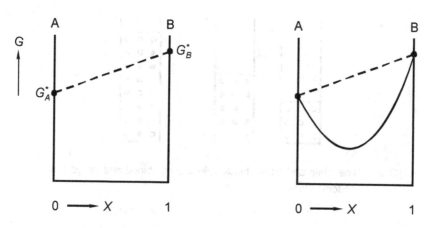

FIG. 2.   Gibbs energy as a function of X; before (left) and after mixing (right)

A convenient way to find the *Gibbs energy change on mixing*, i.e. the distance from the dashed line to the convex curve

$$\Delta_m G(X) = G(\text{mixture}, X) - (1 - X)G_A^* - XG_B^*, \tag{2}$$

is to use the *Boltzmann formula*, to derive the entropy change, $\Delta_m S$, and to state that the enthalpy change, $\Delta_m H$, is zero:

$$\Delta_m G(\text{generally}) = \Delta_m H - T \cdot \Delta_m S \quad (\text{in this case}) \ = -T \cdot \Delta_m S. \tag{3}$$

Indeed, $\Delta_m H = 0$ because an ideal gas is an ensemble of molecules which have no interaction whatsoever.

In terms of the Boltzmann formula ($\leftarrow$106) the entropy change can be found from

$$\Delta_m S = \frac{R}{N_{AV}} \ln \frac{W_b}{W_a}, \tag{4}$$

where R is the gas constant as in Equation (1) and N_{AV} is Avogadro's number. For our purpose W_b and W_a have to represent the number of configurations of the two situations before (a), and after mixing (b). So, if we put W_a equal to one, then W_b has to represent the number of ways in which $(1-X) \cdot N_{AV}$ molecules of A and $X \cdot N_{AV}$ molecules of B can be distributed over N_{AV} positions. Remembering that (the sign) ! is for factorial ($4! = 1 \times 2 \times 3 \times 4$) and knowing that 9 white balls and 15 black balls can be distributed in $24! / 9!15!$ different manners over 24 positions, see Figure 3, we have

$$W_b = \frac{N_{AV}!}{\{(1-X)N_{AV}\}! \ \{XN_{AV}\}!}. \tag{5}$$

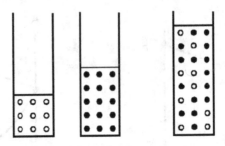

FIG. 3.   Nine white and fifteen black balls are combined and mixed in a
random manner

Making use of *Stirling's approximation* (see Perrot 1998) for $\ln(N_{AV}!)$, which is

$$\ln(N_{AV}!) = N_{AV}\ln N_{AV} - N_{AV}, \tag{6}$$

we obtain, substituting Equation (5) into Equation (4),

$$\Delta_m S = -R\{(1-X)\ln(1-X) + X\ln X\}. \tag{7}$$

With this result the Gibbs energy change on mixing, Equation (3), is given by

$$\Delta_m G = RT\{(1-X)\ln(1-X) + X\ln X\}. \tag{8}$$

The formula for the convex curve in Figure 2, the formula for the Gibbs energy of a mixture of $(1-X)$ mole of ideal gas A and X mole of ideal gas B, now reads

$$G(T,P,X) = (1-X)G_A^{\bullet}(T,P) + XG_B^{\bullet}(T,P) + RT\{(1-X)\ln(1-X) + X\ln X\}; \tag{9}$$

and it can also be written as

$$G(T,P,X) = (1-X)G_A^{\circ}(T) + XG_B^{\circ}(T) + RT\ln P + RT\{(1-X)\ln(1-X) + X\ln X\}. \tag{10}$$

the ideal mixture

The formula for the molar Gibbs energy, Equation (9), plays an important part in thermodynamics. A mixture of A and B in a certain form α (α can be a solid form, liquid, or gas) is referred to as an *ideal mixture* if its Gibbs energy is given by

$$G^{\alpha}(T,P,X) = (1-X)G_A^{\bullet\alpha}(T,P) + XG_B^{\bullet\alpha}(T,P) + RT\{(1-X)\ln(1-X) + X\ln X\}. \tag{11}$$

As an observation, to know the Gibbs energy of an ideal mixture, it is sufficient to know the thermodynamic properties of the pure components A and B, i.e. $G_A^{\bullet}$ and $G_B^{\bullet}$. The rest, the logarithmic term multiplied by RT, is just a matter of arithmetic.

All thermodynamic properties of the ideal mixture are implied in Equation (11) - G is characteristic for T and P, as we know (←107).
The entropy of the ideal mixture is obtained by differentiation with respect to temperature, knowing that $-(\partial G_A^* / \partial T)_P = S_A^*$:

$$S^\alpha(T,P,X) = (1-X)S_A^{*\alpha}(T,P) + XS_B^{*\alpha}(T,P) - R\{(1-X)\ln(1-X) + X\ln X\}. \quad (12)$$

Similarly, the volume of the ideal mixture is given by

$$V^\alpha(T,P,X) = (1-X)V_A^*(T,P) + XV_B^*(T,P). \quad (13)$$

The enthalpy $H = G + TS$ is given by

$$H^\alpha(T,P,X) = (1-X)H_A^*(T,P) + XH_B^*(T,P), \quad (14)$$

and so on.

We may observe that Equation (13) and Equation (14) also hold true for $(1-X)$ mole of A and X mole of B separated from one another, before mixing so to say. In other words, the formation of an ideal mixture out of the pure components is accompanied neither by a heat effect nor a volume effect. In the case of an ideal mixture, the *heat of mixing* and the *volume change on mixing* both are zero.
The chemical potentials of the components A and B of the ideal mixture are (←203: Equations 14, 15; Exc 203:5)

$$\mu_A^\alpha(T,P,X) = G_A^{*\alpha}(T,P) + RT\ln(1-X) \quad (15)$$
$$\mu_B^\alpha(T,P,X) = G_B^{*\alpha}(T,P) + RT\ln X. \quad (16)$$

These two equations can also be written as

$$\mu_A^\alpha = G_A^{*\alpha} + RT\ln X_A \quad (17)$$
$$\mu_B^\alpha = G_B^{*\alpha} + RT\ln X_B; \quad (18)$$

to show that the chemical potential of a substance in the ideal mixture is related to its own mole fraction. It can easily be verified that this property is of general validity: no matter the number of components, the chemical potential of the component i is given by

$$\mu_i = G_i^* + RT\ln X_i. \quad (19)$$

In the ideal mixture the chemical potential of any substance is independent of the mole fractions, with the exception of its own.

The molar Gibbs energy of the ideal mixture of n components is given by

$$G(T,P,\text{all } X_i) = \sum_{i=1}^{n} X_i \mu_i = \sum_{i=1}^{n} X_i \{G_i^*(T,P) + RT \ln X_i\}. \qquad (20)$$

The Gibbs energy of mixing of the ideal mixture is given by RT { (1-X) ln (1-X) + XlnX}. As a consequence there are no heat and volume effects on mixing. The chemical potential of a substance in a mixture is, apart from T and P, a function of its own mole fraction only.

EXERCISES

1. *a Gibbs-Duhem exercise*

A and B are two pure, liquid substances that mix in all proportions.
- Is the statement that the chemical potential of A in the mixture is given by $G_A^* + RT \ln X_A$ sufficient to express that A and B mix ideally?

2. *mathematical analysis of a function*

For the function $f(X) = (1-X) \ln (1-X) + X \ln X$ derive the expressions for the first derivative $f'(X)$ and the second derivative $f''(X)$ and make plots of $f(X)$, $f'(X)$ and $f''(X)$.

§ 205 NON-IDEAL BEHAVIOUR,
EXCESS FUNCTIONS

In the real world of materials the Gibbs energy of mixing is a function - whose complexity differs from case to case - of temperature, pressure and composition. The thermodynamic description of real mixtures starts from the formulae derived for the ideal mixture - the mixing effects of the latter are just standard expressions in temperature and mole fractions.

deviation from ideal-mixing behaviour

In the foregoing section we derived for the *ideal mixture* of $(1-X)$ mole of substance A and X mole of substance B - more precisely for a mixture of $(1-X) \cdot N_{AV}$ particles A and $X \cdot N_{AV}$ particles B - the following formula for its Gibbs energy

$$G(T,P,X) = (1-X)G_A^*(T,P) + XG_B^*(T,P) + RT\{(1-X)\ln(1-X) + X\ln X\}. \qquad (1)$$

One has to fear that the thermodynamic description of a *real system* by means of Equation (1) becomes more and more inaccurate the more the mixture differs from the ideal-gas mixture, the more the particles differ from looking like spheres and the greater the difference in chemical nature between the constituents.
Equation (1) may be good enough for a liquid mixture of argon and krypton or a liquid mixture of n-heptane and n-octane; it certainly will fail for a liquid mixture of silica and magnesium oxide. In the case of the latter, one has the right of saying that a molar mixture consists of $(1-X)$ times 60.085 g of silica and X times 40.304 g of magnesium oxide. Thereafter one has not the right of saying that such a mixture consists of N_{AV} particles, N_{AV} structural units that are capable of changing their positions with respect to one another. In reality SiO_2 and MgO react with each other, giving rise to all sorts of anions whose polymeric nature changes with composition.

No matter the deviation from ideal-mixing behaviour, or rather the extent of that deviation, one has, of course, the right to take the ideal mixture as a reference for the description of a real mixture.
In Figure 1 the Gibbs energy of a *real system* at a given T and P is compared with the Gibbs energy of the *hypothetical ideal system* of the same components under the same conditions. At the bottom of the figure the difference between the two functions is drawn. Obviously that difference is zero at $X = 0$ and $X = 1$, that is to say for the two pure components, the substances A and B.

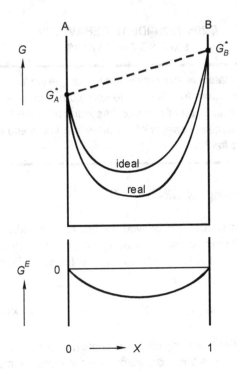

FIG. 1.   For given temperature and pressure, the Gibbs energy of a real
mixture differs from the Gibbs energy of the hypothetical ideal
mixture of the same components. The difference between the two
is the excess Gibbs energy

The difference between real and ideal Gibbs energy is called *excess Gibbs energy*
and it is defined as

$$G^E_{T,P}(X) = G_{T,P}(X) - G^{id}_{T,P}(X),$$ (2)

in which the superscript *id.* is used for the ideal mixture. Generally therefore, the
molar Gibbs energy of a *real mixture* can be formulated as

$$G(T,P,X) = (1-X)G^*_A(T,P) + XG^*_B(T,P) + RT\ LN(X) + G^E(T,P,X) .$$ (3)

$LN(X)$ is a short rotation for the logarithmic term:

$$LN(X) = (1-X)\ln(1-X) + X\ln X .$$ (4)

The corresponding formulae for entropy, enthalpy and volume are ($\leftarrow$107)

$$S(T,P,X) = (1-X)S^*_A(T,P) + XS^*_B(T,P) - R\ LN(X) + S^E(T,P,X);$$ (5)

$$H(T,P,X) = (1-X)H^*_A(T,P) + XH^*_B(T,P) + H^E(T,P,X) ;$$ (6)

$$V(T,P,X) = (1-X)V_A^\bullet(T,P) + X\,V_B^\bullet(T,P) + V^E(T,P,X). \tag{7}$$

The *excess enthalpy* is identical with the *heat of mixing* and the *excess volume* with the *volume change on mixing*. The latter, by the way, is the partial derivative of the excess Gibbs energy with respect to pressure. The *excess Gibbs energy* and the *excess entropy*, on the other hand, are not identical with the *Gibbs energy* and *entropy changes on mixing*.

The *partial excess Gibbs energies*, the *excess chemical potentials* follow from the excess Gibbs energy with help of the recipes:

$$G_A^E \equiv \mu_A^E = G^E - X\frac{\partial G^E}{\partial X} \tag{8}$$

$$G_B^E \equiv \mu_B^E = G^E + (1-X)\frac{\partial G^E}{\partial X}. \tag{9}$$

In its most simple form the excess Gibbs energy is a parabola; and because it has to be zero for $X = 0$ and $X = 1$, it can be represented by

$$G_{T,P}^E(X) = A_{T,P}\cdot X(1-X); \tag{10}$$

or, generally,

$$G^E(T,P,X) = A(T,P)\cdot X(1-X). \tag{11}$$

The excess functions generated by this form are

$$S^E = -\left(\frac{\partial A}{\partial T}\right)_P X(1-X) \tag{12}$$

$$V^E = \left(\frac{\partial A}{\partial P}\right)_T X(1-X) \tag{13}$$

$$H^E = \left\{A - T\left(\frac{\partial A}{\partial T}\right)_P\right\} X(1-X) \tag{14}$$

$$\mu_A^E = A\cdot X^2 \tag{15}$$

$$\mu_B^E = A\cdot(1-X)^2, \tag{16}$$

and so on.

The excess function of the form $A \cdot X\,(1-X)$ is symmetrical with respect to $X = 0.5$ and has a constant second derivative. In reality an excess function may be less symmetrical and to a lesser or greater extent 'sharper' than the form $A \cdot X\,(1-X)$. These aspects can be accounted for by the addition of 'correction terms' such as in

$$G^E = A \cdot X\,(1-X)\left\{1 + B \cdot (1-2X) + C \cdot (1-2X)^2\right\}, \qquad (17)$$

with A, B and C depending on T and P. The parameter A with the dimension of energy per mole expresses the magnitude of the excess function; $0.25\,A$ is the function value for $X = 0.5$. The dimension-less parameters B and C are a measure of the asymmetry and the sharpness of the function, respectively. The expansion, between the curly brackets, in $(1-2X)$ rather than in X, has the advantage that the absolute values of A, B and C do not change when the components of the binary system are interchanged.

NB Expressions of the type of Equation (17) usually are referred to as Redlich-Kister expressions (Redlich and Kister 1948).

an intermezzo on linear contributions

Let's isolate the term, $X \cdot G_B^{*}$, from Equation (3), and subject it to a close inspection as regards the nature of its zero point(s). As always, it has an enthalpy and an entropy part:

$$G_B^{*} = H_B^{*} - TS_B^{*}. \qquad (18)$$

We know that it is possible to put S^{*} on an absolute scale ($\leftarrow$106): in this little digression S_B^{*} is the *absolute entropy* of pure B.

The quantity H_B^{*}, on the other hand, does not have a *natural zero point* ($\leftarrow$109): nothing in the world of B will change if one is going to replace H_B^{*} by $H_B^{*} + a$, where a can have any value, e.g. 10 kJ·mol^{-1}. If one is really going to do so, the term XG_B^{*} changes to $XG_B^{*} + a \cdot X$, and with it the complete Gibbs function, Equation (3).

In a next step we take two forms α and β in which the substances A and B are miscible. The forms α and β can be, e.g. mixed crystalline solid and liquid, respectively. Their Gibbs energies are given, T and P being constant, by

$$G^\alpha = (1-X)G_A^{*\alpha} + XG_B^{*\alpha} + RT\,\mathrm{LN}(X) + G^{E\,\alpha}(X) \qquad (19)$$

$$G^\beta = (1-X)G_A^{*\beta} + XG_B^{*\beta} + RT\,\mathrm{LN}(X) + G^{E\,\beta}(X). \qquad (20)$$

Now, if we are going to change $H_B^{*\alpha}$ to $H_B^{*\alpha} + a$, we are obliged to change, at the same time, $H_B^{*\beta}$ to $H_B^{*\beta} + a$, because of the fact that $\Delta H_B^{*} = H_B^{*\beta} - H_B^{*\alpha}$ is a physical reality. In the case of α = solid and β = liquid that reality is the *heat of melting* of substance B.

As a result so far, nothing in the treatment of equilibria between α and β will change if we replace G^α in Equation (19) by

$$G'^\alpha = G^\alpha + aX, \tag{21}$$

and at the same time G^β in Equation (20) by

$$G'^\beta = G^\beta + aX. \tag{22}$$

The effect of the addition of the linear term is shown in Figure 2: A's chemical potential remains unchanged, while B's chemical potential changes into μ_B +a. The equilibrium equations $\mu_A^\alpha = \mu_A^\beta$ and $\mu_B^\alpha = \mu_B^\beta$ remain unaffected and the same holds true for the abscissae of the points of contact of the common tangent, which represent the mole fractions of the coexisting phases.

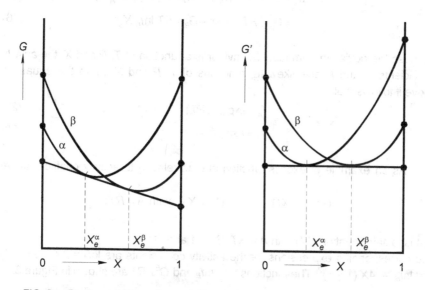

FIG. 2. By the addition of equal linear contributions to the two Gibbs functions, the points of contact can be made to coincide with the minima

The *linear contribution concept* is a valuable tool in phase equilibrium matters. By means of linear contributions sketch drawings of (enlargements of) GX diagrams can be given a lucid appearance - a speaking example is Figure 211:8. By means of linear contributions any double tangent line can be made parallel to the X axis: the complex problem of numerically calculating the positions of the points of contact is reduced to the simple problem of determining the positions of minima ($\rightarrow$213).

activity coefficients

An alternative manner to treat *deviation from ideal-mixing behaviour,* is to start from the expressions for the ideal chemical potentials (*partial Gibbs energies*), rather than from the Gibbs energy (the *integral Gibbs energy*) itself.
The formulae for the chemical potentials of A and B that correspond to Equation (3) are

$$\mu_A = G_A^* + RT\ln(1-X) + \mu_A^E \tag{23}$$

$$\mu_B = G_B^* + RT\ln X + \mu_B^E . \tag{24}$$

The alternative manner is the use of *activity coefficients*: the mole fractions that appear under the logarithms in the 'ideal' expressions for the chemical potentials are multiplied by some kind of 'correction factor' - the activity coefficient:

$$\mu_A = G_A^* + RT\ln f_A \cdot (1-X) = G_A^* + RT\ln f_A \cdot X_A \tag{25}$$

$$\mu_B = G_B^* + RT\ln f_B \cdot X + G_B^* + RT\ln f_B \cdot X_B . \tag{26}$$

Because the deviation from ideal behaviour is a function of T, P and X, the activity coefficients f_A and f_B are, likewise, functions of T, P and X. From the equations above it follows that

$$f_A = \exp(\mu_A^E/RT) \tag{27}$$

$$f_B = \exp(\mu_B^E/RT) . \tag{28}$$

As an example of excess function and its relation to activity coefficients we take

$$G^E = A \cdot X(1-X)\left[1-(1-2X)\right], \text{ with } A = RT;$$

that is to say Equation (17) with $A = RT$; $B = -1$ and $C = 0$.
The corresponding expressions for the activity coefficients are $\ln f_A = 2X^2 \cdot (2X-1)$ and $\ln f_B = 4X(1-X)^2$. The functions $\ln f_A$; $\ln f_B$ and G^E/RT are shown in Figure 3.

activities

The products of *activity coefficient* and mole fraction, $f_A \cdot X_A$ in Equation (25) and $f_B \cdot X_B$ in Equation (26), obviously, like f_A and f_B, are functions of T, P and X. For that reason the products just as well can be replaced by single properties:

$$f_A \cdot X_A = f_A(T,P,X) \cdot (1-X) \Rightarrow a_A(T,P,X) \tag{29}$$

$$f_B \cdot X_B = f_B(T,P,X) \cdot X \Rightarrow a_B(T,P,X). \tag{30}$$

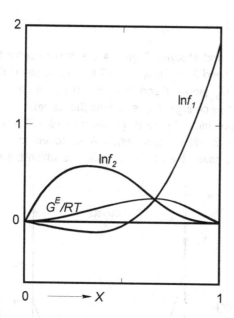

FIG. 3. The *integral* function (excess Gibbs energy divided by RT) and its corresponding two *partial* functions (the logarithms of the activity coefficients of A(1), and B(2))

The properties a_A and a_B, in Equations (29) and (30), are referred to as *activities* - the activity of A, and the activity of B, respectively. In terms of activities, the expressions for the chemical potentials become

$$\mu_A = G_A^* + RT \ln a_A \tag{31}$$
$$\mu_B = G_B^* + RT \ln a_B; \tag{32}$$

and the *integral molar Gibbs energy*

$$G = (1 - X)G_A^* + XG_B^* + RT \left\{ (1 - X)\ln a_A + X \ln a_B \right\}. \tag{33}$$

The relationships between the various properties are depicted in Figure 4. Note that in the case of this figure the distance from G_A^* to μ_A and the distance from G_B^* to μ_B are negative; as a result, the activities a_A and a_B are smaller than 1. Note also, by comparing Figures 2 and 4, that a_A and a_B are independent of *the addition of linear contributions*, i.e. independent of the *choice of zero points*. In other words, a_A and a_B are single-valued characteristics of the system (A+B), accessible to experimental determination. The chemical potentials, on the other hand, can be given any value, by playing with zero points.

activity and standard state

Taking Equation (32) and studying Figure 4 we may observe that, for the given circumstances of T, P and X, the property $RT \ln a_B$ so to say is the 'correction' that has to be applied to G_B^* (same T and P) to obtain μ_B (for other values of X at the same T and P, a_B and μ_B change, G_B^* remaining the same).

G_B^* is the chemical potential of pure B and as a result $RT \ln a_B$ is the change the chemical potential of B will undergo when A is added to (pure) B. In activity terminology pure B is called the *standard state* - to which the correction $RT \ln a_B$ has to be applied.

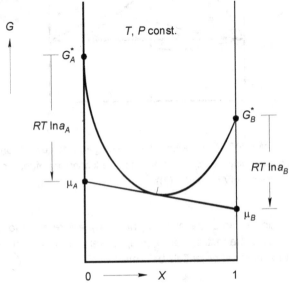

FIG. 4.   *GX* diagram showing the relationships between molar Gibbs energy, chemical potentials and activities

In the case of Figure 4 the mixture of A and B, which is considered, is in the same form (e.g. liquid) as pure B - the mixture and the standard state have the same form. In the case of Figure 5 - think of A = water and B = sodium chloride - the conditions of T and P are such that A is liquid and B is solid. In such a case pure liquid B is not a convenient standard state because its position on the G scale - the position of $G_B^{* \, liq}$ - has to be obtained by extrapolation. The inconvenience can be circumvented by taking pure solid B for standard state: $RT \ln a_B$ is the "correction" to be applied to the molar Gibbs energy of pure *solid* B to obtain the chemical potential of B in the *liquid* mixture of A and B:

$$\mu_B^{liq} = G_B^{* \, sol} + RT \ln a_B' ; \tag{34}$$

see Figure 5.

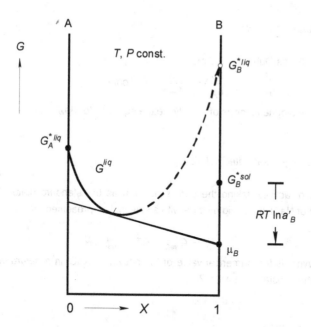

FIG. 5. If one prefers, one can take the pure solid form of a substance as the standard state for its activity in a liquid mixture

---

Non-ideal mixing behaviour can be expressed by means of functions that are based on the formulation of the ideal mixture. The integral excess Gibbs energy is the deviation of the Gibbs energy of mixing from the ideal expression. The description in terms of activities starts from the ideal expressions for the partial Gibbs energies - the chemical potentials

---

EXERCISES

1. *ideal or not?*

For a given mixture at given T and P the excess Gibbs energy is zero over the whole mole fraction range.

- Is this mixture an *ideal mixture* ?

2. *Gibbs-Duhem and activity*

From the Gibbs-Duhem equation

$$\frac{d\mu_A}{d\mu_B} = -\frac{X}{1-X}; \quad T,P \text{ const.}$$

derive its analogue in terms of *activity*, i.e. the equation for da_A/da_B.

3. *NaCl's activity in saturated solution*

In terms of *activity*, taking the pure solid form as the standard state, the chemical potential of NaCl in a liquid mixture with a solvent is represented by

$$\mu_{NaCl}^{liq} = G_{NaCl}^{*sol} + RT \ln a_{NaCl}^{liq}.$$

- What is the numerical value of the activity of NaCl in a *saturated solution* of the substance in water?

4. *equality of activities*

If one wishes to do so, one can use one and the same standard state to define the activities of substance B for *all* phases in which it is present.
- Show that in that case 'equality of chemical potentials', as equilibrium condition, can be replaced by *equality of activities*.

5. *heat of mixing and activity coefficients*

At given T and P two substances A and B have the same form, in which they mix in all proportions.
- Show that the following relation is valid between the heat of mixing and the activity coefficients

$$\Delta_m H = -RT^2 \left\{ (1-X)\left(\frac{\partial \ln f_A}{\partial T}\right)_{P,X} + X\left(\frac{\partial \ln f_B}{\partial T}\right)_{P,X} \right\},$$

X being the mole fraction of B.

6. *a convenient formula*

A convenient formula for expressing the isobaric excess Gibbs energy of binary mixed crystalline materials is

$$G^E(T, X) = A\left(1-\frac{T}{\Theta}\right)X(1-X)[1+B(1-2X)],$$

in which A, B and Θ are system-dependent constants.

- For the system {(1–X) mole of component A + X mole of component B} derive the corresponding expressions for the excess properties S^E, H^E, C_P^E, μ_1^E and μ_2^E and the activity coefficients f_1 and f_2, the subscripts 1 and 2 referring to the components A and B, respectively.

7.   *alcohol is mixed with water – a classroom experiment*

Water is poured into a 100 ml measuring cylinder to the level of 51.3 ml. The temperature of the water in the cylinder is 19.2 °C. Next, avoiding mixing by keeping the cylinder in a nearly horizontal position, absolute alcohol (= waterless, pure ethanol) is added to the cylinder in a gentle manner until - the cylinder being in a vertical position again - the 100 ml mark is reached. The temperature of the alcohol layer is 21.9 °C. In a subsequent step the cylinder is shaken, as a result of which the two liquids mix completely. Directly after mixing the temperature of the homogeneous mixture is 27.3 °C. One hour later the temperature has dropped to 21.4 °C, and the level of the liquid in the cylinder has become 95.8 ml.

- What are the signs of the heat of mixing, and the volume change on mixing?
- For the composition the mixture has, neglecting the heat capacity of the cylinder and ignoring the existence of systematic and random errors, calculate the values of the molar excess enthalpy and the molar excess volume.

	molar mass (g)	density (g cm^{-3})	heat capacity (J K^{-1} mol^{-1})
Alcohol	46.1	0.79	111.5
Water	18.0	1.00	75.3

§ 206 MAGIC FORMULAE

Two simple formulae that account for non-ideal behaviour are shown to have the power of explaining, of predicting the existence of critical points. One of them is the Van der Waals equation. The other is a formula for the Gibbs energy of mixing.

magic formulae

The *ideal gas* and the *ideal mixture* are two *model systems* that are frequently brought into action. The ideal gas is characterized by the *ideal-gas equation,* which is the following relation between molar volume, temperature and pressure

$$V \cdot P = R \cdot T. \tag{1}$$

The R in the equation is the *gas constant;* $R = 8.314472$ $m^3 \cdot Pa \cdot K^{-1} \cdot mol^{-1}$.
The second model system, the ideal mixture, is characterized by a special expression for the *molar Gibbs energy of mixing,* which, by the way, is linear in temperature and independent of pressure. That expression is

$$\Delta_m G(T, P, X) = \Delta_m G(T, X) = RT\{(1 - X)\ln(1 - X) + X \ln X\}. \tag{2}$$

Real gases and *real mixtures* deviate - to a lesser or greater extent and depending on chemical nature and circumstances - from the ideal behaviour expressed by Equations (1) and (2), respectively. For real systems, therefore, the formulae have to be adapted or extended, to include system-dependent parameters - parameters that depend on chemical nature and, in addition, may be functions of the state variables.

The *Van der Waals equation* for gases

$$(V - b)\left(P + \frac{a}{V^2}\right) = RT \tag{3}$$

is meant to take into account (*i*) the proper volume of the molecules, by the system-dependent parameter b, and (*ii*) the energetic interaction between the molecules, expressed by the system-dependent parameter a divided by V^2; or rather multiplied by $(1/V) \cdot (1/V)$, which expresses the probability of an interaction between two molecules.

Ideal-mixing behaviour presupposes a *neutral interaction* between the units (e.g. molecules) of the substances that are mixed (let us say that neutral stands for some kind of mean interaction, symbolized by AB = 1/2 (AA + BB)). In reality, that is to say for real mixtures, the interaction between unlike units is either more attractive or less attractive (more repulsive) than the mean. A simple way to account for the 'extra interaction' is to add a term, which is the product of strength

(Ω) and probability ($X_A \cdot X_B = X(1-X)$) of the interaction between the unlike units (←005; ←205):

$$\Delta_m G = RT\{(1-X)\ln(1-X) + X \ln X\} + \Omega X(1-X). \tag{4}$$

In first approximation the *interaction parameter* Ω is a constant; negative for a net attraction and positive for a net repulsion.

It is obvious, indeed, that Equations (3) and (4) with their *adjustable parameters* will give a better description of reality than will do Equations (1) and (2), respectively. It is also obvious - maybe to a somewhat lesser extent - that neither of the two equations will have the power of giving a correct description of the systems for all experimentally accessible circumstances. The real power, the *magic* of the two equations is that they are the key to the interpretation and understanding of a series of phenomena and properties - the existence of *critical behaviour* in particular.

critical behaviour

The *boiling curve*, the curve that represents the equilibrium states of the equilibrium between liquid and vapour - in the *PT phase diagram* of a *pure substance* - is ending in an abrupt manner, at the *critical point*, see Figure 1.

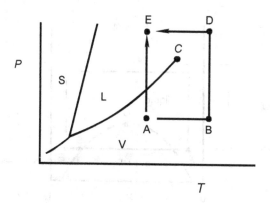

FIG. 1.   *PT* phase diagram of a pure substance. The boiling curve ends
abruptly at the critical point (*C*)

Two routes are depicted along which an amount of substance is transferred from state A, where it is gaseous, to state E, where it is liquid. Along the direct route the system is isothermally compressed and, as we know, at the boiling curve the gaseous form will change completely into the liquid form. There is an experimentally visible transition from vapour to liquid. Along the indirect route, on the other hand, there is no abrupt and clear transition to be seen! The system passes gradually from the gaseous form at A into the liquid form at E, via B where it is gaseous and D where it is said to be *supercritical*.

The continuity of the gaseous and liquid forms which follows from this behaviour is anchored in the Van der Waals equation - can be accounted for by the equation. Without going into detail, we may remark that the coordinates of the critical point, the *critical pressure* P_C and the *critical temperature* T_C follow from the constants a and b:

$$P_c = \frac{a}{27b^2} \; ; \quad T_c = \frac{8a}{27b \cdot R} . \tag{5}$$

The other way round,

$$a = \frac{27R^2T_c^2}{64P_c} \quad \text{and} \quad b = \frac{RT_c}{8P_c} . \tag{6}$$

The *critical volume*, the volume of the system at the critical point, is three times the value of the constant b:

$$V_c = 3b . \tag{7}$$

region of demixing

Phenomena that resemble the ones around the critical point in Figure 1 are observable for (liquid) mixtures that involve a *region of demixing*, see Figure 2.

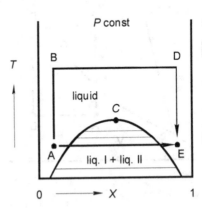

FIG. 2.   *TX* phase diagram of a binary system showing a region of demixing with its critical point

In the case of Figure 2 the system can be transferred from state A to state E by feeding it with the right-hand pure substance, the second component. Along the direct, the isothermal route from A to E, the *binodal* - the boundary of the *region of demixing* - is crossed twice. It can clearly be observed that along the route two liquid layers are formed, that one of the two liquids grows at the expense of the other and that the latter from a certain moment on is exhausted. Along the indirect route, on the other hand, the system undergoes the transfer from A to E without producing the phenomena of demixing and remixing.

This time the observations, read, the existence of a region of demixing and its critical point, can be fully explained in terms of the second *magic formula*, Equation (4). Otherwise expressed, the *continuity of* the *states* A and E is embedded in the formula in case Ω is positive. As was observed above, a positive value for Ω corresponds to a net repulsion. The repulsion, now, may become such that it will give rise to a separation, a *separation into* two different *phases*. The explanation is as follows.

The Gibbs energy of a mixture of $(1-X)$ mole of component A and X mole of component B is, in terms of Equation (4) for the Gibbs energy of mixing, given by

$$G(T,X) = (1-X)G_A^*(T) + X G_B^*(T) + RT\{(1-X)\ln(1-X) + X \ln X\} + \Omega X (1-X). \quad (8)$$

For positive Ω the extra term $\Omega X (1-X)$ is able to give the function the appearance of Figure 3, that is to say to introduce a section where the function is *concave*, i.e. its second derivative being negative (note that the ideal function, for $\Omega = 0$, and also the function with negative Ω are *convex* over the whole range, from $X = 0$, to $X = 1$).

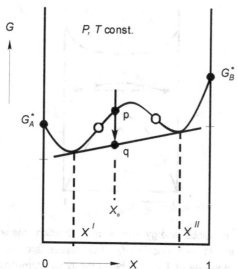

FIG. 3.   Curve of Gibbs energy versus mole fraction having a concave
section. Open circles: points of inflexion

With reference to § 203 we observe that for *overall mole fractions* X_o, with $X' < X_o < X''$, there is a *driving force for phase separation*: the Gibbs energy of a two-phase system (point q) in which the phases have compositions X' and X'' is lower than the Gibbs energy (point p) a single-phase system of composition X_o would have.

As a next step, with reference to the intermezzo in § 205, we know that nothing in the treatment will change if we are going to add a *linear term* to the formula for the Gibbs energy, Equation (8). Or, in other terms, nothing will change when we are going to remove the linear term $(1-X)\,G_A^{\cdot} + X\,G_B^{\cdot}$ and concentrate on the function

$$G'(T,X) = RT\{(1-X)\ln(1-X) + X\ln X\} + \Omega X(1-X),\qquad(9)$$

which is in fact $\Delta_m G$, Equation (4), the Gibbs energy of mixing.

To continue, we will examine the properties of the function G' for positive Ω. The function, to start with, is symmetrical with respect to $X = 1/2$; it is represented in Figure 4, from bottom to top for increasing temperatures. At the bottom, for zero kelvin, just the parabola $\Omega\,X\,(1-X)$; the two pure substances, the two pure components have the lowest possible Gibbs energy for every X_o: there is no tendency to mixing at all. At the top, for high temperatures the $RT\,\{(1-X)\,\ln(1-X) + X\,\ln X\,\}$ term dominates: the function is convex - the components mix in all proportions.

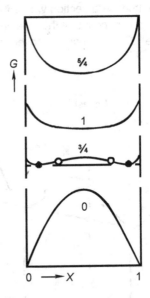

FIG. 4.   Curves of Gibbs energy versus mole fraction: magic formula Equation (4) with positive Ω. The curves are for different, indicated values of T/T_c., T_c being the critical temperature. See also Figure (5)

For the temperature corresponding to the curve 3/4 there will be *demixing* into two phases, the compositions of which being the abscissae of the points of contact of the *double tangent line*. For this special case the points of contact are identical with the minima; exceptionally, therefore, the mole fractions of the coexisting phases follow from $\partial G'/\partial X = 0$. The two *points of inflexion* of the curve, as always, follow from $\partial^2 G'/\partial X^2 = 0$.

On increasing temperature, the contact points and the inflexion points move inwards, and finally coincide all. This happens at the critical temperature; the critical mole fraction follows from $\partial^3 G' / \partial X^3 = 0$. The G' function for the critical temperature is the curve marked 1.

In Figure 5 the set of points of contact and the set of points of inflexion are represented. The set of points of contact, the *binodal*, is the boundary of the region of demixing. The set of points of inflexion is named *spinodal*. The binodal and the spinodal have a common maximum, which is the critical point.

Summarizing in terms of formulae:
the binodal, in this special case, is the solution of

$$\frac{\partial G'}{\partial X} = 0 ; \tag{10}$$

that is to say

$$\frac{\partial G'}{\partial X} = RT \ln \frac{X}{1-X} + \Omega(1-2X) = 0 . \tag{11}$$

The solution is

$$T_{BIN}(X) = \frac{-\Omega(1-2X)}{R \ln \dfrac{X}{1-X}} . \tag{12}$$

The spinodal generally is the solution of

$$\frac{\partial^2 G'}{\partial X^2} = 0 ; \tag{13}$$

which gives rise to

$$\frac{\partial^2 G'}{\partial X^2} = \frac{RT}{X(1-X)} - 2\Omega = 0 \tag{14}$$

and

$$T_{spin}(X) = \frac{2\Omega}{R} X(1-X) . \tag{15}$$

The critical point generally follows from

$$\frac{\partial^3 G'}{\partial X^3} = 0 , \tag{16}$$

giving rise to

$$\frac{\partial^3 G'}{\partial X^3} = -\frac{RT}{X^2(1-X)^2}(1-2X) = 0 . \tag{17}$$

The solution of the equation is (obviously, in view of the symmetry)

$$X_c = 0.5 . \tag{18}$$

§ (206)

Substitution of $X_C = 0.5$ into the spinodal equation, Equation (15), gives the following relation between the parameter Ω and the critical temperature T_C:

$$T_C = \frac{\Omega}{2R}.\qquad(19)$$

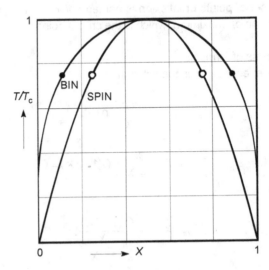

FIG: 5. Binodal and spinodal curves resulting from the magic formula Equation (4) with positive Ω. The circles correspond with those in Figure 4; open: points of inflexion; closed: points of contact of double tangent line

In summarizing, the magic formula, Equation (9), is the simplest thermodynamic formula to account for the phenomenon of demixing. In the chapters ahead the formula is going to play a leading part. To start with, it will be shown that the laws of dilute solutions directly follow from the magic formula and, therefore, in retrospect are a proof of its power. The formula, to end with, is a combination of *Boltzmann's formula* $S = k \ln W$, which gives rise to $RT\{(1-) \ln(1-X) + X \ln X\}$, and a term $\Omega X (1-X)$, which is no more than a little piece of common sense.

The magic formula $RT\{(1-X) \ln(1-X) + X \ln X\} + \Omega X (1-X)$ - a combination of Boltzmann and common sense - is going to play a leading part in the sections ahead. The results, to which it will give rise, are proof of its power.

EXERCISES

1. *volume properties of supercritical carbon dioxide*

For carbon dioxide, CO_2, the coordinates of the critical point are $T_C = 304.2$ K and $P_C = 73.8$ bar.
- For the supercritical state $T = 325$ K; $P = 80$ bar, calculate, in terms of the Van der Waals equation, the numerical values of molar volume, isothermal compressibility and expansion coefficient.

2. *Van der Waals and Helmholtz*

Three PV isotherms are sketched for a system which obeys the Van der Waals equation of state. The isotherms are for $T > T_C$; for $T = T_C$ and for $T < T_C$, T_C being the critical temperature.

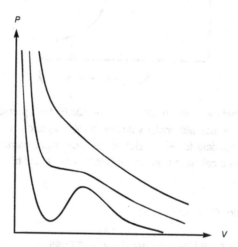

- For the three different cases, sketch the corresponding curves of Helmholtz energy (A) versus volume at constant temperature, remembering that $dA = -SdT - PdV$.

3. *critical coordinates from Van der Waals constants*

Write down the derivation of the Equations (5) and (7) for the relations between the coordinates of the critical point P_C and T_C and critical volume V_C and the constants a and b of the Van der Waals equation.

Note that the critical point is characterized by $(\partial P/\partial V)_T = 0$ and $(\partial^2 P/\partial V^2) = 0$; see foregoing exercise. Note also that these two conditions are equivalent to $(\partial^2 A/\partial V^2) = 0$ and $(\partial^3 A/\partial V^3) = 0$; see next exercise.

4. *minimal Helmholtz energy*

For temperatures below T_C the Van der Waals equation gives rise to a Helmholtz energy-versus-volume function, which looks like the figure which is shown below (see also second exercise).
In an experiment 1 mol of substance, satisfying the function shown, is brought in a vessel of constant volume V_E, kept at $T = T_E$.

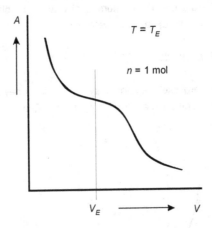

- Explain that the criterion $(dA)_{T,V.} \leq 0$ ($\leftarrow$108:Exc 11) predicts the formation of two phases, say with molar volumes of $(V_E - a)$ and $(V_E + b)$, a and b being positive. Indicate to which relation the two phases stand to the A versus V function and calculate their amounts in terms of a and b.

5. *critical temperature for a model system*

For a model system the Gibbs energy of mixing is given by

$$\Delta_m G = RT\{(1-X)\ln(1-X) + X\ln X\} + A\left(1 - \frac{T}{\Theta}\right)X(1-X) \qquad (20)$$

where A and Θ are constants.
- Find the relation between the critical temperature T_C and the constants A and Θ.
- Is critical behaviour possible for negative A in combination with positive Θ?

6. *spinodal and critical point for a given function*

For the Gibbs energy of mixing function
$$RT\{(1-X)\ln(1-X) + X\ln X\} + AX(1-X)[1 + B(1-2X)]$$
- derive the equation for the spinodal and the equation for the mole fraction of the critical point.

The function recipes of the chemical potentials of solvent and solute in a dilute solution are derived from the magic formula for the Gibbs energy of a binary mixture.

In this section, and the following two, the systems have phases that are either a *pure substance* or a *dilute solution*. The equilibrium conditions, the set of equations from which the equilibrium states can be solved, will be in terms of chemical potentials. First, in this section, the *function recipes* of the *chemical potentials* are derived from the *magic formula* (←206). Next, in § 208, equilibria are considered where the main component, the *solvent* is present in more than one phase. In § 209 the *solute* is partitioned over more than one phase.

First of all we need to specify the systems to be treated. There is a main component, which is the solvent, referred to as substance A. The solute, which is present in small proportions, is substance B. The *composition variable (X)* to be used is the mole fraction of B, which, in the case of dilute solutions, has the property $X \ll 1$.

chemical potentials

The starting point is the magic formula for the Gibbs energy of a mixture of $(1-X)$ mole of A and X mole of B, Equation (206:8)

$$G(T,P,X) = (1-X)G_A^*(T,P) + XG_B^*(T,P) + RT\{(1-X)\ln(1-X) + X\ln X\} + \Omega X(1-X) . \quad (1)$$

At this place we underline the fact that, strictly speaking, the formula is derived for a mixture of $(1-X)$ N_{AV} molecules of A and $X \cdot N_{AV}$ molecules of B, such that A and B do not react or associate or dissociate in each others presence.

At constant T and P the expressions for the chemical potentials are obtained with the help of the general recipes, Equations (203:14,15):

$$\mu_A = G - X\frac{\partial G}{\partial X} ; \quad (2)$$

$$\mu_B = G + (1-X)\frac{\partial G}{\partial X} . \quad (3)$$

On substitution of Equation (1) into Equations (2) and (3) the following expressions are obtained

$$\mu_A = G_A^* + RT\ln(1-X) + \Omega \cdot X^2 \quad (4)$$

$$\mu_B = G_B^* + RT\ln X + \Omega \cdot (1-X)^2 . \quad (5)$$

the solvent

For $X \ll 1$, or in other words for X approaching zero, the influence of the quadratic term ΩX^2 (which is the *excess chemical potential* of A) is negligible. As a result Equation (4) takes the form

$$\mu_A = G_A^* + RT \ln(1 - X) \tag{6}$$

that is to say the form of the chemical potential in the *ideal mixture!*
Next, for $X \to 0$, $\ln(1-X)$ can be safely replaced by the first term of its *power series expansion* (verify this with the help of your pocket calculator!)

$$\ln(1 - X) = -X - 1/2 X^2 - 1/3 X^3. \tag{7}$$

The expression therefore, or the *function recipe* that we are going to use for the *chemical potential of the solvent* in the dilute solution is

$$\mu_A = G_A^* - RTX. \tag{8}$$

the solute

For X approaching zero, the expression for the chemical potential, Equation (5), changes into

$$\mu_B = G_B^* + RT \ln X + \Omega. \tag{9}$$

At this stage the most important thing to observe is that (at constant T, P) the expression contains two constants, namely G_B^* and Ω, of which G_B^* in many cases not even is known. For instance, at the conditions of T and P taken, A may be a liquid and B a solid whose liquid state is far away; see also § 205, around Figure 5. The best thing to do, is to combine the two and to write the expression as

$$\mu_B = G_B^\circ + RT \ln X. \tag{10}$$

The *chemical potential of the solute* in the dilute solution has the property of varying logarithmically with mole fraction. Generally, the mole fraction of substance B, our symbol X, is defined as

$$X = \frac{n_B}{n_A + n_B}; \tag{11}$$

that is to say, amount of B in the mixture divided by the mixture's total amount of substance. In the dilute solutions considered ($n_B \ll n_A$) the influence of n_B in the denominator is negligible and as a result X reduces to amount of B divided by amount of A. The implication of this is that logarithmic in X can be changed to logarithmic in amount of solute.

For practical purposes the composition variable *molality* is frequently preferred. Molality is defined as amount of solute per kg solvent, symbol m. As a result, Equation (10) can be written as

$$\mu_B = G_B^o + RT \ln m; \tag{12}$$

along with the warning that the G_B^o in Equation (12) is another G_B^o than the one in Equation (10). The two G_B^o refer to different *standard states*; see also § 205. The standard state for Equation (12) is the dilute solution with $m = 1$. And to underline the possibility that Equation (12) may not be valid any more in the region of $m = 1$, one can better say that the standard state is a *hypothetical solution* having $m = 1$, see Figure 1.

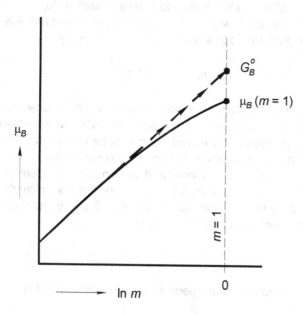

FIG. 1.   Chemical potential of solute as a function of the logarithm of molality

ideal dilute solutions and real dilute solutions

Dilute solutions which obey Equation (6), and as a result also Equation (8), for the chemical potential of the solvent and Equation (10), and its equivalent Equation (12), for the chemical potential of the solute are referred to as *ideal dilute solutions*. Equations (8) and (12) can be taken as a starting point for the thermodynamic description of *real dilute solutions* and the following expressions are met with

$$\mu_A = G_A^* - RT\varphi X; \tag{13}$$

$$\mu_B = G_B^o + RT \ln \gamma_B m. \tag{14}$$

The 'correction factor' φ in Equation (13) is known as the *osmotic coefficient* (the laws based on ideal dilute solutions in cases where the solvent is in two phases are referred to as the *osmotic laws*, see § 208). The factor γ_B in Equation (14) is called *activity coefficient*; not to be confused with other activity coefficients.

Henry's Law and Raoult's Law

When we 'follow' the substance B - in a mixture of (1–X) mole of A + X mole of B - from X = 0 to X = 1, i.e. from high dilution to high concentration, we may observe that
- for very small X, μ_B will behave according to Equation (10);
- for X approaching 1, where B has the status of solvent and A is the solute, μ_B will be given by the analogue of Equation (6), i.e. by

$$\mu_B = G_B^* + RT \ln X. \tag{15}$$

The region where μ_B behaves according to Equation (10) is often referred to as *Henry's Law region*; and μ_B is said to display *Henry's Law behaviour*. The law named after William Henry (1774-1836), which will be derived in §209, deals with the solubility of a gaseous substance B in a liquid substance A as a function of pressure. When the pressure P is increased, i.e. when the pressure of gaseous B is increased, the amount of B in the liquid phase will increase. *Henry's Law* states that, for sparingly soluble B, the mole fraction of B in the liquid phase will be proportional to the pressure exerted by the gas.

$$X_B = k \cdot P. \tag{16}$$

The constant k is a function of temperature and, of course, of the combination of A and B.
The region where μ_B satisfies Equation (15) is referred to as *Raoult's Law region*. One speaks of *Raoult's Law behaviour*. The law named after François-Marie Raoult (1830-1901) deals with the isothermal equilibrium between a liquid phase composed of A and B and a vapour phase likewise composed of A and B. Now, let P_B represent the partial pressure of B in the vapour phase, i.e. the mole fraction of B in the vapour phase multiplied by the pressure indicated by the manometer, and P_B^o the pressure indicated if B would be alone in the space, then, according to *Raoult's Law* , the partial pressure P_B (= $X^{vap} \cdot P$) is given by

$$P_B = X^{liq} \cdot P_B^o. \tag{17}$$

As a remark, for the *ideal mixture* Equation (15) for μ_B is valid over the whole composition range, i.e. for 0 < X < 1. The same, in that case, holds true for the validity of Equation (17), and it implies that the pressure indicated by the manometer obeys the relation

$$P = (1 - X^{liq})\, P_A^o + X^{liq}\, P_B^o: \tag{18}$$

for ideal behaviour, the *liquidus* in the *PX liquid-vapour* equilibrium *diagram* is a straight line, Equation (18).

In spite of the last remark, we will reserve Raoult's Law for dilute solutions. To that end it is expedient to modify it somewhat; starting from Equation (205:24), which is the general expression for B's chemical potential in a mixture (dilute or not):

$$\mu_B = G_B^* + RT \ln X + \mu_B^E, \tag{19}$$

and in which the *excess chemical potential* follows from the excess Gibbs energy, Equation (205:9), as

$$\mu_B^E = G^E + (1 - X)\, \frac{\partial G^E}{\partial X}. \tag{20}$$

From the last equation and the fact that G^E is zero for $X = 1$, it follows that μ_B^E is zero for $X = 1$. Now, from Raoult's Law, in the form of Equation (15), it follows that μ_B^E must be zero in the vicinity of $X = 1$. This characteristic - being zero and staying zero for a while - implies that not only μ_B^E but also its derivative with respect to X has to be zero at $X = 1$. And this is how we are going to formulate Raoult's Law:

$$\left(\frac{\partial \mu_B^E}{\partial X} \right)_{T,P,X=1} = 0. \tag{21}$$

And, returning to A as the solvent,

$$\left(\frac{\partial \mu_A^E}{\partial X} \right)_{T,P,X=0} = 0. \tag{22}$$

Note that this is exactly the case for Equation (4), where μ_A^E is equal to ΩX^2.

At high dilution the chemical potential of the solute depends logarithmically on its mole fraction (Henry's Law behaviour); for a given solute, the potential changes from solvent to solvent. The chemical potential of the solvent, at high dilution, is the same as its potential in ideal mixtures (Raoult's Law behaviour); the solvent's chemical potential is independent of the solute.

EXERCISES

1. *the ideal isothermal vaporus*

    A and B are two substances that mix completely and ideally in the liquid as well as in the gaseous state. For the isothermal equilibrium between liquid and vapour, the relation between the vapour pressure and the composition of the liquid phase is given by Equation (18).
    - Derive the equation for the vaporus: the vapour pressure as a function of the composition of the vapour.

2. *Raoult, Henry, and Gibbs-Duhem*

    Prove that Raoult's Law is a direct consequence of Henry's law, or *vice versa*.
    N.B. This is typically a 'Gibbs-Duhem problem'.

3. *the activity coefficient*

    What is the relation between the activity coefficient γ_B in Equation (14) and Figure 1 for the solution with $m = 1$?

4. *the solute's chemical potential*

    Show that the formula for the chemical potential of the solute in the ideal dilute solution, Equation (12), follows from Henry's Law, Equation (16).

5. *a strange question (?)*

    A solution of 2 mole percent of toluene (component 2) in benzene (component 1) is compared with a solution of 1 mole percent of toluene in benzene.
    - What is the difference for these two solutions between the values of the differential coefficient $(\partial U / \partial n_1)_{S,V,n_2}$ at 50 °C?

6. *a trigonometric excess function*

    Does the excess function $G^E = a_n \sin n\pi X$, where n is an integer, satisfy Raoult's Law in the sense that, for $X \rightarrow 0$, the partial derivative of A's excess chemical potential with respect to mole fraction is equal to zero (Equation (22))?

7. *the ideal dilute solution's quantites*

    For the ideal dilute solution of X mole of solute B in $(1-X)$ mole of solvent A give the expressions for the *partial quantities* G_A, G_B, S_A, S_B, H_A, H_B and V_A and V_B and for the *integral molar quantities* G_m, S_m, H_m and V_m.

    N.B. the subscripts m for molar were abandoned, in fact, by the end of § 203.

§ (207)

Three different kinds of equilibrium are considered where one the two phases is an ideal dilute solution of solute B in solvent A and the other phase pure A. In each case a relationship is derived between the two variables of the set M of variables that are necessary to define the state of the system under the constraints applied.

the solute disturbs the pure solvent's equilibrium

Three cases are examined where, by the addition of a small amount of a foreign substance (the solute), the equilibrium between two phases of a pure substance (the solvent) is disturbed. In all of the three cases, the foreign substance enters in only one of the phases, which invariably is liquid, and lowers the chemical potential of the main component in that phase. Every time the system has just one possibility to repair the inequality in the main component's chemical potentials.

This time, and by way of a change, two types of scheme are used: one for the definition of the system, and the other for the derivation of the relationship between the variables. In the first type of scheme the system is defined, the variables are stated, the equilibrium conditions are formulated; and finally, guided by the number of degrees of freedom, the type of mathematical solution is indicated. In the other type of scheme, the function recipes of the chemical potentials are substituted two times: one time for the situation after the disturbance, and the other time for the situation before – the difference between them yielding the desired relationship.

vapour pressure lowering

The first example considered is the equilibrium between liquid and vapour - studied at *constant temperature* in a vessel-with-manometer; Scheme 1, which is the scheme under here. The pressure indicated by the manometer is a function of the solute's mole fraction in the liquid phase. The lowering of the solvent's chemical potential is responded by the vapour phase in lowering its pressure.

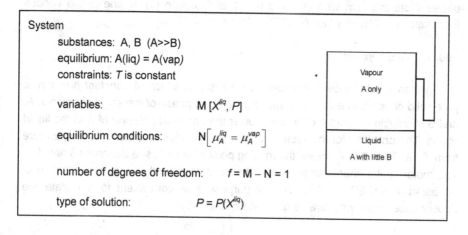

System

    substances: A, B (A>>B)

    equilibrium: A(liq) = A(vap)

    constraints: T is constant

    variables: $\qquad M\,[X^{liq},\,P]$

    equilibrium conditions: $\qquad N\left[\mu_A^{liq}=\mu_A^{vap}\right]$

    number of degrees of freedom: $\qquad f=M-N=1$

    type of solution: $\qquad P=P(X^{liq})$

Vapour

A only

Liquid

A with little B

As follows from Scheme 2, the function recipe of the solvent's chemical potential in the liquid is the one for the *ideal dilute solution*, Equation (207:8); and for the vapour phase the recipe for the *ideal gas*, Equation (108:18).

$$(X \neq 0; P = P) \qquad\qquad G_A^{*liq} - RTX^{liq} = G_A^{o\,vap} + RT\ln P$$

$$(X = 0; P = P_A^o) \qquad\qquad G_A^{*liq} = G_A^{o\,vap} + RT\ln P_A^o$$

$$- RTX^{liq} = RT\ln(P/P_A^o)$$

SCHEME 2

After subtraction and division by RT, the relation found is

$$\ln(P/P_A^o) = -X^{liq}. \qquad (1)$$

Realizing that, in the ideal dilute solution approximation, P will not be far from P_A^o, we may, after writing

$$P = P_A^o + \Delta P, \qquad (2)$$

modify Equation (1) to (NB verify that $\ln(1+\alpha)$ for small α reduces to α; use your pocket calculator)

$$\frac{\Delta P}{P_A^o} = -X^{liq}, \qquad (3)$$

which is equivalent to

$$P = P_A^o(1 - X^{liq}). \qquad (4)$$

From Equation (3) it follows that ΔP, indeed, is negative: there is a *lowering of the vapour pressure* when a non-volatile solute is added to the solvent. The relative lowering of the pressure is equal to the mole fraction of the solute in the liquid phase. Note that Equation (4), and with that Equation (3), is one of the various formulations of *Raoult's Law* ($\leftarrow$207).

freezing point depression

When a substance B is added to a vessel, in which at *constant pressure* a pure solid phase is in equilibrium with a pure liquid phase of the same substance A, and B is soluble in liquid A only, it will lower the chemical potential of A in the liquid phase. The only way for the system to face this situation is to lower its temperature from T_A^o to $T_A^o + \Delta T$, T_A^o being the melting point of pure A; see Schemes 3 and 4. The molar Gibbs energy of pure A in the solid and liquid phases is a function of temperature ($\partial G^*/\partial T = -S^*$). For our purpose it is convenient to formulate the dependence on temperature as given in Equation (5)

§ (208)

System
    substances: A, B (A>>B)
    equilibrium: A(sol) = A(liq)
    constraints: P is constant

    variables: $M\,[X^{liq},\,T]$

    equilibrium conditions: $N\left[\mu_A^{sol}=\mu_A^{liq}\right]$

    number of degrees of freedom: $f = M - N = 1$

    type of solution: $T = T(X^{liq})$

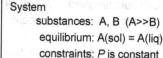

liquid

A with B

solid

A only

SCHEME 3

$(X \neq 0;\ T = T)$ $G_A^{*sol}(T_A^o) - S_A^{*sol}\cdot\Delta T = G_A^{*liq}(T_A^o) - S_A^{*liq}\cdot\Delta T - RTX$

$(X = 0;\ T = T_A^o)$ $\underline{\quad\quad G_A^{*sol}(T_A^o)\quad\quad\quad\quad\quad = G_A^{*liq}(T_A^o)\quad\quad\quad\quad\quad\quad\quad}$

                        $-S_A^{*sol}\cdot\Delta T =$ $-S_A^{*liq}\cdot\Delta T - RTX$

SCHEME 4

$$G^{*\alpha}(T) = G_A^{*\alpha}(T_A^o) - S_A^{*\alpha}\Delta T;\ \alpha = \text{sol, liq} \tag{5}$$

It follows that the change in the equilibrium temperature is given by

$$\Delta T = -\frac{RT}{\Delta_s S_A^*}\cdot X, \tag{6}$$

where $\Delta_s S_A^* = S_A^{*liq} - S_A^{*sol}$ is the *entropy of melting* of A which is, as we know, equal to the *enthalpy of melting* divided by the melting temperature $\Delta_s H_A^* / T_A^o$ and as a result of which Equation (6) can be written as

$$\Delta T = -\frac{RT\cdot T_A^o}{\Delta_s H_A^*}\cdot X. \tag{7}$$

The temperature T in the numerator of Equations (6) and (7) is close to T_A^o and for that reason we will formulate the final result as

$$\Delta T = -\frac{RT_A^o}{\Delta_s S_A^*}\cdot X = -\frac{RT_A^{o\,2}}{\Delta_s H_A^*}\cdot X. \tag{8}$$

Indeed, ΔT is negative: there is a *depression of the equilibrium temperature*. And all other things being equal, the depression of the melting temperature is the greater, the lower the heat of melting of the main component (A).

For the case that the solute is present in each of the two phases, the X in Equation (8) has to be replaced by $\Delta X = X^{liq} - X^{sol}$. A variant of the equation with ΔX is the following *equation for the initial slopes in TX phase diagrams*:

$$\frac{dX^{\beta}}{dT} - \frac{dX^{\alpha}}{dT} = -\frac{\Delta H_A^*}{RT_A^{o2}}. \tag{9}$$

The superscripts α and β are for the low-temperature form and the high-temperature one, respectively.

The expression for the *depression of the freezing point*, Equation (8), usually is referred to as *Van 't Hoff's Law*, after Jacobus Henricus van 't Hoff (1852-1911).

osmotic pressure

The last case considered in this section is a typical example of the inventiveness of matter: in finding a way out, when faced with an *inequality of chemical potentials*.

In 1748 the French Abbé J.A. Nollet carried out an experiment which is more or less as follows. He first filled a pot with a mixture of water and spirit of wine. He sealed the pot with a piece of pig's bladder and next immersed the whole thing in a barrel with water. Much to his surprise the bladder expanded and finally burst. The explanation is as follows. The bladder material is permeable for water but not for the spirit. As a result the spirit has to remain in the pot. The water inside, having a lower chemical potential than the water outside, tries to compensate the difference by driving back the influence - the mole fraction - of the spirit, i.e. by absorbing more and more water from outside.

In a more formal set-up, see Scheme 5, a solution of A and B is inside a vessel, provided with a vertical glass tube and having a *membrane* at the bottom. The membrane is permeable for the solvent A but not for the solute B. Outside the vessel there is just pure A.

The inequality between the chemical potentials of A inside and outside is removed by the *hydrostatic pressure* of the column of liquid in the tube - by the *osmotic pressure* Π. The inside potential will have an extra term $\left(\partial G_A^* / \partial P\right) \cdot \Pi = V_A^* \cdot \Pi$ which compensates the term $-RT\, X^{in}$, see Scheme 6; as a result

$$\Pi = \frac{RT}{V_A^{*\,liq}} \cdot X^{in}. \tag{10}$$

The mole fraction variable X^{in} is, in the case of the dilute solution, equal to the quotient of the amount of solute and the amount of solvent, i.e. n_B / n_A. When this quotient is introduced in Equation (10), the denominator part will be $n_A \cdot V_A^{*liq}$, which is, still in the case of high dilution, the volume of the whole inside part of the system. In this volume V there are n_B mole of solvent. Taking these things into account, Equation (10) changes to

$$\Pi = n_B \cdot \frac{RT}{V}. \tag{11}$$

In Equation (11) we recognize the *ideal-gas equation*!

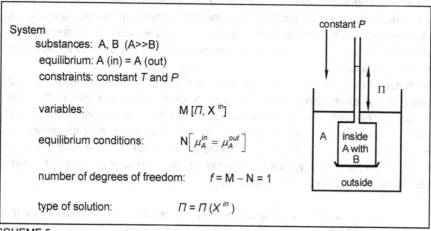

System
 substances: A, B (A>>B)
 equilibrium: A (in) = A (out)
 constraints: constant T and P

 variables: $M[\Pi, X^{in}]$

 equilibrium conditions: $N\left[\mu_A^{in} = \mu_A^{out}\right]$

 number of degrees of freedom: $f = M - N = 1$

 type of solution: $\Pi = \Pi(X^{in})$

constant P

A | inside
A with B

outside

SCHEME 5

	outside	inside
$(X \neq 0; \Pi = \Pi)$		$G_A^{*liq} = G_A^{*liq} - RTX^{in} + V_A^{*liq} \cdot \Pi$
$(X = 0; \Pi = 0)$		$G_A^{*liq} = G_A^{*liq}$
	$0 =$	$-RTX^{in} + V_A^{*liq} \cdot \Pi$

SCHEME 6

The important observation to make is that the *osmotic pressure* caused by n_B mole of solute in a 'liquid space' is equal to the pressure exerted by the same amount of matter 'dissolved' in an empty space, having the same volume as the liquid and being kept at the same temperature. Van 't Hoff's name is firmly associated with the osmotic law, Equation (11).

Looking over the relationships derived in this section, we may ask ourselves to what extent they do depend on the choice of solvent, the choice of solute and their combination. We also might ask ourselves to what extent it is imperative to be sure that the dilute solution is composed, indeed, of $(1-X)\ N_{AV}$ particles of solvent and $X\ N_{AV}$ particles of solute (as an example, if we want to calculate the freezing point depression for 2 g of NaCl dissolved in 100 g of water, do we have to take into account that each 'molecule' of NaCl gives two particles, an ion Na^+ and an ion Cl^-, and is it necessary to take 18 $g \cdot mol^{-1}$ for the molar mass of water, instead of e.g. 36 $g \cdot mol^{-1}$ assuming that water is composed of molecules/particles $(H_2O)_2$?).

As for the choice of substances (and assuming that the mixture really contains $(1-X)\ N_{AV}$ particles A and $X\ N_{AV}$ particles B) the following can be observed. The relative lowering of the vapour pressure resulting from the addition of a soluble but not volatile substance, apparently, is independent of the choice of (combination of) substances. The same holds true for the osmotic pressure, which is as substance - independent as the ideal-gas pressure (for that matter *osmometers* are used to determine the molar masses of the large protein molecules). The freezing point depression depends on the choice of solvent and does not depend thereafter on the choice of solute.

The question about the particles of solvent and solute is a fundamental one - an answer is not easily given. In the case of NaCl in water, one has to take into account the *electrolytic dissociation* of NaCl (the other way round, *freezing point depression* provides information about the degree of association or dissociation of the solute molecules). The degree of association or dissociation of the solvent molecules, on the contrary, cannot be derived from the depression of the solvent's freezing point brought about by a solute (see, e.g, Oonk 1981).

The ensemble of solvent laws, derived above, often are referred to as the *'osmotic laws'*. The fundamental characteristic behind the osmotic laws is the ideal behaviour of the chemical potential of the solvent at high dilution.

Equations have been derived for vapour pressure lowering, freezing point depression and osmotic pressure as a function of solute mole fraction in ideal dilute solutions. In these cases the solvent - whose chemical potential behaves as in ideal mixtures - is present in two phases. In the case of vapour pressure lowering and freezing point depression the equations contain, apart from solute mole fraction, only pure solvent properties pertaining to the same equilibrium. The osmotic-pressure law is identical with the ideal-gas law.

EXERCISES

1. *the solute is present in both phases*

    - How do the relationships derived for the lowering of vapour pressure ($\Delta P/P$), the freezing point depression (ΔT) and the osmotic pressure (Π) change if the solute is not only present in the phase considered so far - and where it has mole fraction X - but also in the other phase with mole fraction Y? (NB

For the case of freezing point depression, the answer has already been given; Equation (9)).

- Can $\Delta P/P$ and ΔT be positive and Π negative?

2. *chemical potentials of A in liquid and vapour versus pressure*

- Make a drawing in which by straight lines the chemical potentials of a pure substance A, for the liquid and the vapour states at constant temperature, in the vicinity of the equilibrium vapour pressure, are given as a function of pressure.
- How has the figure to be changed and in what sense does the equilibrium pressure change if
  - a. the temperature is increased;
  - b. a non-volatile substance is added which is soluble in the liquid?

3. *chemical potentials of H_2O in liquid and vapour versus temperature*

- Make a drawing in which by straight lines the chemical potentials of liquid and gaseous water at 1 atm pressure are represented as a function of temperature in the vicinity of 100 °C.

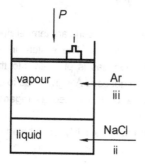

- How has the figure to be changed and in what sense does the equilibrium temperature change, if
  - i. the pressure on the system is increased;
  - ii. salt is added to the system;
  - iii. argon is added to the system?

4. *the molecular formula of a hydrocarbon*

The equilibrium vapour pressure of pure benzene (C_6H_6) is 9954 Pa at 20 °C. The equilibrium pressure over a solution of 2 g of a non-volatile hydrocarbon in 100 g benzene - at the same temperature - is 9867 Pa.

- Find the molecular formula of the hydrocarbon for which it is known that its carbon content is 94.4 wt%.

5.   *the molar mass of the solvent*

The freezing point of pure benzene (C_6H_6) is 5.50 °C at 1 atm; its heat of melting is 126 $J \cdot g^{-1}$. B is a substance which is soluble in liquid benzene and insoluble in solid benzene. The freezing point of a solution of 2 g B in 100 g benzene is 4.93 °C.
   - Calculate B's molar mass.
   - Will the same value be found for B's molar mass, if a fictitious molar mass of 100 $g \cdot mol^{-1}$ is taken for benzene?

6.   *initial slope of solidus*

In § 213 experimental liquidus and solidus temperatures - derived from cooling curves - are given for the system NaBr+NaCl; see Table 213:4. It is known that the cooling curve technique often yields incorrect solidus data (←006; Figure 006:8).
   - Use the liquidus data to calculate the initial slope of the liquidus at the NaCl side, and from it the initial slope of the solidus. NaCl's heat of melting is 28 $kJ \cdot mol^{-1}$. The result must be in agreement with the calculated solidus shown in Figure 213:4.

7.   *the intervention by a foreign gas*

In a vessel-with-manometer, kept at constant temperature, pure liquid substance A is in equilibrium with its (pure) vapour. The system in equilibrium is exposed to an intervention by which a foreign gas is pressed into the vessel (the gas does not dissolve in the liquid and with gaseous A it gives rise to an ideal gas mixture).
   - What kind of imbalance is induced as regards the chemical potentials of liquid and gaseous A?
   - How will the system react to nullify the imbalance?

8.   *an equation for the relative change in partial pressure*

   - For the case considered in Exc 7 - where a foreign gas is pressed into a vessel in which liquid A is in equilibrium with gaseous A - derive an expression for the relation between the relative change in A's partial pressure, $\Delta P_A / P_A^o$, and the pressure exerted by the gas.
   - Show that the relation can be modified to

$$\frac{\Delta P_A}{P_A^o} = n^{gas}\, \frac{V_A^{*liq}}{V^{gas}} \, ,$$

in which V_A^{*liq} is the molar volume of liquid A and n^{gas} and V^{gas} are the amount of gas and the volume occupied by the gas phase, respectively.

Three kinds of equilibrium are considered where the solute and not the solvent is present in two different phases in equilibrium. The theoretical framework is the ideal dilute solution, again.

three cases

Three cases are considered in which the substance B is present in each of two phases in equilibrium under isothermal conditions. The first case is the equilibrium between a gas, which is pure B, and a liquid phase, which is a solution of B in solvent A. In the second example pure solid B is in equilibrium with B dissolved in liquid solvent A. In the third case A and C are two immiscible liquid solvents and B the solute, soluble in A as well as in C. In all of the three cases, the liquid phases are *ideal dilute solutions* with B as the solute.

In contrast to the equilibria in § 208, there is no pure-substance situation free of B - and it means that, in the equilibrium relationships to derive, there will remain a Gibbs-energy property. It has, however, the advantage that, from the relationships found, the dependence on temperature can directly be derived - and for two of the cases the dependence on pressure as well. As always, the dependences on pressure and temperature are related to an enthalpy effect, and a difference-in-volume property, respectively.

solubility of gas in liquid

It is obvious, see Scheme 1, which is the scheme under here, that, when the external pressure P is increased, more and more B will dissolve in the liquid. Inversely when the pressure is released bubbles of gaseous B will escape from the liquid - think of a strong bottle of champagne which is opened.

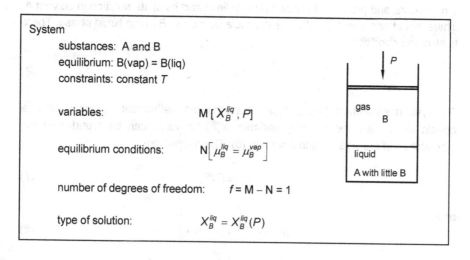

System
    substances: A and B
    equilibrium: B(vap) = B(liq)
    constraints: constant T

    variables: $M\,[\,X_B^{liq},\,P]$

    equilibrium conditions: $N\left[\mu_B^{liq} = \mu_B^{vap}\right]$

    number of degrees of freedom: $f = M - N = 1$

    type of solution: $X_B^{liq} = X_B^{liq}(P)$

P

gas
  B

liquid
A with little B

The recipes for the potentials are, if we take the ideal approximation - *ideal gas* and *ideal dilute solution*

$$\mu_B^{vap} = G_B^{o\,vap} + RT \ln P \; ; \tag{1}$$

$$\mu_B^{liq} = G_B^{o\,liq} + RT \ln X_B^{liq} \; . \tag{2}$$

After substitution of the recipes into the equilibrium condition

$$\mu_B^{liq} = \mu_B^{vap} \; , \tag{3}$$

it directly follows that the result, the relation between the mole fraction of B in the liquid phase and the pressure of the gas, can be written as

$$X_B^{liq} = k \cdot P. \tag{4}$$

The proportionality constant k is given by

$$k = \exp\left(\frac{\Delta G_B^o}{RT}\right) = \exp\left(\frac{G_B^{o\,vap} - G_B^{o\,liq}}{RT}\right) ; \tag{5}$$

and it is clear that k is a function of temperature.
The result is known as *Henry's Law*, and it dates from 1803; see also §207. The constant k is often referred to as *Henry's Law constant*.

solubility of solid in liquid

In this example, see Scheme 2, we consider the equilibrium at constant temperature and pressure between pure solid B and its dilute solution in solvent A. There is just one variable, which is the mole fraction of B in the liquid phase. There is also one condition:

$$\mu_B^{sol} = \mu_B^{liq} \; . \tag{6}$$

The system is invariant ($f = 0$), which means that the mathematical solution will be a certain fixed value for the only variable X_B^{liq}. Its value can be found after the introduction of the μ recipes in Equation (6). The recipes are

$$\mu_B^{sol} = G_B^{*sol}, \tag{7}$$

and

$$\mu_B^{liq} = G_B^{o\,liq} + RT \ln X_B^{liq} \; . \tag{8}$$

System
   substances: A and B
   equilibrium: B (sol) = B (liq)
   constraints: T and P constant

   variables: $M [X_B^{liq}]$

   equilibrium conditions: $N [\mu_B^{sol} = \mu_B^{liq}]$

   number of degrees of freedom: $f = M - N = 0$

   type of solution: X_B^{liq} = numerical value

liquid
A with
little B

solid B

SCHEME 2

The value of the mole fraction of the liquid phase - the *saturated solution* - follows from

$$\ln X_B^{liq} = -\Delta G_B^o / RT,$$ (9)

where

$$\Delta G_B^o = G_B^{o\,liq} - G_B^{*sol},$$ (10)

which at constant T and P is just the difference between two numerical values (to be taken, e.g., from thermodynamic tables: the table for pure solids and the table for solutes in solvents).

partition of solute between two solvents

   We now consider the equilibrium between two different ideal dilute solutions of the solvent B. Again temperature and pressure are kept constant. The system, see Scheme 3, consists of two solvents A and C which are liquid and not miscible with one another. The equilibrium condition is

$$\mu_B^I = \mu_B^{II};$$ (11)

and the recipes for the potentials are

$$\mu_B^I = G_B^{oI} + RT \ln X^I,$$ (12)

$$\mu_B^{II} = G_B^{oII} + RT \ln X^{II}.$$ (13)

System

substances: A, B and C

equilibrium: B(liq I) = B(liq II)

constraints: T and P constant

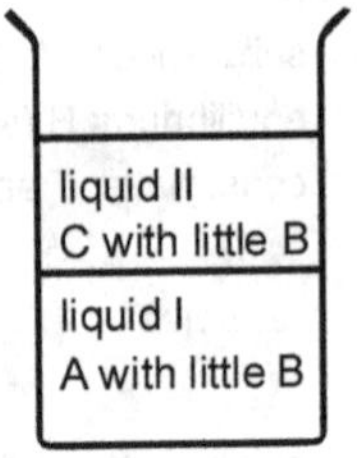
liquid II
C with little B
liquid I
A with little B

variables: $\mathrm{M}\left[X^{I}, X^{II}\right]$

equilibrium conditions: $\mathrm{N}\left[\mu_B^{I} = \mu_B^{II}\right]$

number of degrees of freedom: $f = \mathrm{M} - \mathrm{N} = 1$

type of solution: $X^{I} = X^{I}(X^{II})$

SCHEME 3

The relation between X^{I} and X^{II}, we are looking for, is

$$\ln\frac{X^{II}}{X^{I}} = \frac{-(G_B^{oII} - G_B^{oI})}{RT} = -\frac{\Delta G_B^{o}}{RT}. \tag{14}$$

At constant T and P the right-hand member is a constant: the relation between X^{II} and X^{I} simply is

$$\frac{X^{II}}{X^{I}} = k. \tag{15}$$

The constant k is the *partition coefficient*. The result is referred to as the *partition law* and is named after Walther Hermann Nernst (1864-1941) - *Nernst's partition law.*

influence of temperature

The results obtained for the three kinds of equilibrium all have the same structure in that they generally can be given as

$$\ln k = -\frac{\Delta G_B^{o}(T)}{RT}, \tag{16}$$

where k denotes, in respective order, Henry's Law constant, the mole fraction of the solute in the saturated solution and the partition coefficient. Realizing that

$$\Delta G_B^{o} = \Delta H_B^{o} - T\Delta S_B^{o}, \tag{17}$$

the general result can also be given as, see also § 007,

$$\ln k = -\frac{\Delta H_B^o}{RT} + \frac{\Delta S_B^o}{R}. \tag{18}$$

Neglecting ΔC_P, one can say that Equation (18) gives the dependence of k on temperature; it can, when comparing the values for the temperatures T_1 and T_2, be expressed as

$$\ln \frac{k_1}{k_2} = -\frac{\Delta H_B^o}{R}\left(\frac{1}{T_2} - \frac{1}{T_1}\right). \tag{19}$$

Generally for the change with temperature we can write

$$\frac{d(\ln k)}{dT} = \frac{\Delta H_B^o}{RT^2}; \tag{20}$$

and in the modified form

$$\frac{d(\ln k)}{d(1/T)} = -\frac{\Delta H_B^o}{R}. \tag{21}$$

As remarked before (←110), the last of these equations is an invitation to plot data in a $\ln k$ versus $1/T$ diagram (Clausius-Clapeyron like plot): if the range of data is relatively small a straight line will be obtained. The slope of the straight line is determined by ΔH_B^o; or, the other way round, the slope provides experimental information on ΔH_B^o.

Great care must be exercised as regards the *use of superscripts* - the small open circles. If one likes, one can change from B's mole fraction to its *molality* (←207). In that case the numerical value of ΔS_B^o will change and with it the numerical value of ΔG_B^o. The enthalpy property ΔH_B^o, on the other hand, is insensitive to the change to other units: as long as the solution is an ideal dilute solution its H_B^o - the *partial enthalpy of* the solute - is a constant, i.e. independent of B's mole fraction . In other terms, when $\ln k$ is expressed in mole fraction(s) and plotted against $1/T$ and next expressed in molality(ies) and again plotted against $1/T$, the second plot is shifted with respect to the first such that the two are parallel.

Equations have been derived for the solubility of a pure gas in a liquid, for the solubility of a pure solid in a liquid and for the partition of a solute between two immiscible liquids. The equations have a common structure: the effect - solubility, partition coefficient - is related to the difference in partial Gibbs energy of the solute in the two phases.

EXERCISES

1. *the Kritchevsky-Kasarnovsky equation*

In Nordstrom and Muñoz (1985) one can read on p.77 "Since most gases dissolve in sea water at the surface, the effect of pressure on the solubility is not normally considered to be important. The effect of pressure on the solubility of gases produced in the deep oceans (e.g. He) can be estimated from the *Kritchevsky-Kasarnovsky* (1935) *equation* ln $(k^P/k^o) = (\bar{V}/RT)\cdot P$"
In this equation κ represents Henry's Law constant.

In the ideal gas and ideal dilute solution approximation and in terms of molality of solute B, and after rearranging, the equation can be given as

$$m_B^P = m_B^o \cdot \frac{P}{P^o} \exp\left[-(V_B^o/RT)\cdot P\right].$$ (B)

- What is the meaning of the symbols m_B^P, m_B^o, and V_B^o ?
- Give the derivation of the equation (B).

On several occasions the non-ideal behaviour of the gas is accounted for by replacing - in equilibrium equations like (B) - pressure P by *fugacity f.*

- Would you, to that end, replace each of the two P's of equation (B) by f?

2. *helium in deep ocean water*

The solubility of helium (He, molar mass 4 $g \cdot mol^{-1}$) in water at 25°C is said to be about 1 ml per 100 ml.

- To what molality (m_{He}) does this correspond?
- Up to what molality does helium dissolve in deep ocean water at a depth of 8 km?

NB use the formula given in Exc 1, Equation (B), the partial volume of helium in water (V_{He}) being 30 $cm^3 \cdot mol^{-1}$

3. *calculation of molality and Gibbs energy*

For the substances A (liquid) and B (solid) and for ideal dilute solutions of B in A, data are given for the standard (T = 298.15 K; P = 1 bar; molality of B, m = 1) thermodynamic properties:
$G^{*liq} = -1000$; $G^{*sol} = -2500$; $G_B^o = 3200$; all three expressed in $J \cdot mol^{-1}$.

The molar masses of A and B are 100 and 250 $g \cdot mol^{-1}$, respectively.

- For the saturated solution of B in A (at 298.15 K and 1 bar), calculate a) its molality, and b) its molar Gibbs energy.

4. *simultaneous saturation?*

The substances A and C are liquid at room temperature and virtually immiscible; substance B is a solid - soluble in A as well as in C.
In an experiment B is added little by little to a tube containing A and C; thermodynamic equilibrium is guaranteed during the whole experiment.
- Will the two liquid phases be saturated with B at the same time, or not?

5. *a system for storage of thermal energy*

- Show that the equilibrium between solid Na_2SO_4 and solid $Na_2SO_4 \cdot 10H_2O$ and solution of Na_2SO_4 in water is invariant at constant pressure.
- Using the solubility data given below and valid for 1 atm, determine the temperature of the invariant equilibrium, and estimate the heat effect of the reaction $Na_2SO_4 \cdot 10H_2O \rightarrow Na_2SO_4 + 10H_2O$.

NB Owing to the position of the transition temperature and the considerable heat effect, the system can be used for the *storage of thermal energy*.

$t\,/^{\circ}C$	molality of Na_2SO_4	solid in equilibrium with solution
10	0.629	decahydrate
20	1.331	"
30	2.890	"
40	3.375	anhydrate
50	3.269	"
60	3.179	"

6. *mixing of salt and water*

A person, interested in calorimetry, found a table on the decrease in temperature that goes together with the mixing of equal weights of water and a given salt. Part of the table is NaCl 4 °C; KCl 12 °C; NH_4Cl 14 °C; $NaNO_3$ 9.5 °C; KNO_3 10 °C; and NH_4NO_3 25 °C. The table inspired him to do some simple experiments. First he mixed, in a plastic beaker, 50 g KCl with 50 g water, both having a temperature of 21.0 °C, and observed that the temperature dropped to 8.0 °C. In a second experiment he mixed just 16 g KCl with 50g water, both again 21.0 °C, and observed, to his surprise, that the temperature dropped to 7.7 °C.
- What is your interpretation of the observations?

This time the distinction between solvent and solute is absent. Two-phase equilibria are considered for systems in which two substances of equal 'quality' mix ideally in the mixed phases they are giving rise to. The set of equilibrium equations is solved for three cases: the isothermal and the isobaric equilibrium between liquid and vapour and the isobaric equilibrium between liquid mixture and pure solid.

three examples of ideal equilibria

Three cases of equilibrium between two phases are considered for systems in which the mixed phases are *ideal mixtures* of the two components. In the first case an ideal liquid phase is in equilibrium with an ideal vapour phase at isothermal conditions. For the second case the liquid + vapour equilibrium is considered again, but this time under constant pressure. In the last case the two substances A and B are immiscible when solid; the equilibrium, which is considered, is between liquid mixture and one of the pure solids.

In the first two cases there are two composition variables. The structure of the recipes for the chemical potentials is such that one of the composition variables can easily be eliminated from the two equilibrium equations - providing explicit relationships for the equilibrium curves in the phase diagram.

isothermal liquid + vapor equilibrium

As can be seen in Scheme 1, the mathematical solution - of the set of equations N acting on the variables in the set M - corresponds to two curves in the PX plane.

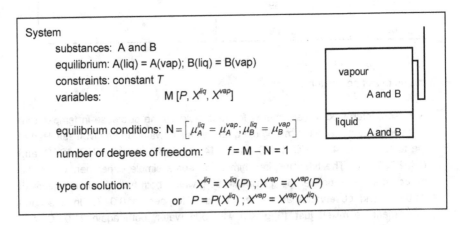

System
    substances: A and B
    equilibrium: A(liq) = A(vap); B(liq) = B(vap)
    constraints: constant T
    variables: $M\ [P, X^{liq}, X^{vap}]$

equilibrium conditions: $N = \left[\mu_A^{liq} = \mu_A^{vap}; \mu_B^{liq} = \mu_B^{vap} \right]$

number of degrees of freedom: $f = M - N = 1$

type of solution: $X^{liq} = X^{liq}(P)$; $X^{vap} = X^{vap}(P)$
        or $P = P(X^{liq})$; $X^{vap} = X^{vap}(X^{liq})$

SCHEME 1

The *liquidus* is one of these curves and it gives the relation between liquid composition and the pressure indicated by the manometer. The other is the *vaporus* and it represents the relation between vapour composition and pressure. Liquidus and vaporus have common points on the axes $X = 0$ and $X = 1$: the vapour pressures of the pure substances, P_A^o and P_B^o, respectively.

In Scheme 2 the situation is sketched which is obtained after the introduction of the function recipes - for A's chemical potentials - in the condition $\mu_A^{liq} = \mu_A^{vap}$.

$$
\begin{array}{ll}
(P = P; X \neq 0) & G_A^{*liq} + RT \ln(1 - X^{liq}) = G_A^o + RT \ln P + RT \ln(1 - X^{vap}) \\[2mm]
(P = P_A^o; X = 0) & \underline{G_A^{*liq} \qquad\qquad\qquad\quad = G_A^o + RT \ln P_A^o} \\
 & RT \ln(1 - X^{liq}) = RT \ln(P \cdot (1 - X^{vap}) / P_A^o)
\end{array}
$$

SCHEME 2

The result of Scheme 2 can be expressed as

$$P \cdot (1 - X^{vap}) = (1 - X^{liq}) \cdot P_A^o. \tag{1}$$

On the same lines, starting from $\mu_B^{liq} = \mu_B^{vap}$, we obtain

$$P \cdot X^{vap} = X^{liq} \cdot P_B^o. \tag{2}$$

The *liquidus equation*, pressure as a function of liquid composition, is obtained simply by adding the Equations (1) and (2)

$$P = P(X^{liq}) = (1 - X^{liq}) \cdot P_A^o + X^{liq} \cdot P_B^o. \tag{3}$$

In the *PX* plane, as follows from Equation (3), the liquidus is the straight line connecting P_A^o and P_B^o, see Figure 1.

The vaporus is obtained by eliminating X^{liq} from Equations (1) and (2); it is a rectangular hyperbola, or rather part of a rectangular hyperbola, represented by

$$P = P(X^{vap}) = \frac{P_A^o \cdot P_B^o}{(1 - X^{vap}) \cdot P_B^o + X^{vap} \cdot P_A^o}. \tag{4}$$

The asymptotes of the hyperbola are the lines $P = 0$ and $X = P_B^o / (P_B^o - P_A^o)$.

The *width of the two-phase region* in the *PX* phase diagram is determined by the relative difference of the vapour pressures of the pure substances.

For $X^{liq} = 0.5$ the width of the two-phase region is

$$(X^{vap} - X^{liq})_{X=0.5} = \frac{P_B^o - P_A^o}{2(P_B^o + P_A^o)}. \tag{5}$$

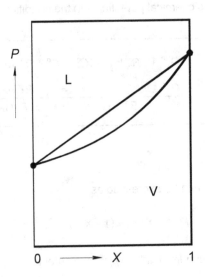

FIG. 1.   Equilibrium at constant temperature between ideal liquid and ideal
vapour mixtures

isobaric equilibrium between liquid and vapour

Just like the isothermal situation, there is one *degree of freedom*

$$f = M\left[T, X^{liq}, X^{vap}\right] - N\left[\mu_A^{liq} = \mu_A^{vap}; \ \mu_B^{liq} = \mu_B^{vap}\right] = 1; \tag{6}$$

and it means that X^{liq} as well as X^{vap} can be expressed in T. In the TX *phase diagram* the representation of $X^{liq}(T)$ and $X^{vap}(T)$ are the liquidus and vaporus curves, respectively.

Again, the equilibrium is self-supporting in the sense that it provides itself the necessary information on the molar Gibbs energies of the pure components (otherwise expressed, the information needed comes from the involved phase change itself - think of heat of vaporization). Neglecting the influence of heat capacities we can formulate the following scheme, Scheme 3.

$$\Delta G_A^*(T) = G_A^{*vap}(T) - G_A^{*liq}(T) = \Delta H_A^* - T\,\Delta S_A^*$$
$$\Delta G_A^*(T_A^o) \qquad\qquad\qquad = \Delta H_A^* - T_A^o \Delta S_A^* = 0$$
$$\Delta G_A^*(T) \qquad\qquad\qquad = -\Delta S_A^*(T - T_A^o)$$

SCHEME 3

The change in Gibbs energy of component A, as follows from Scheme 3, is given by

$$\Delta G_A^*(T) = -\Delta S_A^*(T - T_A^o), \tag{7}$$

in which ΔS_A^* is the entropy of vaporization, the quotient of the heat of vaporization and boiling point temperature. Similarly for component B,

$$\Delta G_B^*(T) = -\Delta S_B^*(T - T_B^o). \tag{8}$$

After substitution of the recipes for the potentials into the equilibrium equations, the following relations are obtained.

$$(1 - X^{liq}) = (1 - X^{vap})\exp(\Delta G_A^* / RT); \tag{9}$$

$$X^{liq} = X^{vap}\exp(\Delta G_B^* / RT). \tag{10}$$

By addition of the last two equations, X^{liq} is eliminated and X^{vap} is obtained as a function of temperature:

$$X^{vap}(T) = \frac{1 - \exp(\Delta G_A^* / RT)}{\exp(\Delta G_B^* / RT) - \exp(\Delta G_A^* / RT)}, \tag{11}$$

which is, in other words, the formula for the vaporus. The formula for the liquidus is obtained by combining Equations (10) and (11):

$$X^{liq}(T) = \frac{1 - \exp(\Delta G_A^* / RT)}{\exp(\Delta G_B^* / RT) - \exp(\Delta G_A^* / RT)}\exp(\Delta G_B^* / RT). \tag{12}$$

Phase diagrams, calculated with Equations (11) and (12) and using Equations (7) and (8), show *lens-shaped two-phase regions*. An example is represented by Figure 2, where the two-phase region is pronouncedly wide: in the middle it extends from $X = 0.20$ to $X = 0.80$.

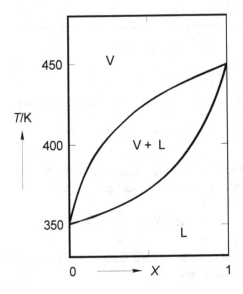

FIG. 2.   Phase diagram depicting the isobaric equilibrium between ideal vapour and liquid phases. Calculated with $\Delta S_A^* = \Delta S_B^* = 11R$

Generally, the width of a two-phase region is determined by two factors; these are the magnitudes of the ΔS^* and the difference between T_A^o and T_B^o. This is clearly shown by the equation

$$X^{liq} = X^{vap} \exp\{\Delta S_B^*(T_B^o - T)/RT\},\tag{13}$$

which is obtained by combining Equations (10) and (8).
The smaller the ΔS^* and $(T_B^o - T_A^o)$ are, the narrower the two-phase region will be.

In the case of metals, for the equilibrium between (mixed-crystalline) solid and liquid (of which the treatment for ideal mixtures is exactly the same), the ΔS^* values are smaller by a factor 10 in comparison with the liquid-vapour values. All other things being equal, the two-phase region of Figure 2 would reduce, at 400 K, to the boundaries $X = 0.466$ and $X = 0.534$ (←004, for *Trouton's rule*, and *Richard's rule*).

For $\Delta S_A^* = \Delta S_B^* = 0$ the two-phase region is infinitesimally narrow (think of a *second-order transition* ←004). And when $T_A^o = T_B^o$ the 'two-phase region' in the TX plane is just the horizontal line between T_A^o and T_B^o - a type of diagram that can be expected when A and B are a *pair of optical antipodes* (Figure 006:3c).

equilibrium between pure solid(s) and liquid mixture

In the foregoing cases each of the two phases in equilibrium is an ideal mixture. This time the case is considered where two substances A and B mix completely and ideally in the liquid state and, at the same time, do not mix at all in the solid state, i.e. do not give rise to *mixed crystalline material*. The solid-liquid phase diagram - at isobaric conditions and excluding the formation of compounds - will then have the appearance of Figure 3. It has two liquidus curves which intersect at the *eutectic point*. The left-hand liquidus represents, for the equilibrium between pure A and liquid mixtures, the composition of the liquid phase as a function of temperature. The right-hand liquidus has the same function for the equilibrium between liquid mixtures and solid B.

At the *eutectic temperature* the liquid mixture with the *eutectic composition* can at the same time coexist with solid A and solid B. In other words, the eutectic point represents the invariant equilibrium between the liquid and the two solid phases - invariant, of course, at fixed pressure.

For the equilibrium between solid B and the liquid mixture, the equilibrium condition is

$$\mu_B^{sol} = \mu_B^{liq}; \tag{14}$$

and the recipes of the potentials are

$$\mu_B^{sol} = G_B^{*sol}; \tag{15}$$
$$\mu_B^{liq} = G_B^{*liq} + RT \ln X. \tag{16}$$

The right-hand liquidus accordingly is the solution of the equation

$$\Delta G_B^* + RT \ln X = 0. \tag{17}$$

In the simplest case, i.e. neglecting C_P influences, Equation (8) for ΔG_B^* can be introduced whereafter the following relation is obtained for liquidus temperature as a function of mole fraction

$$T = \frac{\Delta S_B^* \cdot T_B^o}{\Delta S_B^* - R \ln X}. \tag{18}$$

The analogous formula for the left-hand liquidus is

$$T = \frac{\Delta S_A^* \cdot T_A^o}{\Delta S_A^* - R\ln(1-X)} . \tag{19}$$

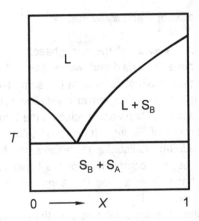

FIG. 3. Simple eutectic TX phase diagram; the eutectic point is the intersection of the two liquidi

From these liquidus formulae it follows that the *depression of the freezing point* for given X increases with decreasing ΔS^*. Substances with low entropies of melting give rise to large depressions of the freezing point (see also § 208). The influence of entropy of melting is shown in Figure 4.

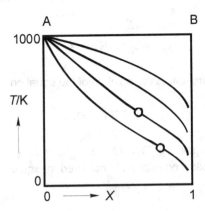

FIG. 4.   Equilibrium between pure solid A and liquid mixtures of A and B. Liquidus curves for different values of entropy of melting: (0.5; 1.0; 2.0; and 4.0) times R. Open circles mark points of inflexion

The formulae derived for equilibrium curves in PX and TX phase diagrams are simple, without exception. As 'diagnostic formulae', they represent a sound starting point for the evaluation of experimental phase diagrams. Apart from a pocket calculator, the properties of the pure substances are needed - and only as far as these properties are related to the phase change studied: phase equilibria are 'self-supporting'.

EXERCISES

1. *deviation from ideal vapour pressure*

For the isothermal equilibrium between ideal vapour mixtures and non-ideal liquid mixtures, for which the deviation from ideal-mixing behaviour is given by $G^E = \Omega X(1 - X)$, the equation for the liquidus is

$$P = (1-X){\cdot}P_A^o \cdot \exp\left\{\frac{\Omega X^2}{RT}\right\} + X{\cdot}P_B^o \cdot \exp\left\{\frac{\Omega(1-X)^2}{RT}\right\}.$$

- Give the derivation of this formula.
- Modify the formula to a *'diagnostic formula'* with the help of which the *interaction parameter* Ω can be read from a PX liquid + vapour phase diagram; that is to say, calculated from $P(X=\frac{1}{2})$ and $P^{id.}(X=\frac{1}{2}) = 1/2\,(P_A^o - P_B^o)$.

2. *like and unlike*

Vapour-pressure data are given below for two systems: i. at 60 °C for a combination of chemically like substances, namely methanol and 1-propanol, and ii. at 100 °C for the unlike combination of isobutylalcohol and toluene (the data can be found in Ohe 1989).

- Use the pressure data, which are expressed in Torr, to calculate the value of Ω; see Exc 1.

A	B	P_A^o	P_B^o	$P(X^{liq}=0.5)$
methanol	1-propanol	151.5	633.8	398.4
isobutylalcohol	toluene	559	570.5	736.9

3. *entropies of vaporization from phase diagram data*

Liquid mixtures of n-pentane (C_5H_{12}) and n-hexane (C_6H_{14}) - intuitively - are expected to be nearly ideal. If so, the isobaric liquid + vapour equilibrium in the system $\{(1-X)\ C_5H_2 + X\ C_6H_{14}\}$ will allow the evaluation of the *entropies of vaporization* of the two substances.

- Use the set of data, which are printed under here, to calculate the entropies of vaporization of n-pentane and n-hexane.

The data, which are valid for $P = 1$ bar, are from Tenn and Missen (1963).

$t\ /^oC$	X^{liq}	X^{vap}
35.6	0	0
41.2	0.253	0.103
43.6	0.342	0.167
47.5	0.478	0.250
51.0	0.600	0.350
54.5	0.690	0.461
56.8	0.755	0.542
59.3	0.801	0.620
61.1	0.845	0.688
63.1	0.892	0.769
64.2	0.912	0.812
68.0	1	1

4. *point of inflexion in ideal liquidus*

The 'ideal' liquidus curve for the equilibrium between pure solid A and liquid mixture of A and B can be written as

$$X(T) = \exp\{\Delta S\ (T - T_A^o\ )\ /RT\}, \text{ where } X = X_A$$

- Show that for $0 < X \leq 1$ there will be a *point of inflexion* if $(\Delta S_A^{\cdot} /R) \leq 2$.
- Verify the positions of the points of inflexion in Figure 4.

5. *solubility of anthracene in benzene*

Estimate the solubility of anthracene ($C_{14}H_{10}$) in liquid benzene at 25 °C.

	melting point	heat of melting
C_6H_6	5.5 °C	9.9 kJ·mol^{-1}
$C_{14}H_{10}$	217 °C	28.8 kJ·mol^{-1}

6. *from TX to PX*

Use the data and the result of Exc 3 to predict/calculate the *PX* liquid + vapour phase diagram in the system { $(1-X)C_5H_{12} + X C_6H_{14}$ } at 60°C.

7. *narrow two-phase regions?*

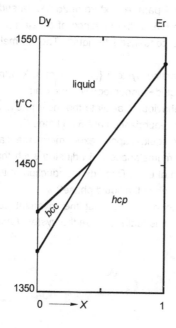

The figure which is shown resembles a published phase diagram on the lanthanide system Dy + Er. It looks more like an *EGC diagram* (such as Figure 211:13) than a true phase diagram in which the two -phase regions have a 'visible' width.
*   Perform a calculation in order to find out if the diagram - from a 'visibility point of view' - is wrong or not.
Entropies of melting are about (4/3) times *R*.

8. *heteroazeotrope*

Toluene (boiling point 383.78 K; heat of vaporization 33.18 $kJ \cdot mol^{-1}$) and water (373.15 K; 40.866 $kJ \cdot mol^{-1}$) are virtually immiscible.

- Estimate the temperature at which a heterogeneous mixture of the two substances will start to boil (temperature of *heteroazeotrope*).

9. *the ortho and para forms of H_2*

Hydrogen (H_2) exists in two forms, which are ortho-hydrogen (o-H_2, nuclear spins are parallel) and para-hydrogen (p-H_2, spins antiparallel). In the presence of a catalyst (active coal) a rapid transformation between the two forms can be realized, as a result of which the equilibrium composition can be determined as a function of temperature. At zero K the equilibrium composition is 100% para; at 298.15 K it is 25% para. The normal boiling point of *equilibrium hydrogen* (e-H_2) is 20.28 K, the composition being 99.8% para. A mixture of 25% para and 75% ortho is referred to as *normal hydrogen* (n-H_2). In the absence of a catalyst the interconversion is so slow that n-H_2 can be condensed to liquid. The normal boiling point of n-H_2 is 20.40 K.

- For the non-reacting system $\{(1-X)$ para $+ X$ ortho$\}$ calculate the isobaric ($P = 1$ atm) liquid+vapour equilibrium diagram, assuming ideal gas and ideal mixing behaviour. Calculate the mole fractions to four decimal places. The entropies of vaporization are 5.36 times R.
- In a virtual TX liquid+vapour experiment, the catalyst being absent, the experimental circumstances are adjusted such that the vapour phase has the composition of e-H_2. Does de introduction of the catalyst have an effect on the composition of the liquid phase?
- In the presence of the catalyst the invariant equilibrium between solid, liquid, and vapour is realized. Give the *system formulation* for this case.

The correspondence between Gibbs energy and phase diagram is studied by means of qualitative, graphical representations of isobaric-isothermal cross-sections of the Gibbs energy space.

non-ideal systems

In the last three sections of this volume systems are considered where at least one of the forms, in which the two components mix, deviates from ideal-mixing behaviour. Again, a restriction is made to isothermal or isobaric conditions. And even so, and for most of the cases, the equilibrium equations do not lead to explicit formulae for the relation between mole fraction variables and temperature or pressure. In other terms, for given temperature or pressure the compositions of the coexisting phases have to be calculated in a non-analytical manner, that is to say geometrically or numerically. The geometric approach, with a key role for *Gibbs energy versus mole fraction diagrams*, is the subject of this section. In the following section the problem-of-two-composition-variables is reduced to a 'one-mole-fraction-problem'; that is to say, a problem that can be tackled in an analytical manner. The numerical approach, at the level of a pocket calculator, is the subject of the last section.

As a matter of fact, the Gibbs energy versus mole fraction diagram, the *GX diagram* is a pre-eminent tool for the understanding of binary equilibrium. Sketches of *GX* diagrams are extremely useful vehicles for thinking and talking about phase diagrams. Moreover and most importantly, the *GX* diagram surpasses the notions of ideality and non-ideality: it is free from any theoretical model whatever. Therefore, the true purpose of this section is to demonstrate the power of *GX* diagrams.

Gibbs energy versus mole fraction diagrams

Let α and β be two forms, such as liquid and vapour, in which the two substances A and B, system $\{(1 - X)$ mole of A $+ X$ mole of B$\}$, are completely miscible. Then, the *molar Gibbs energies* of mixtures of A and B in each of the two forms are generally represented by

$$G^\alpha(T,P,X) = (1-X)G_A^{*\alpha}(T,P) + XG_B^{*\alpha}(T,P) + RT\,LN(X) + G^{E\alpha}(T,P,X) + C\cdot X\,;\quad (1)$$

$$G^\beta(T,P,X) = (1-X)G_A^{*\beta}(T,P) + XG_B^{*\beta}(T,P) + RT\,LN(X) + G^{E\beta}(T,P,X) + C\cdot X\,.\quad (2)$$

The first two members at the right-hand side of Equations (1) and (2) represent the Gibbs energy contributions of the pure components A and B before mixing. The third term

$$RT\,LN\,(X) \equiv RT\{\,(1-X)\ln(1-X)+X\ln X\,\}, \tag{3}$$

which represents the Gibbs-energy effect of *ideal mixing*, is the same for α and β. The deviation from ideal-mixing behaviour is given by the fourth term; the *excess Gibbs energy*, which is different for α and β. The last term, where the value of C is arbitrary, but necessarily the same for α and β, is referred to as *the linear contribution* (←205).

Whenever, for a given T and P, the two curves representing the two Gibbs functions have one or more points in common, there will be the possibility of equilibrium between phases (←203; in particular Figure 203:3). Equilibrium between and α-phase of composition X_e^α and a β-phase of composition X_e^β; and only so if the *overall composition* of the system (X^o) satisfies the condition $X_e^\alpha < X^o < X_e^\beta$. In the GX diagram, the compositions of the coexisting phases are given by the points of contact of the *common tangent line*. In terms of *chemical potentials*, the equilibrium compositions of the phases are the solution of the set N of equilibrium conditions (←203)

$$N\left[\mu_A^\alpha = \mu_A^\beta;\ \mu_B^\alpha = \mu_B^\beta\right]. \tag{4}$$

Purely mathematically (i.e. if one would ignore the existence of chemical potentials), as follows from Figure 1, the two conditions (two signs of equality) for the points of contact of the common tangent line are

$$N\left[\left(\frac{\partial G^\alpha}{\partial X}\right)=\left(\frac{\partial G^\beta}{\partial X}\right)=\frac{\Delta_e G}{\Delta_e X}\right]. \tag{5}$$

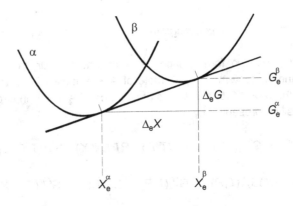

FIG. 1.   Towards a 'more mathematical' formulation of the two conditions for the common tangent

The operator Δ_e has the function of indicating that the difference is taken between the value of a property in β and the value of the same property in α, and such that α and β have taken their equilibrium compositions. In the following the operator Δ without the subscript 'e' is used for a difference property such that α and β have the same composition.

Under *isobaric conditions* the difference between $G_A^{*\beta}$ and $G_B^{*\alpha}$ can be given as

$$G_A^{*\beta}(T) - G_A^{*\alpha}(T) \equiv \Delta G_A^*(T) = \Delta G_A^*(T = T_r) + \int_{T_r}^{T} \left(\frac{\partial \Delta G_A^*}{\partial T} \right)_P dT, \qquad (6)$$

where T_r is a reference temperature. If T_r is taken as T_A^o, the temperature at which for pure A there is equilibrium between a phase α and a phase β, then the first term at the right-hand side of Equation (6) will vanish, because

$$\Delta G_A^*(T = T_A^o) = 0. \qquad (7)$$

Hence, and because $(\partial \Delta G_A^* / \partial T)_P = -\Delta S_A^*$,

$$\Delta G_A^*(T) = -\int_{T_A^o}^{T} \Delta S_A^*(T) \, dT \approx -\Delta S_A^*(T - T_A^o); \qquad (8)$$

similarly for pure B

$$\Delta G_B^*(T) \approx -\Delta S_B^*(T - T_B^o). \qquad (9)$$

equilibrium between liquid and vapour

As a first example of the relation between Gibbs energy cross-sections and phase diagram, the case is considered where β = vapour and α = liquid, and where the excess parts of the Gibbs functions are negligible (ideal-mixing behaviour in vapour as well as in liquid), see Figure 2. At temperatures above B's boiling point (T_B^o) the G–curve for the vapour is entirely below the one for the liquid: for all compositions the whole amount of matter is gaseous. Similarly, at temperatures below T_A^o the G-curve for the liquid is below the one for the vapour: the whole amount of matter is liquid, no matter the composition of he system. For temperatures between T_A^o and T_B^o the two curves intersect, and, for overall compositions between the points of contact, there will be equilibrium between liquid and vapour. The matter divides itself over a liquid and a gaseous phase, respecting the *lever rule*. Note that in Figure 2 the distances between the G-curves for $X = 0$ are indicated. These distances are given by Equation (8).

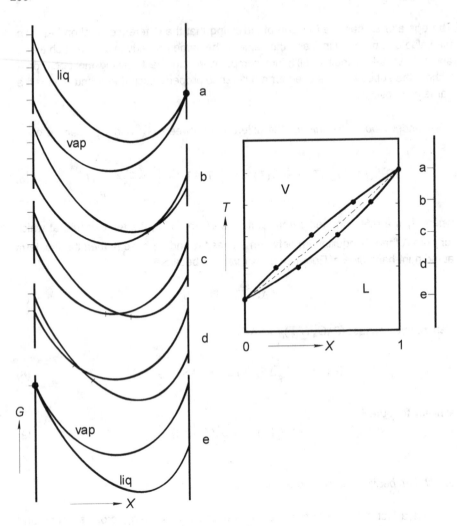

FIG. 2.   Relation between Gibbs-energy-versus-mole-fraction cross-
sections and phase diagram for the isobaric equilibrium between
liquid and vapour

In the case of Figure 3, and compared with Figure 2, the *G*-curve of the liquid
has a greater curvature - caused by a negative excess Gibbs energy, characteristic
of an attractive interaction between A and B. The phase diagram has a maximum, a
stationary point (←005) where the *liquidus* and *vaporus* touch one another.

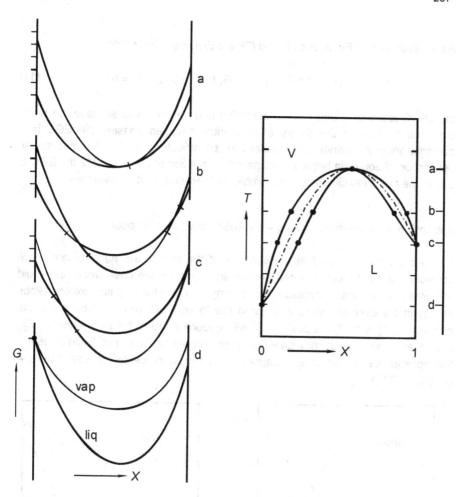

FIG. 3.   Isobaric equilibrium between liquid and vapour. Relation between
           Gibbs-energy-versus-mole-fraction   cross-sections   and   phase
           diagram-with-maximum

the equal-G curve

The *liquidus* and *vaporus* curves in the TX plane are the loci of the points of
contact of the common tangent lines drawn to the G-curves; see Figures 3 and 4.
The dash-dotted curves, which are drawn in the phase diagrams, are the
equal-G curves, EGC for short. In a TX or PX diagram, the EGC is the locus of the
points of intersection of the G-curves. Or, in other words, the EGC is the solution of
the equation

$$\Delta G(T, X) = 0. \tag{10}$$

After substitution of Equations (1) and (2), the EGC equation reads

$$(1-X)\,\Delta G_A^*(T) + X\,\Delta G_B^*(T) + \Delta G^E(T,\,X) = 0;\qquad(11)$$

the LN(X) terms of α and β have cancelled one another, and so have the linear contributions $C{\cdot}X$. In the theory of equilibrium between phases, the EGC is a powerful tool: it permits a considerable simplification of the thermodynamic description of equilibria between two mixed, or solution phases. In §212 the EGC is one of the means to circumvent the problem-of-two-composition-variables.

equilibrium between three phases – two liquid phases and vapour

As a next step, and as an example of the use of the *equal-G curve*, we consider the equilibrium between vapour and liquid for the case where the liquid mixtures have a positive excess Gibbs energy. Just like the negative excess Gibbs energy in the case of Figure 3 is giving rise to an EGC, which is above the line connecting T_A^o with T_B^o, a positive G^E will produce an EGC below the line $T_A^o\,T_B^o$. Moreover, a positive G^E may give rise to equilibrium between two liquid phases - the appearance in the phase diagram of a *region of demixing*; see Figure 4 ($\leftarrow$ 206;$\rightarrow$ 212).

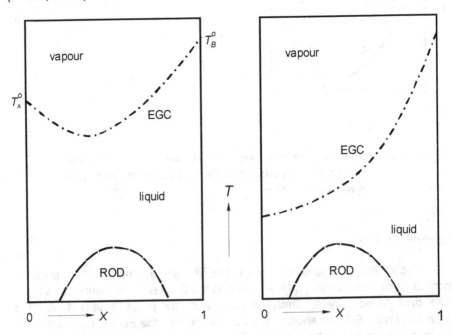

FIG.4. Isobaric equilibrium between liquid and vapour. Positive excess Gibbs energies of liquid mixtures i) make that the equal-G curve is bent down, and ii) give rise to a region of demixing

§ (211)

At this place we consider the case where a large positive G^E makes that the (liquid + vapour) equilibrium interferes with the region of demixing. If so, there is room for a situation of *equilibrium between three phases* ($L_I + L_{II} + V$), for which the *system formulation* is

$$f = M\left[T, X^{L_I}, X^{L_{II}}, X^{vap}\right] - N\left[\mu_A^{L_I} = \mu_A^{L_{II}} = \mu_A^{vap}; \; \mu_B^{L_I} = \mu_B^{L_{II}} = \mu_B^{vap}\right] = 0. \quad (12)$$

And from this expression it follows that the isobaric equilibrium between the three phases is invariant: a unique situation with fixed values for the four quantities of the set M, say, indicated as T_3, $X_3^{L_I}$, $X_3^{L_{II}}$, X_3^{vap}, see Figure 5 with its *three-phase equilibrium line*.

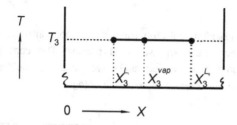

FIG. 5.   Three-phase equilibrium line connecting the points that represent the compositions of the phases

From the three points on the line, three pairs of *two-phase equilibrium curves* are emanating. One pair for the equilibrium ($L_I + L_{II}$); another pair for ($L_I + V$); and the third pair for ($L_{II} + V$). In Figures 6 and 7 it is shown how the phase diagram straightforwardly follows from the combination of *binodal* (the boundary of the region of demixing) and equal-G curve. The important things to keep in mind are i) the three-phase equilibrium line has two points on the binodal, and ii) the EGC invariably is situated between liquidus and vaporus - for ($L_I + V$) and ($L_{II} + V$).

Sketches of the *GX* sections for the three-phase equilibrium temperatures, corresponding to Figures 6 and 7, are shown in Figure 8. The fact that *linear contributions* do change nothing in the world of phase equilibria, makes that the common tangent lines can be given no matter what slope - e.g. slope zero as is done in the case of Figure 8.

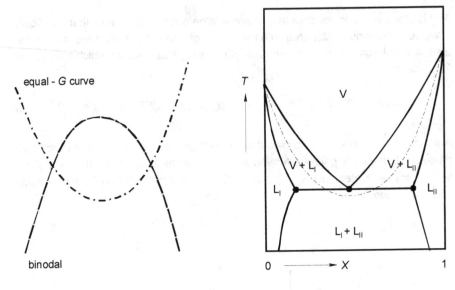

FIG. 6.   The equilibrium between liquid and vapour interferes with the equilibrium between two liquid phases. The correspondence between diagram with equal-*G* curve and binodal (left) and phase diagram with three-phase equilibrium (right)

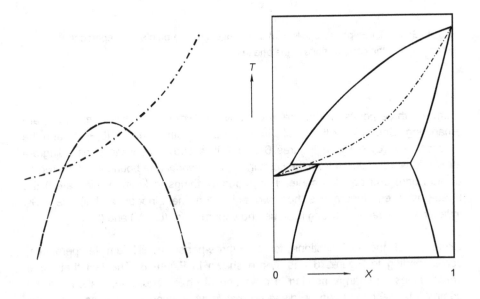

FIG. 7.   Analogous to Figure 6 - with the distinction, however, that this time the EGC's minimum is absent

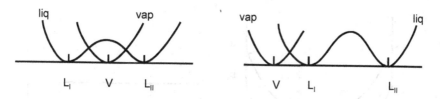

FIG. 8.   Sketches of the *GX* sections for the three-phase equilibrium
temperatures in Figure 6 (left) and Figure 7 (right)

types of phase diagram

The phase diagrams Figures 2, 3, 6, 7, and also the one with a minimum in
the (L + V) two-phase region, implied in Figure 5 left-hand side, are the basic types
of *TX* phase diagram in the case of two mixed forms. Not only for the equilibrium
between liquid and vapour, but also for the equilibrium between liquid and mixed
crystalline solid, and any other combination of two forms in which the components
are miscible. The issue of *types of phase diagram* is readdressed in the following
section.

solids of fixed composition in equilibrium with liquid

The Figures 9, 10, and 11 pertain to systems where solids of fixed
composition are in equilibrium with liquid mixtures. In the case of Figure 9 the solid
phases just are the component solids; the phase diagram is the *simple eutectic
phase diagram*. The symmetrical phase diagram in Figure 10 is characteristic of
systems where a *pair of optical antipodes* gives rise to a *racemic compound*, a
racemate.

In Figure 11 the relative positions of the Gibbs energies are such that the 1:1
compound is a so-called *incongruently melting compound*: on heating the
compound changes into (liquid P + solid B) at the three-phase (compound + liquid
P + solid B) equilibrium temperature (isothermal section b). In the *GX* diagram for
section c, a (dashed) line can be drawn for the *metastable equilibrium* between
solid B and liquid; the point of contact corresponds to a point on the *metastable
extension* of the (L + S$_B$) liquidus in the phase diagram.

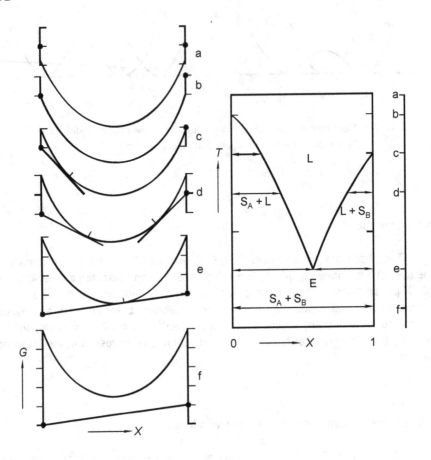

FIG. 9. Isobaric equilibrium between liquid mixtures and solids of constant composition. Relation between GX sections and phase diagram for the case of a simple eutectic: the only solid phases are the pure component ones

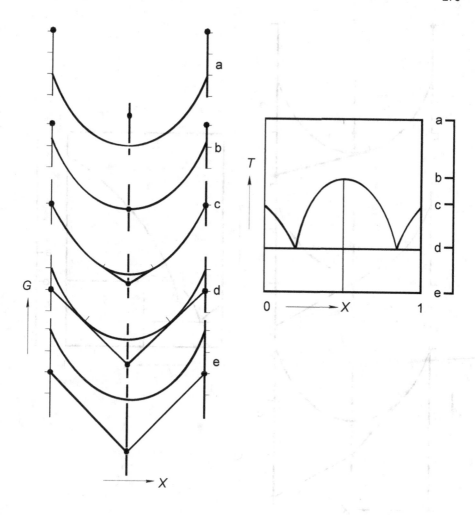

FIG. 10. Isobaric equilibrium between liquid mixtures and solids of constant
composition. The latter are the two pure component solids and a 1:1
compound which fully dissociates on melting. The relative positions of
the G-points for the solids in the GX sections are such that the
compound is congruently melting

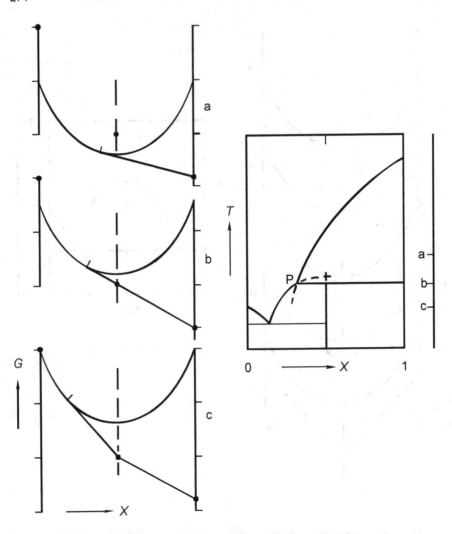

FIG. 11. Analogous to Figure 10 - will the distinction, however, that this time the relative positions of the G-points make that the compound is incongruently melting

crossed isodimorphism

Finally, Figures 12 and 13 are for systems where the molecules of A and B can replace one another in the crystal lattice; the two components, however, having different crystal structures. The phase diagram in Figure 12 shows a *eutectic type of three-phase equilibrium*, and the one in Figure 13 a *three-phase equilibrium of the peritectic type*. Owing to the fact that the phase diagrams in Figures 12 and 13 can be regarded as the stable result of two, each other crossing, *solid-liquid loops*, the term *crossed isodimorphism* is used to refer to this situation.

§ (211)

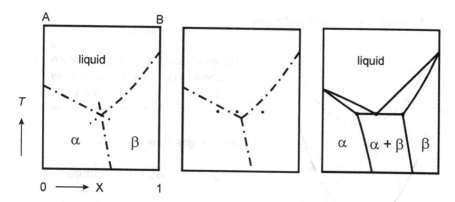

FIG. 12. From EGC diagram to phase diagram with a eutectic three-phase
equilibrium. Two, each other crossing (solid + liquid) Equal-G
curves invariably involve a phase diagram showing incomplete
solid-state miscibility

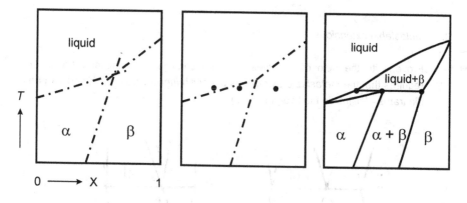

FIG. 13. The peritectic analogue of Figure 12

*The qualitative GX diagram is a powerful tool for grasping the ins and outs of binary
phase diagrams. Unlike theoretical and empirical models, it has the advantage of
being free from any limitations.*

EXERCISES

1. *validation of linear contributions*

For the case of Figure 1, the equilibrium conditions in N, Equation (5), are satisfied for
$X^\alpha = X_e^\alpha$ and $X^\beta = X_e^\beta$.

- Prove that the two conditions are satisfied for the same X values, when equal
  linear contributions are added to each of the two G functions; i.e. when G^α is
  replaced by $G'^\alpha = G^\alpha + CX^\alpha$; and similarly for G^β.

§ (211)

2. *from G-curves to phase diagram*

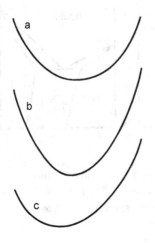

Using transparent paper, or otherwise, construct (liquid + vapour) phase diagrams for a number of combinations of the *G*-curves given here, and labeled a, b, and c.

Some suggestions

PX or TX	liquid	vapour
PX	a	c
PX	b	c
TX	a	a

3. *metastable extensions*

Prove, with the help of *G*-curves, that the condition of minimal Gibbs energy requires that the *metastable extensions of equilibrium curves* - in *PX* and *TX* phase diagrams - fall inside the two-phase regions.

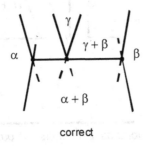

correct

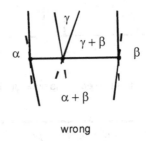

wrong

4. *isothermal solid+vapour equilibrium*

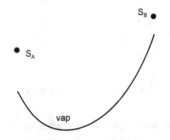

The system is considered in which, at constant temperature, two pure solid components, either individually or together, are in equilibrium with a gaseous mixture of the two components. For the relative positions of the Gibbs energies shown here, construct the *PX* phase diagram.

5. *phase diagram for given ROD and EGC*

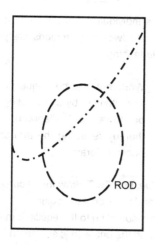

Construct the *TX* (liquid + vapour) phase diagram corresponding to the figure with the boundary of a closed region of demixing (ROD) in the liquid state and the equal-*G* curve for the liquid to vapour transition.

ROD

6. *from EGC diagram to phase diagram*

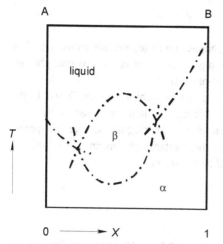

A   B

liquid

T

β

α

0 ⟶ *X*   1

The two substances A and B are isomorphous in form α. From a certain moment (temperature) on, and for intermediate compositions, the form β makes its appearance (is stabilized).

- Transform this EGC diagram into the corresponding phase diagram.
- Make sketches of the *GX* sections for the two three-phase equilibrium temperatures.

7. *phase diagram for EGC and two ROD's*

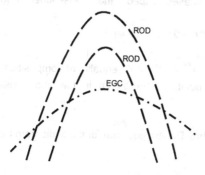

ROD

ROD

EGC

For the purely hypothetical case implied in the sketch, the region of demixing (ROD) in the solid state is below the ROD in the liquid, the solid-liquid equal-*G* curve having a maximum inside the ROD's.

- Make a sketch of the corresponding *TX* phase diagram.

§ (211)

8.   *overlapping two-phase regions*

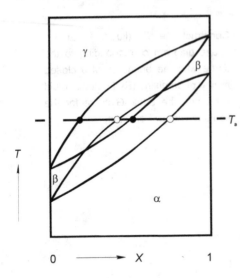

The individually correct $(\alpha + \beta)$ and $(\beta + \gamma)$ two-phase regions are partly overlapping.

- Work out the consequences of this situation by constructing one or more *GX* diagrams, and thereby revealing the true stable phase diagram.

Clue. For $T = T_a$ draw the *G*-curve for β and mark on it the points corresponding to the equilibrium with α and the one with γ; and so on.

9.   *the system formulation for a symmetrical binary system*

Generally, the equilibrium between four phases in a binary system is invariant. This is e.g. the case for the equilibrium between two liquid phases and two solid phases, the latter being the two pure solid components A and B.

- For the case that A and B are a pair of optical antipodes, D and L, the equilibrium between the four phases is monovariant, due to symmetry.
- To demonstrate this, first, construct the *GX* diagram for the four-phase situation and, next, write down the system formulation such that the conditions are given in terms of G and $(\partial G/\partial X)$.

10.   *azeotropy and Gibbs-Duhem*

Along the liquidus and the vaporus in the *TX* plane, the liquid and the vapour phase, in combination, satisfy the equilibrium conditions in chemical potentials; and, on its own, each of the two phases respects the Gibbs-Duhem equation ($\leftarrow$ 203):

$$(1 - X^\alpha)d\mu_A^\alpha + X^\alpha d\mu_B^\alpha + S^\alpha dT = 0; \qquad \alpha = \text{liq, vap.}$$

- Demonstrate that $dT = 0$, for $X^{liq} = X^{vap} = X$: equality of composition goes together with a stationary point, an extremum in the phase diagram ($\leftarrow$ Exc 003:2).

Clue. To get started, subtract the Gibbs-Duhem equation for the liquid from the one for the vapour.

On the basis of a mathematical model for the Gibbs energy, or rather the excess Gibbs energy, formulae are derived for a variety of binary phase-diagram characteristics. In essence, the procedure is a deepening of the magic formula approach put forward in § 206.

aim and approach

The aim of this section is to examine how binary phase diagrams reflect the deviation from ideal mixing behaviour, read, the excess Gibbs energy as a function of the system variables. This is realized by using simple models, models that can be looked upon as extensions of the *magic formula* ($\leftarrow$206)

$$G^E = \Omega X(1-X). \tag{1}$$

One of these models is the *ABΘ* model, in which Ω is made a linear function of both temperature and mole fraction.

To realize this goal, and as already stated ($\leftarrow$211), the problem-of-two-composition-variables is reduced to a one-mole-fraction-problem. More precisely, for those classes of systems where each of the two phases-in-equilibrium has a variable composition, the treatment is directed either to the *spinodal* (*TX* systems) or to the *equal-G curve* (*TX* and *PX*). For the classes of *TX* systems, where only one of the two phases in equilibrium has a variable composition, on the other hand, there is one equilibrium equation, the solution of which is the relation between phase composition and temperature. Accordingly, for all of the *TX* systems examined, there is one condition, one equation to be solved - an equation of the type

$$\delta G'' = 0, \tag{2}$$

in which δ represents an operator, acting on a Gibbs-energy-related property G''. As a next step, and in order to set up formulae that give temperature as an explicit function of temperature, all Gibbs energies are taken as linear functions of temperature. This comes down to neglecting heat capacities, as a result of which the corresponding enthalpies and entropies are independent of temperature, depending only on composition:

$$G(T, X) = H(T, X) - T\,S(T, X) \Rightarrow G(T, X) = H(X) - T\,S(X). \tag{3}$$

The (didactic) advantage is in the fact that the solution of Equation (2), defining a characteristic curve in the *TX* phase diagram, directly can be written down as an explicit relation:

$$T(X) = \delta H''/\delta S'' \tag{4}$$

the ABΘ model

The extended magic formula for the excess Gibbs energy, referred to as the *ABΘ* model, has the form

$$G^E(T, X) = A\left(1 - \frac{T}{\Theta}\right)X(1 - X)[1 + B(1 - 2X)], \tag{5}$$

in which A, B, and Θ are system-dependent constants. The constant A is a measure of the magnitude of the excess function; it has the dimension of energy. The dimensionless B is a measure of the asymmetry of the function with respect to $X = 0.5$. The dependence on temperature is reflected by Θ, a parameter with the dimension of temperature.

Before putting the *ABΘ* formula into action, it is useful to give a short survey of its characteristics and thermodynamic implications. To start with, cross-sections of the excess function are shown in Figure 1, for $A = 18200$ J·mol^{-1}; $B = 0.2$; and $\Theta = 2565$ K - numerical values that go well with the mixed crystalline state in $\{(1 - X) \text{ NaCl} + X \text{ KCl}\}$.

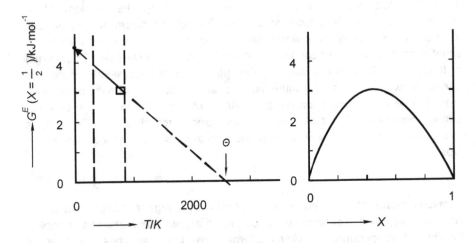

FIG. 1.   Cross-sections of the function $G^E(T,X)$, Equation (5), for $A = 18.2$ kJ·mol^{-1}; $B = 0.2$; and $\Theta = 2565$ K. Left: for $X = 0.5$, as a function of temperature. Right: for $T = 875$ K, as a function of mole fraction

The expressions for the excess enthalpy and excess entropy functions are

$$H^E(X) = AX(1 - X)[1 + B(1 - 2X)], \quad \text{and} \tag{6}$$

$$S^E(X) = (A/\Theta)X(1 - X)[1 + B(1 - 2X)]. \tag{7}$$

The quotient of H^E and S^E is just Θ, a property which is often referred to as *compensation temperature*: at $T = \Theta$, the H^E and S^E compensate one another, in the sense that G^E becomes zero, or rather goes through zero - changing sign. *Enthalpy-entropy compensation* is one of the guiding principles in experimental thermodynamics. The reason is in the fact that it is often observed, that the quotient of enthalpy change (ΔH) and entropy change (ΔS), for a given class of systems, is system-independent. As an example, for the class of common-ion alkali halide mixed crystals, having the NaCl type of structure, it is observed that the parameter A (Equation 5) is system-dependent, whereas Θ is not. The latter has appeared to be system-independent: the class of systems is characterized by a uniform value, which is $\Theta = 2565$ K (van der Kemp et al. 1992).

For the system $\{(1 - X) A + X B\}$ the full expression for the (molar) Gibbs energy is

$$G(T, X) = (1 - X)G_A^*(T) + XG_B^*(T) + RT\{(1 - X)\ln(1 - X) + X \ln X\} +$$
$$+ A\left(1 - \frac{T}{\Theta}\right)X(1 - X)[1 + B(1 - 2X)] \qquad ; \qquad (8)$$

and the expressions for the chemical potentials are

$$\mu_A(T, X) = G_A^*(T) + RT \ln(1 - X) + X^2 \cdot A\left(1 - \frac{T}{\Theta}\right)[1 + B(3 - 4X)]; \qquad (9a)$$

$$\mu_B(T, X) = G_B^*(T) + RT \ln X + (1 - X)^2 \cdot A\left(1 - \frac{T}{\Theta}\right)[1 + B(1 - 4X)]. \qquad (9b)$$

The expressions for the (integral) entropy and enthalpy functions are (still neglecting heat capacities)

$$S(X) = (1 - X)S_A^* + XS_B^* - R\{(1 - X)\ln(1 - X) + X \ln X\} +$$
$$+ (A/\Theta)X(1 - X)[1 + B(1 - 2X)] \qquad ; \qquad (10)$$

$$H(X) = (1 - X)H_A^* + XH_B^* + AX(1 - X)[1 + B(1 - 2X)]. \qquad (11)$$

the Gibbs function has points of inflexion

The Gibbs energy expression

$$G(T, X) = (1 - X)G_A^*(T) + XG_B^*(T) + RT\{(1 - X)\ln(1 - X) + X \ln X\} + \Omega X(1 - X) \quad (12)$$

made its appearance in §206 as the second of two *magic formulae* - part of its magic being that it accounts for the existence of critical points of mixing.

§ (212)

If the asymmetry parameter B is set at zero, the findings of §206 can be summarized as follows (see also Figure 2):

- For negative values of $A(1-T/\Theta)$ the Gibbs function is entirely convex; it means that demixing is excluded.
- For $A(1-T/\Theta) > 2RT$ the function is partly concave - the function has two points of inflexion. The situation of minimal Gibbs energy is realized by an equilibrium between two phases, the compositions of which are given by the points of contact of the double tangent line (see hereafter, the phases being labeled I and II).
- The critical temperature of mixing, T_C, makes its appearance for $A(1-T/\Theta) = 2RT$:

$$T_C = \frac{A}{A/\Theta + 2R} \qquad (13)$$

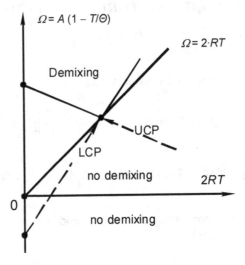

FIG. 2.   In terms of the model $G^E = \Omega X(1-X)$, with $\Omega = A(1-T/\Theta)$, there will be demixing whenever $\Omega > 2 \cdot RT$. The model accounts for the existence of regions of demixing with upper critical points (UCP), as well as for regions of demixing with lower critical points (LCP)

As follows clearly from Figure 2, the model not only accounts for *upper critical points* but also for *lower critical points*. The lower critical points are for regions of demixing having a minimum.

[(*besides*) By virtue of the *linear contribution* property (← 205), Equation (8) may be reduced to

$$G'(T,X) = RT\{(1-X)\ln(1-X) + X\ln X\} + A\left(1-\frac{T}{\Theta}\right)X(1-X)[1+B(1-2X)]. \quad (14)$$

For $B = 0$, the G' function is symmetrical with respect to $X = 0.5$, and the determination of the compositions of the coexisting phases, X'_e and X''_e, is a first derivative problem. The situation can be formulated as

$$f = M[T, X'] - N[(\partial G'/\partial X)_T] = 0, \tag{15}$$

because there is one independent composition variable (as $X'' = 1 - X'$).
For this case the operator δ, Equation (2), is the first partial derivative with respect to mole fraction. And, according to Equation (4), the solution of the equation under N simply is,

$$T_{BIN}(X) = \frac{A(1-2X)}{R\ln\left(\dfrac{1-X}{X}\right) + \dfrac{A}{\Theta}(1-2X)} \quad ; \quad (B = 0) \tag{16}$$

it is the formula for the *binodal*, the boundary of the region of demixing.]

spinodal

The *spinodal*, which in the TX plane is the locus of the points of inflexion, is the solution of the equation

$$\frac{\partial^2 G}{\partial X^2} = 0. \tag{17}$$

According to Equation (4), the solution can be written as

$$T_{SPIN}(X) = \frac{\partial^2 H / \partial X^2}{\partial^2 S / \partial X^2}, \tag{18}$$

and after substitution of the second derivatives, and rearranging, it takes the form

$$T_{SPIN}(X) = \frac{2A \cdot X(1-X)[1+3B(1-2X)]}{R + \dfrac{2A}{\Theta}X(1-X)[1+3B(1-2X)]}. \tag{19}$$

Notwithstanding the fact that the spinodal does not give the exact position of the coexisting phases, it has a key position in the phenomenon of demixing. In the present context, the spinodal defines the extension of the region of demixing over the temperature scale; and its extremum coincides with the critical point.
The mole fraction of the critical point follows from $dT_{SPIN}/dX = 0$: its value (X_C) is the solution of the equation

$$(18\,B)X^2 - (2 + 18B)X + (1 + 3B) = 0, \tag{20}$$

§ (212)

and subsequently, the critical temperature (T_C) is found by substitution of X_C in Equation (19). Clearly (read, from Equation (20) it follows that), the position of the critical point on the X scale is fully determined by the value of B.

two Gibbs functions have one or two points of intersection - the equal-G curve

Two idealized cases are considered where two substances A and B are miscible in two different forms α and β, and where α phases are in equilibrium with β phases. The first is the isothermal equilibrium between a non-ideal liquid mixture, whose volume is neglected, and an ideal mixture of ideal gases (*idealized treatment*). The second case is the isobaric equilibrium between a non-ideal form α (the low-temperature form) and an ideal form β (the high-temperature form).

The fully ideal, isothermal and isobaric equilibria have been treated already ($\leftarrow$210), making use of *chemical potentials*. This time the treatment is based on *equal-G curves*.

In the idealized treatment of isothermal liquid + vapour equilibria, the molar volume of the liquid is neglected with respect to the molar volume of the vapour. It means that, because of $(\partial G / \partial P)_T = V$, the variable P does not appear in the G function of the liquid. The two G functions now are ($\leftarrow$ 204, 205)

$$G^{vap}(P,X) = (1-X)G_A^{o\ vap} + XG_B^{o\ vap} + RT\ln P + RT\{(1-X)\ln(1-X) + X\ln X\} \quad (23)$$

$$G^{liq}(X) = (1-X)G_A^{*liq} + XG_B^{*liq} + RT\{(1-X)\ln(1-X) + X\ln X\} + G^{E\ liq}(X). \quad (24)$$

At the temperature selected for the isothermal equilibrium, the G^o and G^* properties are related by the equilibrium pressures of the pure components, P_A^o and P_B^o, as ($\leftarrow$208)

$$G_A^{*liq} = G_A^{o\ vap} + RT\ln P_A^o; \quad (25a)$$

$$G_B^{*liq} = G_B^{o\ vap} + RT\ln P_B^o. \quad (25b)$$

The equal-G curve (EGC) is the solution of the equation

$$G^{vap}(P,X) - G^{liq}(X) = 0: \quad (26)$$

$$\ln P_{EGC}(X) = (1-X)\ln P_A^o + X\ln P_B^o + \frac{G^{E\ liq}(X)}{RT}. \quad (27)$$

In the $\ln P$ vs X diagram the EGC is obtained by adding to the *zero line*, which is

$$\ln P_{ZERO}(X) = (1-X)\ln P_A^o + X\ln P_B^o, \quad (28)$$

the excess Gibbs energy of the liquid mixtures divided by RT.

§ (212)

In the example, Figure 3, the excess Gibbs energy is negative the EGC is below the zero line. The other way round, from the experimental diagram it can be read that $G^{E\ liq}$ ($X = 0.6$) is about - 0.22 RT = −564 J·mol⁻¹.

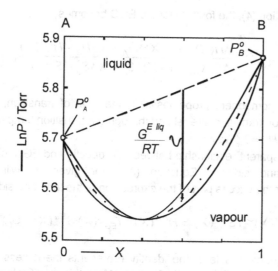

FIG. 3. Isothermal equilibrium between liquid and gaseous mixtures of chloroform (A) and acetone (B) at T = 308.35 K. (Apelblatt et al. 1980)

For isobaric equilibria between a non-ideal, low-temperature form α and an ideal, high-temperature form β the molar Gibbs energies are

$$G^{\beta}(T,X) = (1-X)G_A^{*\beta}(T) + XG_B^{*\beta}(T) + RT\{(1-X)\ln(1-X) + X\ln X\};\qquad (29)$$

$$G^{\alpha}(T,X) = (1-X)G_A^{*\alpha}(T) + XG_B^{*\alpha}(T) + RT\{(1-X)\ln(1-X) + X\ln X\} \\ + G^{E\alpha}(T,X) \qquad (30)$$

The equal-G curve (EGC) is the solution of the equation

$$\Delta G(T,X) \equiv G^{\beta}(T,X) - G^{\alpha}(T,X) = 0, \qquad (31)$$

which, after substitution of Equations (29) and (30), reads

$$\Delta G(T,X) = (1-X)\Delta G_A^{*}(T) + X\Delta G_B^{*}(T) - G^{E\alpha}(T,X) = 0, \qquad (32)$$

or, in terms of enthalpy and entropy,

$$(1-X)(\Delta H_A^* - T\Delta S_A^*) + X(\Delta H_B^* - T\Delta S_B^*) - (H^{E\alpha}(X) - TS^{E\alpha}(X)) = 0. \qquad (33)$$

In line with Equation (4), the formula for the EGC becomes

$$T_{EGC}(X) = \frac{\Delta H(X)}{\Delta S(X)} = \frac{(1-X)\Delta H_A^* + X\Delta H_B^* - H^{E\alpha}(X)}{(1-X)\Delta S_A^* + X\Delta S_B^* - S^{E\alpha}(X)}, \qquad (34)$$

where the pure component properties ΔH^*, the heat of transition, and ΔS^*, the entropy change on transition, are related through the transition temperature; e.g. for component A, $\Delta H_A^* = T_A^o \cdot \Delta S_A^*$.

To obtain a transparent relationship between G^E directly and EGC, one can realize that the left-hand side of Equation (33), for given X, will be zero at $T = T_{EGC}(X)$. For the excess part of the expression at the left-hand side:

$$H^{E\alpha}(X) - TS^{E\alpha}(X) \Rightarrow H^{E\alpha}(X) - T_{EGC}(X) \cdot S^{E\alpha}(X) \equiv G_{EGC}^{E\alpha}(X), \qquad (35)$$

in which the right-hand side of the identity represents the excess Gibbs energy along the EGC as a function of mole fraction. When this modification is introduced in Equation (33), the solution of the equation takes the form

$$T_{EGC}(X) = T_{ZERO}(X) - \frac{G_{EGC}^{E\alpha}(X)}{(1-X)\Delta S_A^* + X\Delta S_B^*}, \qquad (36)$$

where

$$T_{ZERO}(X) = \frac{(1-X)\Delta H_A^* + X\Delta H_B^*}{(1-X)\Delta S_A^* + X\Delta S_B^*}. \qquad (37)$$

In the TX diagram the distance from zero line to EGC is given by (minus) G^E divided by a system property, namely the weighted entropy of transition of the pure components. One could say that in the TX case the entropy change is acting as a scale factor, whereas in the PX (liquid + vapour) case that factor is RT, the T of which is set by the investigator. Figure 4 is an illustration; it is the TX counterpart of Figure 3: a lowering of pressure goes together with an elevation of temperature ($\leftarrow$ 005). Generally, the zero line is close to the straight line between T_A^o and T_B^o, it coincides with that line if $\Delta S_A^* = \Delta S_B^*$, and also when $T_A^o = T_B^o$.

In a further simplification the subscript EGC of G^E, in Equations (35) and (36), is dropped. It comes down to neglecting the change of G^E with temperature in the α to β transition range. Or, in terms of Equation (5), and T_m representing the mean temperature of the transition range,

$$G_{EGC}^E(X) \Rightarrow G^E(T = T_m, X) = A\left(1 - \frac{T_m}{\Theta}\right) X(1-X)[1+B(1-2X)] \qquad (38)$$

and, in shorter form,

$$G_{EGC}^E(X) \Rightarrow X(1-X)[\omega + \omega'(1-2X)] \qquad (39)$$

The simplification is exemplified in Figure 1: within the small rectangle the change of the excess Gibbs energy with temperature is virtually negligible.

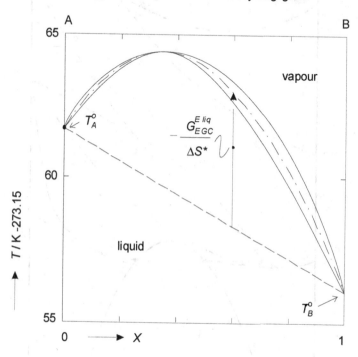

FIG. 4.   Isobaric equilibrium between liquid and gaseous mixtures of chloroform (A) and acetone (B) at $P = 760$ Torr (Reinders and de Minjer 1940)

types of phase diagram

An important observation to be made is that *type of excess behaviour* is not synonymous with *type of phase diagram*. In other terms, two systems showing the same excess behaviour may have quite different phase diagrams. The equal-G curve is a pre-eminent vehicle to make this clear – as is illustrated by Figure 5. The five equal-G curves in Figure 5 all correspond to ideal liquid mixtures, in combination with solid mixtures having $G^{E\,sol}(X) = X(1-X) \cdot 640\,R \cdot K$, the entropies of melting invariably being equal to $8R$. In all five cases the critical temperature of mixing is $T_C = 320$ K.

§ (212)

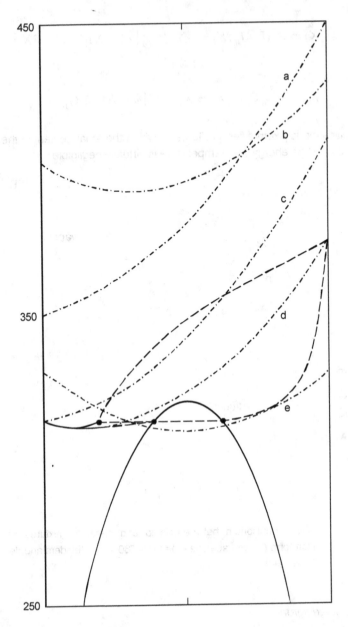

FIG. 5. The five equal-G curves, a, b, c, d, and e, all are for the combination of ideal liquid mixtures and solid mixtures which have, in all five cases, the same excess Gibbs energy, involving the region of demixing shown. In spite of the equality in mixing properties, each case corresponds to a type of phase diagram which is different from the others. For the EGC marked d the phase diagram is completed by dashed curves; it displays a minimum and, separated from it, a peritectic three-phase equilibrium

In Figure 5, there are two *types of equal-G curve*: type (0), absence of extremum (a·, and c·), and type (–), presence of minimum (b·, d·, and e·). In three cases (c·, d·, and e·) the solid + liquid equilibrium interferes with the region of demixing (ROD), i.e. with the solid + solid equilibrium:

e·   the EGC minimum is inside the ROD ; the phase diagram will show a eutectic three-phase equilibrium; type of phase diagram [e];

d·   the EGC minimum is outside the ROD; following the EGC from left to right, first a minimum (in EGC which coincides with minimum in phase diagram) and then a peritectic three-phase equilibrium; the phase diagram (dashed lines) has type code [–p];

c·   type (0) of EGC; phase diagram with peritectic three-phase equilibrium, [p]·

The equal-G curves a· and b· are too far above the ROD to involve a three-phase equilibrium:

a·   type (0) of EGC; phase diagram without extremum, [0];

b·   type (–) of EGC; phase diagram with minimum, [–].

As an observation, to have a type of phase diagram [e] it is necessary that part of the EGC is inside the ROD; for the types [p] and [–p] this is not strictly necessary.

The findings are summarized in the scheme which is shown, and which also includes the possibilities for a negative excess Gibbs energy. The Roman numerals were given by Bakhuis Roozeboom, who originally addressed the issue (Bakhuis Roozeboom 1899).

ω	EGC	Three-phase equil ?	Type of diagram	BR
0	(0)	N	[0]	I
–	(0)	N	[0]	I
–	(+)	N	[+]	II
+	(0)	N	[0]	I
+	(0)	Y	[p]	IV
+	(–)	N	[–]	III
+	(–)	Y	[e]	V
+	(–)	Y	[–p]	

SCHEME: Types of phase diagram for $G^{E\ sol}(X) = \omega \cdot X(1-X)$, $G^{E\ liq}$ being negligible

Bakhuis Roozeboom used *G*-curves for his derivation - such as is the subject of § 211. It explains why in his classification [–p] is missing. Besides, there is not much 'space' for a system to produce a type [–p] diagram.

miscibility in one phase, immiscibility in the other.

If, under isobaric conditions, the substances A and B show miscibility in only one of the two phases in equilibrium, then there will be only one composition variable, and one condition will be needed to establish the relation between temperature and composition. In the simplest case A and B are miscible in β, and phase α is either pure A or pure B. Taking β = liquid and α = solid (pure A), the system formulation is

$$f = M[T, X] - N\left[\mu_A^{sol} = \mu_A^{liq}\right] = 1.$$
(40)

The equilibrium condition can be written as

$$\Delta G_A(T, X) = 0;$$
(41)

and in view of Equation (4), the solution of the equation, the liquidus temperature as a function of X can be written as the quotient of ΔH_A and ΔS_A (which are the changes on melting of A's partial enthalpy and A's partial entropy):

$$T_{LIQ}(X) = \frac{\Delta H_A(X)}{\Delta S_A(X)} = \frac{\Delta H_A^* + X^2 \cdot A[1 + B(3 - 4X)]}{\Delta H_A^* / T_A^o - R\ln(1 - X) + X^2(A/\Theta)[1 + B(3 - 4X)]}.$$
(42)

The expression after the second equality sing is according to the $AB\Theta$ model. In order to avoid confusion: the variable X in Equations (40)-(42) is not A's, but B's mole fraction in the liquid phase!

the presence of a compound

As a last, instructive example, the case is considered where a 1:1 compound on melting dissociates into a liquid mixture of its constituents:

$$\text{AB (sol)} \to \text{A (liq)} + \text{B (liq)}.$$
(43)

In the TX phase diagram, the equilibrium involved in the melting reaction, corresponds to two two-phase fields that are bounded by the line $X = 0.5$ and the liquidus curve. The latter has a maximum at T_m, the melting point of AB; see Figure 211:10.
In terms of chemical potentials, the equilibrium condition for the 'chemical' reaction is

$$\mu_{AB}^{sol} = \mu_A^{liq} + \mu_B^{liq}.$$
(44)

From the GX cross-section, Figure 6, it is read that the condition of minimum Gibbs energy is given by

$$G_{A_{0.5}B_{0.5}}^{sol} = 0.5\left(\mu_A^{liq} + \mu_B^{liq}\right), \text{ or } G_{AB}^{sol} = \mu_A^{liq} + \mu_B^{liq}.$$
(45)

§ (212)

In Equations (44) and (45) μ_{AB}^{sol} and G_{AB}^{sol} are for the amount of AB composed of 1 mol A + 1 mol B.

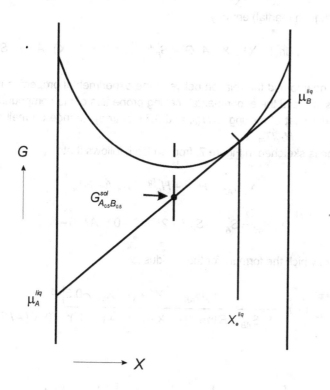

FIG. 6.   GX section for equilibrium between 1:1 compound, wich dissociates on melting, and liquid mixture

To avoid excessive writing, and realizing that in the central part of compositions the influence of parameter B is weak, the equilibrium description is worked out for the parameters A and Θ, putting $B = 0$.
With

$$\Delta_R \mu_i \equiv \mu_A^{liq} + \mu_B^{liq} - G_{AB}^{sol} \tag{46}$$

the liquidus temperature as a function of liquid composition, X, according to Equation (4), is given by

$$T_{LIQ_{AB}}(X) = \frac{\Delta_R H_i}{\Delta_R S_i} = \frac{H_A^{liq} + H_B^{liq} - H_{AB}^{sol}}{S_A^{liq} + S_B^{liq} - S_{AB}^{sol}} \cdot \tag{47}$$

The change in (partial) enthalpy is given by

$$\Delta_R H_i = H_A^{*liq} + X^2 \cdot A + H_B^{*liq} + (1-X)^2 \cdot A - H_{AB}^{sol} \tag{48}$$

and the change in (partial) entropy by

$$\Delta_R S_i = S_A^{*liq} - R\ln(1-X) + X^2(A/\Theta) + S_B^{*liq} - R\ln X + (1-X)^2 \cdot A/\Theta - S_{AB}^{sol} \tag{49}$$

The trick is now to find the relation between the experimental properties in the last two expressions and the experimental melting properties of the compound AB. The latter are AB's heat of melting $\Delta_m H_{AB}$, and AB's entropy change on melting $\Delta_m S_{AB}$, where $\Delta_m S_{AB} = \Delta_m H_{AB}/T_m$.
The situation is sketched in Figure 7, from which it follows that

$$\Delta_m H_{AB} = H_A^{*liq} + H_B^{*liq} + 0.5\,A - H_{AB}^{sol} \tag{50}$$

$$\Delta_m S_{AB} = S_A^{*liq} + S_B^{*liq} + 2R\ln 2 + 0.5(A/\Theta) - S_{AB}^{sol}, \tag{51}$$

as a result of which the formula for the liquidus is

$$T_{LIQ_{AB}}(X) = \frac{\Delta_m H_{AB} + \left\{ X^2 + (1-X)^2 - 0.5 \right\} A}{\Delta_m S_{AB} - R\ln 4X(1-X) + \left\{ X^2 + (1-X)^2 - 0.5 \right\}(A/\Theta)}. \tag{52}$$

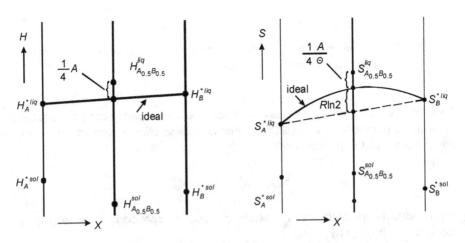

FIG. 7.   Enthalpy and entropy diagrams for the equilibrium between 1:1 compound, which dissociates on melting, and liquid mixtures of which the excess Gibbs energy is given by $\Omega\, X\, (1 - X)$, with $\Omega = A\, (1 - T/\Theta)$

concluding remarks and back to section 005

Certainly, every material has a heat capacity, and its negligence in matters of equilibrium between phases is permitted only as long as it has a minor influence. In practice, the influence of heat capacity is limited every time that there is a large difference between the enthalpies/entropies of the phases in equilibrium. The other way round, it is also true that there are phenomena that can only be attributed to a dominating role of heat capacity. An example is found in the re-entrant behaviour of bcc iron ($\leftarrow$004). And closed regions of demixing ($\leftarrow$211:Exc 5) go together with large *excess heat capacities*, just like the 're-entrant region of demixing' appearing in Exc 4.

For those cases where the influence of (excess) heat capacity is (assumed to be) limited, the formulae derived above are of great value - as *diagnostic formulae* - for a first inspection and thermodynamic analysis of experimental phase diagram data.

With or without heat capacities, whenever the forms of a system and their Gibbs functions are defined in terms of the variables, it is possible to calculate the system's phase diagram. The other way round, information about the system's Gibbs functions can be derived from its phase diagram - it is one of the issues of the next section. And one of the observations to be made is that, although, the *way from given functions to phase diagram* is unique, the way back, *from phase diagram to Gibbs function*, is not. This 'inconvenience' is related to the fact that the Gibbs functions are defined in terms of variables that all are independent, whereas the phase diagram is a representation of the manner in which the variables depend on one another ($\rightarrow$Exc 13).

At the end of this section it is worth while to look back at section 005 with its non-thermodynamic interpretation of the occurrence of extrema and regions of demixing. The interaction between molecules A and B was said to have a neutral character or a non-neutral one - either attractive or repulsive.

At this stage we know that neutral character corresponds to (virtually) ideal mixing; and it is clear that *attractive extra interaction* is a primitive image of negative excess Gibbs energy, and *repulsive extra interaction* of positive excess Gibbs energy. That being the case, we also know now that the excess Gibbs energy is composed of an excess enthalpy term and an excess entropy one - and that the interplay between enthalpy and entropy, as a rule, is such that excess Gibbs energy and excess enthalpy have the same sign. In the case of NaCl+KCl both excess enthalpy and excess entropy are positive: the enthalpy effect is partly 'compensated' by the entropy effect; as a result, the excess Gibbs energy is also positive. Partial compensation is also the case for chloroform+acetone, the three excess properties being negative ($\rightarrow$Exc 5). In the case of ethanol+water, on the other hand, it seems that the enthalpy effect is 'overcompensated' by the entropy effect: the *TX* (liquid+vapour) phase diagram has a minimum ($\leftarrow$Exc 006:6); at room temperature the *PX* (liquid+vapour) phase diagram just fails to have a maximum (Ohe 1989); and the heat of mixing at room temperature is negative ($\leftarrow$Exc 205:7).

Simple mathematical models, for the deviation from ideal mixing, permit the formulation of equations for the typical curves in binary phase diagrams - such as the spinodal, equal-G curve, and liquidus. In spite of its simplicity, both from a mathematical and a physical point of view, the approach is capable of explaining the major part of the phenomenology of binary equilibria, and capable also of providing an opening to the thermodynamic interpretation of phase diagrams.

EXERCISES

1.   *spinodal and binodal for a trigonometric excess function*

The basis for this exercise is a G function with a purely hypothetical, trigonometric type of excess part: $G(T, X) = RT\{(1 - X)\ln(1 - X) + X\ln X\} + C \cdot \sin 2\pi X + A \cdot X$, where C is a temperature and mole fraction independent parameter, whose value is set at $1000 \ J \cdot mol^{-1}$.

- Derive the formula for the spinodal, and make a graphical representation of it; take $0 \le (T/K) \le 1000$.
- For T = zero kelvin, determine the X values of the two binodal points (this can be done in the usual graphical manner; note however, that these values follow from $\tan 2\pi X = 2\pi X$).
- Guided by the results already obtained, estimate the course of the binodal. From the estimated binodal, read the X values for $T = 0.75 \ T_C$.
- For $T = 0.75 \ T_C$ calculate - to three decimal places - the X values of the coexisting phases (the binodal points) by means of the linear contribution method. In this method, the value of A, in $A \cdot X$, is changed until the double tangent line becomes horizontal.

    Procedure: for $A = 0$ calculate the G values for the X-values read from the estimated binodal; from the difference between the two G values calculate the starting value for A, and with that A calculate a series of G values around the estimated positions, in steps of $\Delta X = 0.005$; from the difference in G values of the two minima, calculate the correction to be applied to A and execute a new cycle in steps of $\Delta X = 0.001$; and so on ($\rightarrow$213).

2.   *small solid-state solubility*

Sodium chloride (A) and rubidium chloride (B) both have the NaCl-type of structure. Owing to the difference in size between the Na$^+$-ions (ionic radius Na$^+$ 1.02 Å) and the Rb$^+$-ions (1.52 Å), however, their miscibility at room temperature is poor - and certainly lower than for the combination of NaCl and KCl (K$^+$-ion 1.38 Å).

- For the equilibrium between two phases (I) and (II), sharing the same form, and such that $X_B^I \rightarrow 0$ (I is almost pure A) and $X_A^{II} \rightarrow 0$ (II is almost pure B), show, starting from $N\left[\mu_A^I = \mu_A^{II}; \ \mu_B^I = \mu_B^{II}\right]$, that the $AB\Theta$ model provides the relationships

$$\ln X_B^I = -A(1 - T/\Theta)(1 + B)/RT, \quad \text{and}$$
$$\ln X_A^{II} = -A(1 - T/\Theta)(1 - B)/RT, \quad \text{(see Oonk 2001)}$$

- To get an idea, calculate, for $T = 298.15$ K, the solubilities in one another of NaCl and KCl, and NaCl and RbCl.

	A /J·mol^{-1}	B	Θ /K
NaCl + KCl	18200	0.2	2565
NaCl + RbCl	42000	0.2	2565

3. *conditions imposed on excess enthalpy and entropy*

Guided by Figure 2, which pertains to the excess Gibbs energy function $G^E(T,X) = A (1 - T/\Theta) X (1 - X)$, figure out which conditions are imposed on A (representative of excess enthalpy, heat of mixing) and also on A/Θ (representative of excess entropy) for a system to have an upper (or, lower) critical point.

4. *'re-entrant region of demixing'*

For a hypothetical binary mixture the non-linear part of the Gibbs energy (which is in fact the Gibbs energy of mixing) is given by
$$G'(T,X) = RT\{(1 - X)\ln(1 - X) + X\ln X + X(1 - X)(a + bT + cT^2)\},$$
where $a = 5$; $b = -0.016$ K^{-1}; and $c = 0.00002$ K^{-2}.

- Find out, guided by Figure 2, the nature of the system's demixing behaviour in the temperature range from 200 to 600 K. Construct the phase diagram. Use temperature steps of 50 K.
- Do both S^E and H^E change sing on passing the given temperature range ($\leftarrow$ foregoing Exc)
- Derive the expression for the system's *excess heat capacity*.

5. *indirect evaluation of heat of mixing*

The two phase diagrams, Figure 3 and 4, along with the heats of vaporization of the pure components (31.4 kJ·mol^{-1} for chloroform, and 31.3 kJ·mol^{-1} for acetone), permit the (or rather, a rough) determination of the system's heat of mixing - due to the fact that they pertain to different (mean) temperatures.

- For each of the two diagrams, calculate for $X = 0.4$, which is close to the composition of the extremum, the value of $G^{E\ liq}$, being valid for the prevailing temperature. Next, from the two $G^{E\ liq}$ values, calculate the value of $H^{E\ liq}$ ($X = 0.4$).

NB By direct (= calorimetric) measurement, for $T = 323.15$ K, the heat of mixing, $H^{E\ liq}(X) = X(1 - X)\{-6962 - 1732(1 - 2X) + 1077(1 - 2X)^2\}$ Jmol^{-1}; (Morcom 1965).

6.   *isothermal EGC for change from solid to liquid*

Just write down, or derive the EGC expression for the isothermal (PX) equilibrium between mixed crystals, having negligible excess volumes, and ideal liquid mixtures. All second-, and higher-order derivatives of the Gibbs energies may be neglected.

7.   *types of phase diagram*

Mixed crystals, having $G^E(X) = \omega\,X(1-X)$, and ideal liquid mixtures share six different types of phase diagram: [0], [+], [–], [e], [p], and [–p].
  - Show that, by the extension of the excess function to Equation (39), i.e. $G^E(X) = X(1-X)[\omega + \omega'(1-2X)]$, the number of possible types increases from six to eleven (that is to say, if the occurrence is excluded of points of horizontal inflexion).

8.   *the liquidus formula for a compound*

Derive the expression for the liquidus curve pertaining to the equilibrium between compound A_mB_n, which is fully dissociating on melting, and liquid mixture, whose excess Gibbs energy is given by $G^E(X) = \omega \cdot X(1-X)$.
The case has been studied by Kuznetsov et al. (1975). The authors write the solution as

$$R\ln\left[\left\{\frac{(1-X)(m+n)}{m}\right\}^m\left\{\frac{X(m+n)}{n}\right\}^n\right]+$$
$$+\frac{\omega}{T}\left\{m\cdot X^2+n(1-X)^2-\frac{m\cdot n}{m+n}\right\}=\Delta H_{A_mB_n}\left(\frac{1}{T_m}-\frac{1}{T}\right)$$

where $\Delta H_{A_mB_n}$ is the enthalpy of melting for the amount of compound giving (m mol A + n mol B).

9.   *equations for the isothermal liquid + vapour equilibrium*

In the idealized treatment of isothermal binary liquid+vapour equilibria the equilibrium pressure (P) and equilibrium composition of the vapour phase (X^{vap}) can be given as explicit functions of the composition of the liquid phase (X). The formulae are

$$P(X) = (1-X)\,P_A^o\,\exp(G_A^E/RT) + X\,P_B^o\,\exp(G_B^E/RT)$$

$$X^{vap}(X) = [P(X)]^{-1}X\,P_B^o\,\exp(G_B^E/RT)$$

  - Starting from the two equilibrium conditions in terms of chemical potentials, write down the derivation of the two formulae (consult § 210 for the fully ideal case; see also Exc 210:1).

§ (212)

10. *the pressure above which mixed crystals are stable*

For the *isothermal* transition at $T = 1043$ K from liquid to mixed-crystalline solid in the system $\{(1-X) \text{ NaCl} + X \text{ KCl}\}$ calculate the position of the EGC in the PX plane; say, in order to have an idea of the pressure above which the material will be solid no matter its composition ($\leftarrow$Exc 6).

Suppose that the only data available are the isobaric phase diagram, at $P = 1$ bar, with its minimum at $T = 933$ K and $X = 0.5$, along with the melting properties of the pure components of the system. For convenience, assume that the difference in excess Gibbs energy between liquid and solid can be given by $\Delta A X(1-X)$, and such that ΔA is independent of temperature.

- Complete this exercise by making a sketch drawing of the stable part of the phase diagram.

	T/K	$\Delta S^{\bullet}/(J\cdot K^{-1}\cdot mol^{-1})$	$\Delta V^{\bullet}/(cm^3\cdot mol^{-1})$
NaCl	1073.8	26.22	7.55
KCl	1043	25.20	7.23

11. *heat effects derived from phase diagrams*

Use the phase diagrams in Figure 005:5 to estimate the heats of melting of optically active (D-, L-), and racemic (DL-) IPSA.

Next, give the heat effect of the (racemization) reaction, in which solid DL is formed out of 1 mol solid D and 1 mol solid L.

12. *a rule of thumb for minimum azeotropes*

The maximum *azeotrope* in Figure 5 is due to the fact that the difference between the boiling points of the pure components is small enough to make that the negative excess Gibbs energy of the liquid is capable of producing an equal-G curve with a maximum. Positive excess Gibbs energies are capable of producing *minimum azeotropes* - that is to say, as long as they are small enough to preclude (the interference of) a region of demixing, i.e. the occurrence of a *heteroazeotrope*. As a rule of thumb, minimum azeotropes are not found if the difference between the boiling points of the components is greater than 0.2 times the boiling point (in kelvin) of the lower boiling component.

- Demonstrate the rule, assuming that the excess Gibbs energy is given by Equation (1), and that the entropies of vaporization are in accordance with Trouton's rule ($\leftarrow$004).

13. *unique and not unique*

NaF and RbF are completely miscible when liquid. Their solid-state miscibility is virtually negligible, the *TX* (solid+liquid) phase diagram being eutectic. The change from solid to liquid has been studied by Holm (1965); his experimental data for the liquidus at the NaF side are the first two columns of the table. The temperatures in the third column are calculated liquidus temperatures, based on the values of 3.185 R and 0.5 R for NaF's entropy-of-melting and heat-capacity-change-on-melting, respectively, and putting the parameters of the *ABΘ* model to $A = 780$ J mol^{-1}; $B = 1.18$; and $\Theta = $ infinite.

X_{RbF}	t /°C	t /°C
0.0000	994.5	
0.0526	974.1	973.7
0.1085	951.4	951.6
0.1914	919.4	918.6
0.3005	873.0	873.2
0.3725	841.8	841.3
0.5107	771	772.9
0.5994	723	721.9
0.6769	670	670.7

- In order to appreciate the significance of the concluding remarks formulated above, extend the table with two more columns of calculated liquidus temperatures: 1) ignoring the change in heat capacity, and using the values given for *A*, *B*, and *Θ*; and 2) ignoring again heat capacity, and putting $A = -6155$ J mol^{-1}; $B = 0$; and $\Theta = 817$ K.

In the preceding section the problem of solving the equilibrium equations in terms of chemical potentials was circumvented by using the concepts of spinodal and equal-G curve. This 'shortcoming' is repaired: by showing how the solutions of the equations - the mole fractions of the coexisting phases - can be calculated with the help of a pocket calculator provided with conditional steps. And as a link to Volume II, some elementary examples of phase-diagram analysis are presented, showing how phase-diagram data can be used to calculate excess Gibbs energies.

linear contributions

For equilibria between two phases in a binary system where the phases have variable compositions, the equilibrium compositions of the phases can be determined by means of a programmable pocket calculator provided with conditional steps.

Geometrically, the problem corresponds to the location of the points of contact of the common tangent line to two Gibbs functions. By the use of linear contributions ($\leftarrow$205) the problem can be reduced to finding two minima, such that the two minima are on the same level.

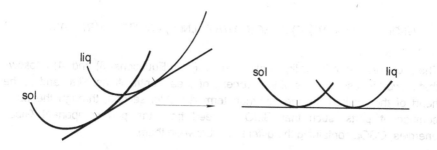

FIG.1. By means of linear contributions, the points of contact of the common tangent line can be made to coincide with minima

The computational procedure is worked out for two cases of isobaric (*TX*) equilibria. The first, the calculator routine ROD, is for regions of demixing. The second, the routine LOOP, is for the case where the phases are in different forms; denoted by solid (superscript 'sol') and liquid (superscript 'liq'). The computations are made in terms of Gibbs energy divided by RT, where R is the gas constant. Two routines, REDLIQ and REDSOL, are used for the calculation of the excess functions, given by the Redlich-Kister expression.

Gibbs energy functions

Generally, the molar Gibbs energy of a homogeneous mixture of $(1 - X)$ mole of A and X mole of B, under isobaric conditions, as a function of temperature and composition is expressed as

$$G(T, X) = (1 - X)G_A^*(T) + XG_B^*(T) + RT\,\mathrm{LN}(X) + G^E(T, X), \tag{1}$$

where

$$\mathrm{LN}(X) = (1 - X)\ln(1 - X) + X\ln X. \tag{2}$$

In the case of ROD there is one G function, and it is indicated by GLIQ. In ROD, GLIQ is not convex over the whole composition range: it corresponds to a separation into two phases; their compositions are given by the points of contact of the double tangent line. In the case of LOOP there are two G functions: GLIQ for the high-temperature form (liquid); and GSOL for the low-temperature form (solid). When these two functions intersect, a common tangent line can be drawn; the two points of contact correspond to the compositions of the coexisting phases.

After division by RT, and the incorporation of linear contributions, the Gibbs functions GLIQ and GSOL are given by

$$\mathrm{GLIQ} = \mathrm{LN}(X) + G^{E\,liq}(T, X)/RT + AX; \tag{3}$$

$$\mathrm{GSOL} = -\left[(1 - X)\Delta G_A^*(T) + X\Delta G_B^*(T)\right]/RT + \mathrm{LN}(X) + G^{E\,sol}(T, X)/RT + AX. \tag{4}$$

The Δ operator stands for liquid minus solid. From Equations (3) and (4) it follows that a two-fold use is made of the property of linear contributions. First, and as the heart of the method, by the common term AX. And second, through the pure-component parts; such that GLIQ is freed from the pure-component Gibbs energies, GSOL containing the differences between them.

The numerical values of $G_i^*(T)$ $(i= A, B)$ follow from [see Equation (109:13)].

$$\mathrm{GIT} = G(T) = H(T_r) - T\,S(T_r) + C_P\left[T - T_r - T\ln\left(\frac{T}{T_r}\right)\right], \tag{5}$$

where T_r is set to T_i $(i=A, B)$, the temperature at which the pure component changes from the low-temperature to the high-temperature form.

The excess parts of GLIQ and GSOL are given by REDLIQ and REDSOL:

$$\text{REDLIQ} = X(1-X)\left[\frac{g_1^{liq}(T)}{RT} + \left(\frac{g_2^{liq}(T)}{RT}\right)(1-2X) + \left(\frac{g_3^{liq}(T)}{RT}\right)(1-2X)^2 + \left(\frac{g_4^{liq}(T)}{RT}\right)(1-2X)^3\right]; \quad (6)$$

$$\text{REDSOL} = X(1-X)\left[\frac{g_1^{sol}(T)}{RT} + \left(\frac{g_2^{sol}(T)}{RT}\right)(1-2X) + \left(\frac{g_3^{sol}(T)}{RT}\right)(1-2X)^2 + \left(\frac{g_4^{sol}(T)}{RT}\right)(1-2X)^3\right]. \quad (7)$$

The main calculator routine, for ROD as well as for LOOP, is TXD. In TXD the value of the linear contribution parameter A is varied, until the points of contact coincide with minima. In this manner, the complex search for the points of contact has been reduced to locating the positions of two minima; carried out by the routine MINGIB. A complete listing of the routines, written in RPN (Reverse Polish Notation), is given at the end of this section. The definitions of the input registers are summarized in Table 1. Additional registers, addressed by the routines, are 06; 07; 36-42; and 46-48. In TXD, flag 07 directs the calculations to ROD (both phases described by GLIQ); and flag 08 directs the calculations to LOOP (one phase described by GLIQ, the other by GSOL). The routine DELMU takes care of the calculation of the ΔG^* by means of GIT.

Table 1:   Use of input registers

00	X (mole fraction)
01	X (for liquid, or 'left-hand phase')
02	X (for solid, or 'right-hand phase')
03	A (parameter linear contribution)
04	$\Delta Gmin$ (tolerance in the difference between the levels of the two minima)
05	ΔX (step size)
10	R (gas constant = 8.314472)
11	Component A's melting- or transition point
12	B's melting/transition point
13	A's enthalpy of melting/transition
14	B's enthalpy of melting/transition
15	T (thermodynamic temperature)
16	A's entropy of melting/transition
17	B's entropy of melting/transition
18	A's change in heat capacity on melting/transition
19	B's change in heat capacity on melting/transition
21-24	the four parameters in REDLIQ
31-34	the four parameters in REDSOL

a hypothetical sample system

To give an idea of the working of ROD and LOOP, a hypothetical system is taken. Its liquid mixtures are ideal; its Gibbs energies are linear in temperature; and its solid state excess Gibbs energy is given by the (*ABΘ* type of) expression

$$G^E(T, X) = 20000X(1-X)\left(1 - \frac{T}{2000 \text{ K}}\right)[1 + 0.2(1 - 2X)] \text{ J·mol}^{-1}. \qquad (8)$$

The melting points of the components A and B are set at 900 K and 1100 K, respectively; and their entropies are both equal to three times the gas constant; and by these properties the heats of melting of the components are fixed to 3*R* times their melting point.

Prior to the calculations by ROD and LOOP, the positions are calculated of the system's spinodal and solid-liquid equal-G curve (EGC) (←212). The two curves, shown in Figure 2, are used to make a proper choice of the starting values for the mole fractions of the coexisting phases.

region of demixing

For the boundary positions of the region of demixing, i.e. for the mole fractions X' and X'' of the coexisting phases I and II, the procedure is as follows; explained for $T = 600$ K.

One Gibbs function is needed, and it is fully defined by the values of the four parameters in REDLIQ, Equation (6). Their values are 2.80635 (→reg. 21); 0.56127 (→reg. 22); 0 (→reg. 23); and 0 (→reg. 24). At 600 K, the X values of the spinodal are about 0.15 and 0.66; it means that the X values to be calculated are in the intervals 0<X<0.15 and 0.66<X<1. The starting values, therefore, can safely be set at $X' = 0.08$ (→reg. 01) and $X'' = 0.83$ (→reg. 02).

A first ROD cycle is carried out with steps $\Delta X = 0.01$ (→reg. 05). A simple rule for the tolerance is $\Delta Gmin = 0.1\Delta X$; accordingly 0.001 is stored in register 04. The result found by ROD is 0.04 for the left-hand mole fraction (in reg. 01) and 0.85 for the right-hand one (in reg. 02). Besides, the less logical choice of starting values 0.20 and 0.60 yields 0.05 (in reg. 01) and 0.86 (in reg. 02)!

For the second cycle the step size (in reg. 05) and the tolerance (in reg. 04) are set at 0.001 and 0.0001, respectively, and ROD is restarted. This time the calculated mole fractions are 0.044 and 0.851. The third cycle yields 0.0445 and 0.8513.

As a remark, if X' approaches zero and/or $X'' \to 1$, the program may come to a stop due to log zero. If that happens, the program should be restarted with a smaller step size in mole fraction.

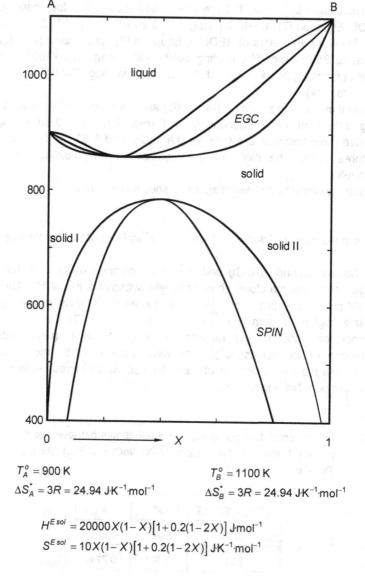

$T_A^o = 900 \text{ K}$ $T_B^o = 1100 \text{ K}$

$\Delta S_A^* = 3R = 24.94 \text{ J·K}^{-1}\text{·mol}^{-1}$ $\Delta S_B^* = 3R = 24.94 \text{ J·K}^{-1}\text{·mol}^{-1}$

$$H^{E\,sol} = 20000X(1-X)[1+0.2(1-2X)] \text{ J·mol}^{-1}$$
$$S^{E\,sol} = 10X(1-X)[1+0.2(1-2X)] \text{ J·K}^{-1}\text{·mol}^{-1}$$

FIG.2. Calculated TX phase diagram with solid-liquid loop, including the equal-G curve (EGC); and region of demixing, including the spinodal (SPIN)

solid-liquid loop

The equal-G curve for the solid to liquid change has a minimum, and the mere fact implies that a good guess can be made of the positions of the solidus and liquidus curves; the latter in particular. At $T = 950$ K, e.g., the liquidus is in the vicinity of $X = 0.50$; and the solidus in the vicinity of $X = 0.80$.

For a calculation at T = 950 K the input numbers are the four parameters in REDSOL, Equation (7): 1.32932 ($\rightarrow$reg. 31); 0.26586 ($\rightarrow$reg. 32); 0 ($\rightarrow$reg. 33); 0 ($\rightarrow$reg. 34); the parameters of REDLIQ, Equation (6), all are zero ($\rightarrow$reg. 21-24); temperature 950 ($\rightarrow$reg. 15); melting points 900 ($\rightarrow$reg. 11), 1100 ($\rightarrow$reg. 12); entropies of melting 24.94 ($\rightarrow$reg. 16,17); heats of melting 22450 ($\rightarrow$reg. 13), and 27440 ($\rightarrow$reg. 14).

For the first cycle, step size 0.01 ($\rightarrow$reg. 05) and tolerance 0.001 ($\rightarrow$reg. 04); and starting values of mole fractions 0.50 ($\rightarrow$reg. 01), and 0.80 ($\rightarrow$reg. 02). Calculated mole fractions are 0.53 for the liquid, and 0.83 for the solid phase. After three cycles the mole fractions calculated are 0.5261, and 0.8276, respectively.

The complete calculated phase diagram is shown in Figure 2.

analysis of the region of demixing in sodium chloride + potassium chloride

Sodium chloride (NaCl) and potassium chloride (KCl), both having the NaCl-type of crystal structure, show complete subsolidus miscibility. Their solid-liquid TX phase diagram is like Figure 2. Experimental data pertaining to the solid-state region of demixing are shown in Table 2. The data stem from isothermal equilibration experiments of long duration (annealing). The compositions of the coexisting phases were determined by means of X-ray diffraction; via the cell parameter of the cubic cell, which increases to the extent of 10% from pure NaCl to pure KCl.

Table 2:   Experimental data pertaining to the equilibrium between two solid phases, I and II, in the system $\{(1-X)$ NaCl + X KCl$\}$; Bunk and Tichelaar (1953)

$(T/K) - 273.15$	X^{I}	X^{II}
309	0.021	0.889
391	0.061	0.771
447	0137	0.634
466	0.195	0.542
472	0.191	0.521
488	-	0.335

Generally, the thermodynamic analysis of region-of-demixing data starts with the choice of a model for the excess Gibbs energy. For the case at hand, a convenient choice is the expression

§ (213)

$$G^E(T, X) = X(1-X)\{g_1(T) + g_2(T)(1-2X)\},\tag{9}$$

and such that the coefficients are linear functions of temperature,

$$g_i = h_i - Ts_i, \quad i = 1, 2 \quad .\tag{10}$$

The $AB\Theta$ model is a special case, having $h_1/s_1 = h_2/s_2$; as a result of which the quotient of g_1 and g_2 does not change with temperature.

In terms of the model defined by Equation (9), the expressions for the chemical potentials of the components A (here NaCl) and B (KCl) are ($\leftarrow$212)

$$\mu_A = G_A^\cdot + RT\ln(1-X) + X^2\{g_1 + g_2(3-4X)\};\tag{11}$$

$$\mu_B = G_B^\cdot + RT\ln X + (1-X)^2\{g_1 + g_2(1-4X)\}.\tag{12}$$

The compositions of the coexisting phases I and II, are the solution of the set N of equations

$$N = N\left[\mu_A' = \mu_A''; \ \mu_B' = \mu_B''\right]\tag{13}$$

For a given T, the equilibrium compositions of the phases can be calculated for given g_1 and g_2. The other way round, if for a given T the compositions are known, the same two equations in N, can be used to calculate the two unknowns g_1 and g_2. The two equations from which g_1 and g_2 can be solved are

$$\{(X'')^2 - (X')^2\}g_1 + \{(X'')^2(3-4X'') - (X')^2(3-4X')\}g_2 = RT\ln(1-X') - RT\ln(1-X'') \quad (14a)$$

$$\{(1-X'')^2 - (1-X')^2\}g_1 + \{(1-X'')^2(1-4X'') - (1-X')^2(1-4X')\}g_2 = RT\ln X' - RT\ln X'' \quad (14b)$$

When applied to the experimental data in Table 2, the values for g_1 and g_2 are obtained that are shown in Table 3.

The calculated g_1 values (indeed) show a nearly linear dependence on temperature. By least squares, the relationship found is

$$g_1 = (29787 - 24.509\, T/K)\, \text{J·mol}^{-1} \quad .\tag{15}$$

In addition, the quotient of g_1 and g_2, as follows from the numbers in Table 3, has a fairly constant value of about 0.26. These facts demonstrate that the calculated result goes well with the $AB\Theta$ model. And such that $A = 29787$ J·mol^{-1}; $B = 0.26$; and $\Theta = 29787/24.509$ K $= 1215$ K. With these values, the equilibrium mole fractions - calculated for the experimental temperatures - are the ones in the last two columns of Table 3. As an observation, the mean difference, each time taken as an absolute number, between experimental and calculated mole fractions is

0.006. This figure is rather representative of the experimentally attainable precision.

Table 3:　Result of thermodynamic analysis applied to the data in Table 2, which are again shown in the first three columns of this table. Columns 4 and 5 give the calculated values of Equation (9)'s g_1 and g_2, expressed in $J mol^{-1}$; and column 5 their quotient. Columns 6 and 7 give the mole fraction values calculated in terms of the overall result of the analysis

T/K	x'	x''	g_1	g_2	g_2/g_1	x'	x''
582.15	0.021	0.889	15478	3832	0.25	0.020	0.887
664.15	0.061	0.771	13554	3568	0.26	0.063	0.769
720.15	0.137	0.634	12268	3097	0.25	0.140	0.624
739.15	0.195	0.542	11661	3024	0.26	0.194	0.543
745.15	0.191	0.521	11402	3399	0.30	0.220	0.509

mathematical versus physical significance

　　　　The above thermodynamic analysis of the region of demixing in NaCl + KCl led to the linear expression, given by Equation (15), for the leading coefficient g_1 of the excess Gibbs energy. In terms of Equation (10), one could say that g_1 is composed of an 'enthalpy part' h_1 of 29787 $J mol^{-1}$, and an 'entropy part' s_1 of 24.509 $J K^{-1} mol^{-1}$. This observation immediately suggests the question to what extent do the two numbers reflect the real situation; the real excess enthalpy and the real excess entropy of the system.

Real excess enthalpies, to start with, are accessible by the methods of calorimetry. For NaCl + KCl the excess enthalpy has been determined, in an indirect manner, by means of solution calorimetry; see e.g. Bunk and Tichelaar 1953. The value found for the leading coefficient h_1 is 17.7 kJ mol^{-1}. Excess entropies, rather inaccessible to experimental determination, are found by combining excess enthalpies with excess Gibbs energies derived from phase equilibria. For NaCl + KCl the leading coefficient s_1 of the excess entropy, calculated that way, has the value of 6.90 $J\cdot K^{-1}\cdot mol^{-1}$ (Van der Kemp et al. 1992). The real g_1, as a result, is

$$g_1 = (17700 - 6.90T\,/K)\ J\cdot mol^{-1}. \tag{16}$$

This real g_1, and the other one, Equation (15), are shown in Figure 3, labeled 'phys' and 'math', respectively. Equation (16) has physical significance: it represents the real property. Equation (15) has a mathematical rather than a physical significance: it allows, along with the result obtained for g_2, a computational reproduction of the phase diagram to the extent that the experimental data are reproduced within their uncertainties.

An important observation to make, is, that the two lines in Figure 3 intersect at $T = 686$ K, the centre of gravity of the experimental data. The mathematical result, therefore, has the physical significance of including, for the mean temperature of the data, the real excess Gibbs energy property. Generally, and in general terms, TX phase diagram data are capable of providing the true Gibbs energy function for their mean temperature, but not its precise dependence on temperature.

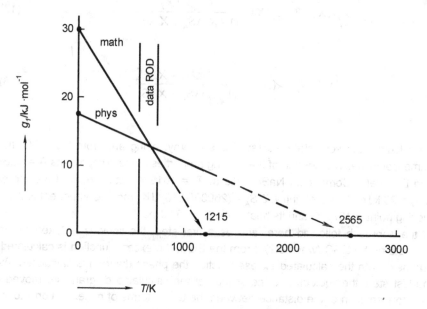

FIG.3.   For the mixed crystalline solid state in the system NaCl + KCl, the dependence on temperature of the leading parameter g_1 of the excess Gibbs energy, Equation (9). The line labelled 'phys' represents the real property; the line marked 'math' is the outcome of the elementary thermodynamic analysis described in the text

the solid-liquid loop of NaBr + NaCl

Before the arrival of the differential methods of microcalorimetry, cooling curves played an important part in the experimental determination of *TX* solid-liquid phase diagrams. For systems showing subsolidus miscibility, cooling curves yield reliable liquidus temperatures, but fail to provide reliable information on solidus temperatures (←006, Figure 8). A speaking example of a cooling-curve result is the phase diagram reported by Gromakov and Gromakova (1953) for {(1 − X) NaBr + X NaCl}, Figure 4 and Table 4.

Notwithstanding the above statement on the reliability of solidus temperatures read from cooling curves, it is quite well possible to derive the true solidus from the position of the liquidus. And, at the same time, obtain information about the excess Gibbs energy. This is worked out here in an elementary manner, i.e. neglecting any heat capacities, and assuming that the liquid mixtures of the system are ideal.

The connection between the phase diagram and the solid-state excess Gibbs energy is provided by the formula for the equal-*G* curve (←212):

$$T_{EGC}(X) = T_{ZERO}(X) - \frac{G_{EGC}^{E\,sol}(X)}{(1-X)\Delta S_A^* + X\,\Delta S_B^*};$$ (17)

$$\text{with } T_{ZERO}(X) = \frac{(1-X)\Delta H_A^* + X\Delta H_B^*}{(1-X)\Delta S_A^* + X\Delta S_B^*}.$$ (18)

To calculate the solid-state excess Gibbs energy, along and from the EGC, the numerical values are needed of the melting properties of the components A = NaBr and B = NaCl. Component NaBr melts at $T_A^o = 1015$ K, with a heat of melting of $\Delta H_A^* = 26$ kJ·mol^{-1}; accordingly, $\Delta S_A^* = (26000/1015)$ J·K^{-1}·mol^{-1}. Component NaCl 's melting point is 1073 K, and its heat of melting 28 kJ·mol^{-1}.

In the approach followed here, and as a first step, the liquidus is taken as an estimate of the EGC (why not?). From the EGC, the excess function is calculated; and next, with the calculated excess function the phase diagram is calculated. As the last step, the liquidus and the solidus of the calculated diagram are moved - with conservation of the distance between the two in terms of mole fraction - to let the calculated liquidus coincide with the experimental one.

In the first step of the approach thought up, and for each of the nine intermediate liquidus points, taken for EGC points, the value of the excess Gibbs energy is calculated; third column in Table 4. Next, the calculated values are divided by $X(1 − X)$, to give the numbers in the fourth column of the table. The latter, by linear least squares, produce the expression

$$G_{EGC}^{E,sol}(X) = X(1-X)\{2135 + 429(1-2X)\} \text{ J·mol}^{-1}.$$ (19)

With this excess function and the known melting properties of the components, the routine LOOP is put into action, to calculate, for the nine isothermal sections the mole fractions of the implied liquidus (fifth column, Table 4) and solidus points (sixth column). The last but one column of the table displays the solidus mole fractions obtained by the shift in the last step of the procedure. The calculated solidus is shown in Figure 4: the 'optimized' solid-liquid loop is four times less wide than the experimental data would suggest.

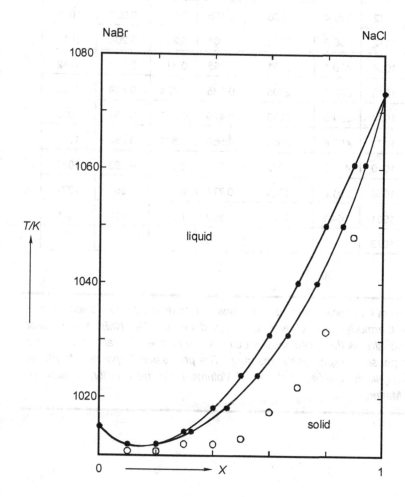

FIG. 4.   The system NaBr + NaCl. Solid-liquid phase diagram. Drawn curves with dots represent the experimental liquidus and the calculated solidus. Open circles represent experimental solidus temperatures as read from cooling curves

Table 4: The system {(1–X) NaBr + X NaCl}. Experimental data (first two, and last column); and course of thermodynamic analysis as described in the text, the last but one column giving the optimized solidus points for the experimental temperature. All numbers are in SI units

X	T^{liq}	G^E	$G^E/X(1-X)$	X^{liq}	X^{sol}	X^{sol}	T^{sol}
0.0	1015						
0.1	1012	228.4	2538	0.105	0.101	0.096	1011
0.2	1012	380.5	2378	0.201	0.205	0.204	1011
0.3	1014	479.1	2281	0.295	0.317	0.322	1012
0.4	1018	529.1	2205	0.385	0.424	0.439	1012
0.5	1024	524.9	2099	0.479	0.533	0.554	1013
0.6	1031	493.6	2056	0.568	0.631	0.663	1015
0.7	1040	410.0	1952	0.670	0.732	0.762	1022
0.8	1050	301.6	1885	0.774	0.825	0.851	1032
0.9	1061	164.1	1823	0.883	0914	0931	1048
1.0	1073						

The contents of this section should be an incentive to do elementary calculations in the field of thermodynamic analysis of phase diagrams. The NaBr + NaCl case clearly demonstrates the usefulness of doing some simple calculations in the pre-publication phase of experimental research. The professional approach to phase-diagram analysis is a substantial part of Volume II on the Equilibrium Between Phases of Matter.

EXERCISES

1. *Chanh's data set*

Small changes in the data for the region of demixing in a given system can have a great influence on the separation of g into h and s. This becomes clear when the analysis of the region of demixing in NaCl + KCl, as described above, is repeated for the set of data published by Nguyen-Ba-Chanh (1964).

T/K	x'	x''
473	0.015	0.938
573	0.031	0.907
623	0.049	0.866
673	0.070	0.814
698	0.099	0.764
723	0.121	0.702
748	0.171	0.594
763	0.224	0.497

- First, for the data triplets calculate the values of g_1 and g_2. Next, by least squares, fit the parameters h_1 and s_1 in $g_1 = h_1 - Ts_1$ to the values calculated for g_1. And, if you like, use the result to refine the findings formulated above under *mathematical versus physical significance*.

2. *excess function constants from critical coordinates*

Knowing that, at the critical temperature T_C of a region of demixing, the second and the third derivatives of the Gibbs energy with respect to mole fraction are zero for the critical composition X_C, formulate the two recipes with the help of which the excess coefficients g_1 and g_2 in Equation (9) can be calculated from the values of T_C and X_C. The two recipes have the form $g_i = a_i R\, T_C$, where a_i is an expression in X_c; you know already that for $g_2 = 0$, $a_1 = 2$.
   - Apply the two formulae two times to the coordinates of the critical point in NaCl + KCl, which to this end should be determined with the help of the data in Table 2, and for the second time, to the data given in the preceding exercise.
   - For each of the two cases, calculate the uncertainties in g_1 and g_2 resulting from an uncertainty of 0.02 in the mole fraction of the critical point.

§ (213)

3.   *analytical check*

Analytically, the mutual thermodynamic consistency of the computed mole fractions of the solid phase, as given in the last but one column of Table 4, and the mole fractions of the liquid phase, in the first column, can be checked as follows.

- First, introduce, in the equilibrium condition for the chemical potentials of B = NaCl, the recipes of the two members, taking into account that (*i*) the liquid mixtures are ideal, (*ii*) the solid mixtures, approximately, respect Equation (19), and (*iii*) the melting properties of NaCl are given.
- Next, from the resulting equation, solve the mole fraction of the liquid phase for given temperature and mole fraction of solid phase.

4.   *isothermal liquid+vapour equilibrium*

Write a routine LVPX, in RPN or otherwise, with the help of which, in terms of the idealized model ($\leftarrow$212; Exc 212:10), with $G^{Eliq}(X) = X(1 - X)\{ g_1 + g_2 (1 - 2X)\}$, for a given temperature, the equilibrium vapour pressure and the composition of the vapour phase can be calculated as a function of the liquid phase's composition.

5.   *elementary analysis of isothermal liquid+vapour data*

The data in the table pertain to the isothermal, T = 353.15 K, equilibrium between liquid and vapour in the system $\{(1 - X)$ n-heptane $+ X$ benzene$\}$ (Brown 1952).
In terms of the idealized model, and the liquid-state excess Gibbs energy expressed by Equation (9), the data can be used to evaluate the numerical values of the constants g_1 and g_2 - in a primitive manner, and according to two different methods, which are the *total vapour pressure method* and the *EGC method*.
By the first, the equilibrium vapour pressure as a function of liquid composition is used to assess the values of the parameters of the excess function, and such that the difference between calculated and experimental pressure is minimized (Barker 1953). By the EGC procedure, the parameters are derived from the position of the EGC in the lnP,X phase diagram (Oonk and Sprenkels 1971; Oonk et al. 1971; Oonk 1981).

- According to the total vapour pressure method, the calculations can be carried out as follows. First, for the liquid composition 0.4857, for $g_2 = 0$, determine, by means of LVPX ($\leftarrow$Exc 4), the value for g_1 for which the experimental pressure of 650.16 Torr is calculated. With that value and $g_2 = 0$, calculate for all of the compositions the corresponding vapour pressures. Compare the calculated pressures with the experimental ones, and thereby fix the sign of g_2. Next, in a series of calculations in which g_2's value is varied, and every time, for given g_2, the value of g_1 is adjusted such that for $X = 0.4857$ the experimental pressure is calculated, find the values of g_1 and g_2 that give the best reproduction of the set of experimental pressures - the values that yield the lowest value for the sum of the squares of the difference between calculated and experimental pressure.

- Owing to the fact that the two-phase region is rather narrow, the position of the EGC virtually coincides with the curve defined by $X = 0.5(X^{liq} + X^{vap})$. For nine values of X, in steps of 0.1, calculate from the EGC the value of G^{Eliq}, Equation (212:27). Following the route outlined above for the NaBr+NaCl case, and leading to Equation (19), calculate the values of the two constants.
- For each of the two results, calculate, for each of the nine experimental mole fractions of the liquid phase, the corresponding vapour pressure and the composition of the vapour phase. Calculate also, for the two results the values of Δ_X, and Δ_P, which are the mean of the absolute difference between calculated value and experimental one, for vapour composition and pressure, respectively.

P /Torr	X^{liq}	X^{vap}
428.28	0.0000	0.0000
476.25	0.0861	0.1729
534.38	0.2004	0.3473
569.49	0.2792	0.4412
613.53	0.3842	0.5464
650.16	0.4857	0.6304
679.14	0.5824	0.7009
708.78	0.6904	0.7759
729.77	0.7842	0.8384
748.46	0.8972	0.9149
758.44	1.0000	1.0000

6. *an application of LOOP*

As a follow-up of Exc 212:10, use the routine LOOP to calculate, for the system {(1 − X) NaCl + X KCl}, the compositions of the solid and liquid phases at $P = 0.00$ kbar and $T = 1043$ K. For convenience, assume that the liquid mixtures are ideal, so that, as a result, the excess Gibbs energy of the solid mixtures is given by $13000\, X\,(1 − X)$ J·mol^{-1}.

§ (213)

7. *an application of ROD*

For the system $\{(1 - X)\ NaCl + X\ RbCl\}$, calculate by means of the routine ROD, the compositions of the coexisting solid phases for $T = 788$ K, which is the temperature of the eutectic three-phase equilibrium.

First, use the equations and the data in Exc 212:2 to assess the starting values of the mole fractions to three decimal places. Next, use ROD to obtain, in three steps, the result to five decimal places.

8. *unilateral region of demixing*

Systems, like $\{(1 - X)\ SiO_2 + X\ MgO\}$, whose components show a great difference in physico-chemical nature, quite often have liquid-state regions of demixing that occupy a unilateral position in the TX phase diagram. Purely mathematically, a unilateral region of demixing can be accounted for by an excess function of the type

$$G^E = B\ (1 - X)^n X,$$

in which B is a positive parameter and the exponent n a number greater than 1. The maximum of the function is at $X = 1/(n+1)$.
- For B independent of temperature derive the expression for the spinodal, and show that the mole fraction of the critical point is given by
  $$X_C = [(2n + 1) - (2n^2 - 1)^{0.5}]/(n + 1)^2.$$

9. *unilateral ROD in SiO_2 + MgO*

In this exercise, which is a follow-up of Exc 8, use is made of data by Hageman and Oonk (1986) for the region of demixing in the SiO_2+MgO system. The coordinates of the critical point are $T_C = 2240$ K, and, from a thermodynamic analysis of the complete set of data, $X_C = 0.156$. A mixture, having an overall mole fraction   of
$X = 0.125$, equilibrated at 1987 K and subsequently quenched to low temperature, showed phases having mole fractions $X' = 0.022$, and $X'' = 0.410$.

The purpose of this task is to use the experimental data to assess the values of the parameters (n, h, and s) in $G^E = B\ (1 - X)^n X$, with $B = h - Ts$. The following route is suggested:
- From the critical data the value of n is calculated, and also the value of B at the critical temperature.
- From the data for 1987 K, the value B has at that temperature is calculated. NB For given n only one condition/equation is needed to calculate B; out of the possibilities, a convenient condition is the equality of first derivatives, which is one of the double tangent conditions, Equation (211:5).

- The result for 1987 K is checked in a graphical manner; by drawing the double tangent in a plot of function values around the compositions found for the two phases. First, the parameter A of the linear contribution in Equation (3) is adjusted such, that the function values are the same for experimental equilibrium compositions.
- The values of h and s are derived from the two values calculated for B.

10. an amusing side-effect

An amusing side-effect of the use of higher-order excess functions in thermodynamic phase-diagram analysis is the (unexpected) appearance of uncommon types of phase diagram. As an example, the function

$$G^E(T,X) = X(1 - X)[\, A(T) + C(T)\,(1 - 2X)^2],$$

which is symmetrical with respect to the equimolar composition, can give rise to a spinodal curve that has two maxima separated by a minimum, type [+ − +].

- By logic, and drawing GX sections, discover the phase equilibrium diagram that goes together with the strange spinodal.

APPENDIX

The routines printed here are meant for a Hewlett-Packard RPN scientific pocket calculator.

```
LBL "ROD"; CF 08; SF 07; XEQ 'TXD'; RTN.

LBL "LOOP"; CF 07; SF 08; XEQ "DELMU"; XEQ "TXD"; RTN.

LBL "DELMU"; RCL 11; STO 36; RCL 13; STO 37; RCL 16; STO 38; RCL 18; STO 39;
XEQ "GIT"; STO 06; RCL 12; STO 36; RCL 14; STO 37; RCL 17; STO 38; RCL 19; STO
39; XEQ "GIT"; RCL 10; RCL 15; ·; +/−; ST: 06; : ; STO 07; RTN.

LBL "GIT"; 0; STO 40; FS? 01; XEQ 02; XEQ 01; RCL 40; RTN; LBL 02; RCL 15; RCL 36;
: ; LN; RCL 15; ·; RCL 36; +; RCL 15; −; RCL 39; ·; ST− 40; RTN; LBL 01; RCL 37; RCL 38;
·; −; ST+ 40; RTN.

LBL "TXD"; LBL 01; RCL 01; STO 00; XEQ "GLIQ"; STO 46; RCL 02; STO 00; FS? 07;
XEQ "GLIQ"; FS? 08; XEQ "GSOL"; STO 47; RCL 46; RCL 47; −; RCL 02; RCL 01; −; : ;
ST+ 03; RCL 01; STO 00; SF 05; XEQ "MINGIB"; XEQ "GLIQ"; STO 46; RCL 00; STO 01;
CF 05; RCL 02; STO 00; FS? 07; SF 05; FS? 08; SF 06; XEQ "MINGIB"; FS? 05; XEQ
"GLIQ"; FS? 06; XEQ "GSOL"; STO 47; CF 06; RCL 00; STO 02; RCL 46; RCL 47; −; ABS;
RCL 04; X<=Y?; GTO 01; BEEP; RTN.
```

LBL "MINGIB"; LBL 04; FS? 05; XEQ "GLIQ"; FS? 06; XEQ "GSOL"; STO 48; RCL 05; ST+ 00; FS? 05; XEQ "GLIQ"; FS? 06; XEQ "GSOL"; RCL 48; X>Y?; GTO 04; LBL 05; FS? 05; XEQ "GLIQ"; FS? 06; XEQ "GSOL"; STO 48; RCL 05; ST− 00; FS? 05; XEQ "GLIQ": FS? 06: XEQ "GSOL"; RCL 48; X>Y?; GTO 05; RCL 05; ST+ 00; RCL 00; RTN.

LBL "GLIQ"; XEQ "LNX"; RCL 00; RCL 03; ·; +; XEQ "REDLIQ"; +; RTN.

LBL "GSOL"; XEQ "LNX"; RCL 00; RCL 03; ·; +; XEQ "REDSOL"; +; RCL 00; RCL 07; ·; +; RCL 41; RCL 06; ·; +; RTN.

LBL "LNX"; 1; RCL 00; -; ENTER; LN; ·; RCL 00; ENTER; LN; ·; +; RTN.

LBL "REDLIQ";1; RCL 00; −; STO 41; RCL 00; −; STO 42; 3; Y^X; RCL 24; ·; RCL 42; X^2; RCL 23; ·; +; RCL 42; RCL 22; ·; +; RCL 21; +; RCL 00; ·; RCL 41; ·; RTN.

LBL "REDSOL";1; RCL 00; −; STO 41; RCL 00; −; STO 42; 3; Y^X; RCL 34; ·; RCL 42; X^2; RCL 33; ·; +; RCL 42; RCL 32; ·; +; RCL 31; +; RCL 00; ·; RCL 41; ·; RTN.

SOLUTIONS OF EXERCISES

LEVEL 0

001 *equilibrium*

001:1 *vessels with water in thermal contact*
It is plausible to assume that the heat involved in changing the temperature of a quantity of water by one degree is proportional to its mass and more or less constant between the given temperatures, say C units per kg, so that
$10 \cdot (t_e - 50) \cdot C + 5 \cdot (t_e - 25) \cdot C = 0$.
Equilibrium temperature $t_e = 41\ \frac{2}{3}\ °C$

001:2 *vessels with water in thermal contact*
The equilibrium temperature (46.4 °C) differs considerably from $41\ \frac{2}{3}\ °C$ (foregoing exc). Apparently the property C is different for different substances.
$10 \cdot (t_e - 50) \cdot C_{H_2O} + 5 \cdot (t_e - 25) \cdot C_{H_2SO_4} = 0$

$C_{H_2SO_4} = 0.34\, C_{H_2O}$

001:3 *sulphuric acid poured into water*
The heat of mixing (of two different substances) comes into action; the final temperature is not the same.

002 *variables*

002:1 *the air's pressure at the Puy de Dôme*
725 Torr and 639 Torr, respectively.

002:2 *pressure at top and bottom*
101329 Pa.

002:3 *Fahrenheit's temperature scale*
$$\frac{t'}{°F} = \frac{t}{°C} \times \frac{9}{5} + 32 \qquad -17.8\ °C \qquad 35.6\ °C$$

002:4 *two phases and their amounts of two substances*

added together		in phase L_I		in phase L_{II}	
$n(A)$	$n(C)$	$n(A)$	$n(C)$	$n(A)$	$n(C)$
10	10	6	2	4	8
5	5	3	1	2	4
6	4	4.8	1.6	1.2	2.4
2	8	0	0	2	8

002:5 *ethanol and water saturate a space*

$P = (300/760) \cdot 101325 \ Pa$

$V = 50 \cdot 10^{-3} \ m^3$ $n = \dfrac{P \cdot V}{RT} = 0.722 \ mol$

$T = 333.15 \ K$

ethanol mole fraction in vapour - in equilibrium with liquid at 60 °C and 300 Torr - is about 0.57.

amount of ethanol $=$ $0.57 \cdot 0.722 \ mol = 0.41 \ mol$

amount of water $=$ $0.43 \cdot 0.722 \ mol = 0.31 \ mol$

003   *the rules of the game*

003:1 *three variables subjected to two conditions*

There is one independent variable, because of $M - N = 3 - 2 = 1$. Taking X as the independent variable, then

$$\begin{cases} Y(X) = -2X \\ Z(X) = \quad X \end{cases}$$

003:2 *phase diagram or not?*

The phase diagram is supposed to be the graphical representation of the mole fractions of pairs of coexisting phases as a function of T (or P). Hence, for every point of the vaporus there has to be a corresponding point on the liquidus. In the case of figure a) this obvious rule is violated. In the case of b) the rule is respected; however, for thermodynamic reasons, as will be seen later on, the two curves are allowed to make contact in an extremum only.

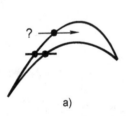   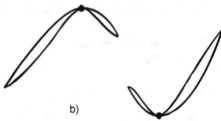

a)   b)

003:3 *derivation of lever rule*

from "law of conservation of substance B"

i.e. overall amount of B = amount of B in α + amount of B in β

$$X^\circ \{n(\alpha) + n(\beta)\} = X^\alpha \cdot n(\alpha) + X^\beta \cdot n(\beta)$$

003:4 *a system formulation*

In addition to the three liquid phases I, II and III there, obviously, is a vapour phase V

$$M = M\left[P, X_B^I, X_C^I, X_B^{II}, X_C^{II}, X_B^{III}, X_C^{III}, X_B^V, X_C^V \right]$$

$$N = N\left[\mu_A^I = \mu_A^{II} = \mu_A^{III} = \mu_A^V, \ \mu_B^I = \mu_B^{II} = \mu_B^{III} = \mu_B^V, \ \mu_C^I = \mu_C^{II} = \mu_C^{III} = \mu_C^V \right]$$

$$f = M - N = 9 - 9 = 0$$

003:5 *amounts of three phases out of three substances*
$n(\alpha) = 0.846$ mol ; $n(\beta) = 1.538$ mol ; $n(\gamma) = 0.615$ mol

003:6 *the experimental advantage of a small vapour phase*
In liquid + vapour equilibrium experiments with low vapour pressures and carried out such that the vapour phase occupies but a small fraction of the space, the equilibrium composition of the liquid phase is 'forced' to be equal to the overall composition, X° (before the experiment adjusted by the investigator), within experimental uncertainty. As an (extreme) example, if $X^\circ = 0.0500$ and $X^{vap} = 0.99$ and $n(liq) = 10^5 \, n(vap)$, then $X^{liq} =$ (calculated value) 0.0499906.
As a result: the answer is $X^{liq} = 0.20$ and $P = 287.7$

003:7 *naphthalene is added little by little to toluene*

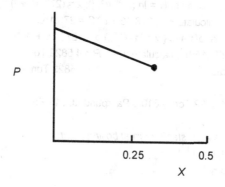

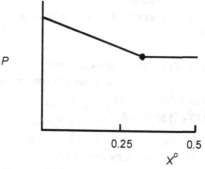

003:8 *does an empty place matter?*
Each new open place makes that the number of variables will be one less, just like the number of conditions; in the end f remains equal to $c - p + 2$.

004 **pure substances**

004:1 *the position of phase symbols*
fields: β top; γ right; α bottom left
increasing entropy: α, β, γ

004:2 *zero Celsius and zero Celsius*

$$\Delta T = \frac{T \cdot \Delta V}{Q} = \Delta P$$

apply SI units

$$= \frac{273\,K \cdot (-1.63 \cdot 10^{-6}\,m^3 \cdot mol^{-1})}{6008\,J \cdot mol^{-1}}\,(611 - 101325)\,Pa$$

$$= 0.0075\,K \qquad \text{the answer is 7 mK}$$

004:3 *water's triple point pressure*

A priori:

the data indicate that the result will be close to 0 °C

and $(4.579/760) \cdot 101325\,Pa = 610.5\,Pa$

Calculations:

- lin. least sq. of pairs $\{y = \ln(P^{sol}/P^{liq});\ x = (273.15 + t)^{-1}\}$
  for $y = 0$, calculate $x \rightarrow 273.15 + t/°C = 273.159$
- lin. least sq. of pairs $\{y = \ln(P^{sol});\ x = (273.15 + t)^{-1}\}$
  for $x = (273.159)^{-1}$, calculate $y \rightarrow P = 4.5822\,Torr$
  or P^{liq} data $\qquad\qquad \rightarrow P = 4.5822\,Torr$

result:

$t = 0.01$ °C; $P = 4.5822$ Torr $= 610.9$ Pa; rounded: 611 Pa.

004:4 *carbon dioxide's metastable normal boiling point*

$$Q^{liq \rightarrow vap} \approx 17\,kJ \cdot mol^{-1}$$
$$T_{nbp} \approx 185\,K$$

004:5 *the substance water under high pressure*

liquid $\rightarrow$ solid at 100°C: at about 2.5 GPa

triple point (II + V + ℓ):   in vicinity of (-19 °C; 0.32 GPa)

see also (Fletcher 1970)

004:6 *a rule to be respected by metastable extensions*

$\qquad\qquad\qquad\qquad\qquad\qquad \rightarrow$ for spontaneous change the shaded field is

- at the α side of $(\alpha + \beta)$: $\beta \rightarrow \alpha$
- at the β side of $(\beta + \gamma)$: $\gamma \rightarrow \beta$ } $\gamma \rightarrow \alpha$
- at the γ side of $(\alpha + \gamma)$: $\alpha \rightarrow \gamma$ } $\otimes$

$\qquad\qquad\qquad\qquad \otimes$ this is the absurdity!

004:7 *a phase diagram acts as a thermobarometer*

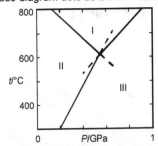

- Triple point at 620 °C and 0.55 GPa
- Equilibrium (I+III): $dt/dP = 530\,K \cdot GPa^{-1}$.
- Rules for equilibrium lines (metastable extensions)
- III at high-pressure side

004:8 *superposition of stable and metastable*

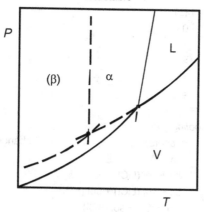

004:9 *Antoine's equation*

A = 15.7576   B = 2408.66   C = 62.060

with these values the following pressures, expressed in Torr, are calculated

289.15   433.56   759.94   1267.95   2025.98

the mean absolute difference in experimental and calculated pressure is just 0.03 Torr (and partly due to round-off effects)!

004:10 *supercritical fluid*

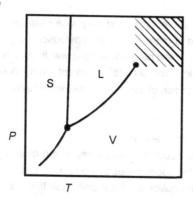

004:11 *iron: the heat effect of magnetic change*

rounded to an integer in $kJ \cdot mol^{-1}$: 8; heat needed 35.6 kJ

004:12 *boiling water altimeter*

Combine $d(\ln P)/dT$ (from the Clapeyron equation) with $d(\ln P)/dh$ (from the barometric formula ←002); heat of vaporization $41 kJ \cdot mol^{-1}$.

005   *binary and ternary systems*

005:1 *a unary diagram made to look like a binary one*

triple point   →   eutectic type of three-phase equilibrium

boiling curve   →   two-phase region like region of demixing in Figure 3

005:2 *the amounts of the phases during an experiment*

	$n(\alpha)$	$n(\beta)$	$n(L)$
i	=	0	0
ii	<	0	>
iii	<	>	>
iv	0	<	>
v	0	0	=

005:3 *phase diagram and cooling curve*

temp. (interval)	phase(s)	change in temp. is
600 to 515 °C	liq	fast
515 to ~ 350 °C	liq + sol (LiCl)	slow
at 350 °C	liq + sol (LiCl) + sol (KCl)	zero
350 to 200 °C	sol (LiCl) + sol (KCl)	fast

005:4 *a reciprocal system*

Two variables:

X (the fraction of the cations that are K^+)

Y (the fraction of the anions that are Br^-)

a composition square, of which the vertexes are occupied by the pure substances:

NaCl ($X= 0$; $Y = 0$); KCl (1; 0); KBr (1; 1); NaBr (0; 1)

005:5 *increasing repulsive interaction and the phase diagram*

The change to liquid is more and more "postponed" to higher temperatures: the A liquidus (except for its initial part) is moving upwards, and the eutectic point is moving up to B's melting point. From a certain "moment" on the A liquidus will be interrupted by a (liquid + liquid) region of demixing - together involving a monetectic three-phase equilibrium.

005:6 *overlapping two-phase regions*

The two three-phase equilibrium temperatures are given by the intersections of the (β + L) solidus and the upper solvus of the (α + β) field. The (α + L) field is between the two three-phase equilibrium lines; and such that the (metastable parts of) the (β + L) liquidus and the (α + β) lower solvus are inside the field.

005:7 *the construction of ternary phase diagrams*

R = racemate = compound AB; Q = quasiracemate = compound AC

section	stable solids	single-phase fields	two-phase fields	invariant triangles
i)	C	L	C + L	
	R		R + L	
ii) [*)	(A)	L	(A + L)	
	(B)		(B + L)	
	C		C + L	C L Q
	Q		Q + L	
	R		R + L	
iii)	A	2 times L	A + L	A L R
	B		B + L	B L R

C	C + L	C L R
Q	Q + L	C Q R
R	R + L	L Q R

- temperature is just above or just below m.p. of A and B

005:8 *the appearance of an incongruently melting compound*

The sequences of the (two -) and single-phase fields are, when considered from high to low temperature and from left to right (assuming that B's melting point is below the peritectic temperature):

L

(A + L) L

(A + AB) (AB + L) L

(A + AB) (AB + L) L (L + B)

(A + AB) (AB + B)

005:9 *ternary compositions having a constant ratio of the mole fractions of two components*

AR : X_B for Q; SB: X_A for Q

AP : X_B for P; PB; X_A for P

Ratios: AP : PB = RP : PS

$$= (AP - RP) : (PB - PS)$$

$$= AR : SB \quad QED$$

005:10 *cyclohexane with aniline - mixing and demixing*

Temperature 19.5 °C - the milky aspect of tube 1.

Initial slopes (negligence of solid-state solubility), given as dT/dX: at cyclohexane (C) side 245 K; at aniline (A) side 56 K.

The phase diagram, from low to high temperature:

at −11 °C three-phase equilibrium solid C + liquid (X = 0.93) + solid A;

at −2 °C three-phase equilibrium (lower boundary of region of demixing) solid (C) + liquid (X = 0.035) + liquid (X =0.88);

critical point (top of region of demixing) at 31.2 °C (X_C = 0.43);

the liquid+vapour two-phase region is bent downwards (repulsive aspect of interaction between C and A), but does not give rise to a minimum.

006 distribution and separation

006:1 *a room saturated with water*

From $P \cdot V = n \cdot R \cdot T$ it follows n = 255.53 mol;

This amount corresponds to 4.603 kg;

The increase is 5.6% per degree.

006:2 *solubility of potassium permanganate in boiling water*

Just as an observation,

by linear regression of ln$m$ as a function of $(t + 273.15)^{-1}$, the solubility at 100 °C is calculated as 3.92 mol·kg^{-1}.

006:3 *the space needed to remove an impurity*
$V \geq 196 \text{ dm}^3$

006:4 *extraction and clever use of solvent*
One time with 1 kg :   0.010 mol of S remain in water;
two times with 0.5 kg: 0.0056 mol of S remain in water.

006:5 *distillation with a fractionating column*
In terms of mole fraction, and X_{az} for the azeotropic composition.

	a	b	c	d	e
first drop	$X = 0$	$X = 0$	$X = 0.5$	$X = X_{az}$	$X = 0$
residue	$X = 1$	$X = 1$	$X = 0.5$	$X = 1$	$X = X_{az}$

006:6 *single-step distillation of wine*
In spite of the need of liquidus data for $X > 0.9$, the answer seems to be "yes".

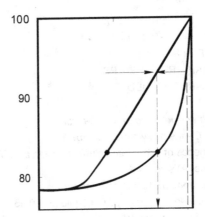

006:7 *vacuum distillation*
$$\ln\left(\frac{P_2}{P_1}\right) = -\frac{\Delta H}{R}\left(\frac{1}{T_2} - \frac{1}{T_1}\right)$$

T_1 = n.b.p; P_2 = 15 Torr; P_1 = 760 Torr (= 1 atm); $\Delta H = 11 \cdot R \cdot T_1$
p-xylene $T_2/K =$ 303
2–nitro–p–xylene   379
4–methylbenzaldehyde   352.

006:8 *distillation with steam*
TX diagram: it has the type of Figure 005:4.
- The distillate, with overall composition $X = 0.045$, contains 21.22 mol of water
- water layer 21.13 mol; aniline layer 1.09 mol
- 21.13 mol of LI contains 0.148 mol of aniline: 15% of the yield!

006:9 *a congruently crystallizing compound*

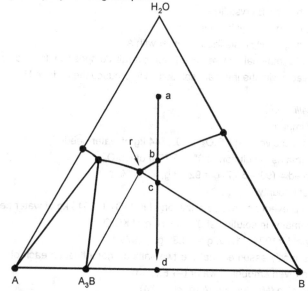

- Overall composition: a→b→c→d
  b→c: B crystallizes
  c→d: A_3B and B crystallize together;
  mother liquor: a→b→r
- 2 mol $A_{0.75}B$: 1 mol B; or, 0.5 mol A_3B : 1 mol B

NB It makes no difference whether or not the solid material is continuously removed (see next exc).

006:10 *an incongruently crystallizing hydrate*

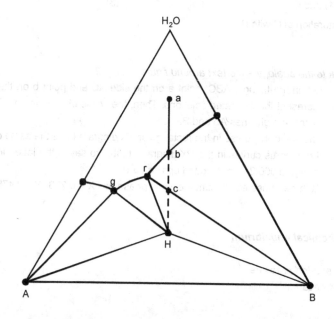

- Mother liquor $a \rightarrow b \rightarrow r \rightarrow g$
  - $b \rightarrow r$: B crystallizes
  - $r \rightarrow g$: H crystallizes
  - at g   H crystallizes together with A
- Solid material not removed: for overall composition from c to H, solid B reacts with the invariant solution (r) to produce the hydrate H.

006:11 recrystallization

i) 2 g of impurity
- to dissolve 98 g $KClO_3$:   0.1744 kg of water needed;
- remains in solution at 0°C: 5.77 g of $KClO_3$;
  yield = (98 – 5.77) g = 92.23 g;   94%

ii) 10 g of impurity
- to prevent any recrystallization of $KMnO_4$: 0.3544 kg of water needed;
- remains in solution at 0°C:11.70 g of $KClO_3$
  yield = (90 – 11.70) g = 73.30 g;   (87%)

NB. It is tacitly assumed that the two salts do not influence each others solubility in a given amount of water ($\rightarrow$ Exc 13)

See also (Mulder and Verdonk 1984).

006:12 a saturation problem

A and C become saturated at the same time (!),
as a result, after the addition of 1.2 g of C:
solid phase B   0.2 g
liquid phase A   contains 0.8 g of B
liquid phase C   contains 0.2 g of B
saturation of A with B $: \mu_B^A = \mu_B^{sol}$ (1)
distribution of B over A and C: $: \mu_B^A = \mu_B^C$ (2)
from (1) and (2): $: \mu_B^C = \mu_B^{sol}$ (3)
(3): saturation of C with B

006:13 back to the analogy – the text around figures 1 and 2

- Let, in the triangle ABC, point a on the side AC and point b on the side BC represent the saturated solutions. Then, the whole phase diagram is defined by the straight lines Ab and Ba.
- The solubility curves in the rectangular YX diagram have the same course as the vaporus curves in the PX diagram; that is to say, if the latter, in contrast to Figure 006:2, is not plotted upside down.
  At three-phase equilibrium the liquid phase has $X = 0.273$; $Y = 0.478$.

007   chemical equilibrium

007:1   the exerted pressure

$P = 14$ decabar.

007:2 *the equilibrium constant of the ammonia equilibrium*
Extrapolated value $6.50 \cdot 10^{-3}$ atm^{-1} = $6.42 \cdot 10^{-3}$ bar^{-1}
NB The square of this value corresponds to the K defined by Equations (9) and (10).
The value of the ΔG° property becomes ΔG° (T = 723.15 K) = 61 kJ·mol^{-1}; P° = 1 bar.

007:3 *Clausius-Clapeyron plot of Horstmann's data*
About 175 kJ·mol^{-1}

007:4 *the equipment to be used*
- vessel-with-manometer, because there is only one degree of freedom:

$$X_{HCl} = \frac{1}{(2+\alpha)}; \quad X_{NH_3} = \frac{(1-\alpha)}{(2+\alpha)}; \quad X_{N_2} = \frac{0.5\alpha}{(2+\alpha)}; \quad X_{H_2} = \frac{1.5\alpha}{(2+\alpha)}$$

$$f = M[T,P,\alpha] - N\left[\mu_{NH_4Cl} = \mu_{NH_3} + \mu_{NHCl}; \; \mu_{NH_3} = 0.5\,\mu_{N_2} + 1.5\,\mu_{H_2}\right] = 3 - 2 = 1$$

007:5 *three pure substances taking part in a reaction*
Pure substances: there are no composition variables.
$$f = M[T,P] - N[\mu_A + \mu_B = \mu_C] = 2 - 1 = 1 :$$

only one of the variables T and P can be chosen in an arbitrary manner.
The answer is NO!

007:6 *Professor Denbigh's example from the zinc smelting industry*
Three phases
$$f = M\left[T,P,X_{Zn}^{vap},X_{CO}^{vap}\right] - N\left[\mu_{Zn}^{vap} + \mu_{CO}^{vap} = \mu_{ZnO} + \mu_C; \; \mu_{Zn}^{vap} + \mu_{CO_2}^{vap} = \mu_{ZnO} + \mu_{CO}^{vap}\right] = 2$$
"c" = $f + p - 2 = 3$;   Zn , CO, CO$_2$
Four phases
$$f = M[\text{same}] - N\left[\text{same plus } \mu_{Zn}^{liq} = \mu_{Zn}^{vap}\right] = 4 - 3 = 1$$
"c" = 3;   Zn , CO, CO$_2$
see also (Denbigh 1955)

LEVEL 1

101   *differential expressions*

101:1   *the ideal gas*

$$P = \frac{RT}{V} \qquad \left(\frac{\partial P}{\partial T}\right)_V = \frac{R}{V} \qquad \left(\frac{\partial P}{\partial V}\right)_T = -\frac{RT}{V^2}$$

$$\frac{\partial^2 P}{\partial T^2} = 0 \qquad \frac{\partial^2 P}{\partial V^2} = 2\frac{RT}{V^3} \qquad \frac{\partial^2 P}{\partial V\, \partial T} = -\frac{R}{V^2} = \frac{\partial^2 P}{\partial T\, \partial V}$$

$$T = \frac{PV}{R} \qquad \left(\frac{\partial T}{\partial P}\right)_V = \frac{V}{R} \qquad \left(\frac{\partial T}{\partial V}\right)_P = \frac{P}{R}$$

$$\frac{\partial^2 T}{\partial P^2} = 0 \qquad \frac{\partial^2 T}{\partial V^2} = 0 \qquad \frac{\partial^2 T}{\partial V\, \partial P} = \frac{1}{R} = \frac{\partial^2 T}{\partial P\, \partial V}$$

101:2   *the Van der Waals gas*

$$P(V,T) = \frac{RT}{(V-b)} - \frac{a}{V^2}$$

$$dP = \left(\frac{\partial P}{\partial V}\right)_T dV + \left(\frac{\partial P}{\partial T}\right)_V dT = \left\{-\frac{RT}{(V-b)^2} + \frac{2a}{V^3}\right\} dV + \left(\frac{R}{(V-b)}\right) dT$$

101:3   *integration of different expressions along different routes*

expression	route	change
(A)	(1)	71
(A)	(2)	71
(B)	(1)	25
(B)	(2)	39

(A), the calculated change of which is independent of the route followed, corresponds to the total differential of a function Z of X and Y:

$$Z(X,Y) = X^2 Y^3 + \text{constant}$$

(B) is not the total differential of a function Z of X and Y.

101:4 *a simplistic method*

$$V(T+dT, P+dP) = \frac{R(T+dT)}{(P+dP)} = \frac{R(T+dT)}{P\left(1+\dfrac{dP}{P}\right)}$$

$$= \frac{R(T+dT)}{P}\left(1-\frac{dP}{P}\right) = \frac{RT}{P} + \frac{R}{P}dT - \frac{RT}{P^2}dP - \ldots dTdP$$

$$dV = V(T+dT, P+dP) - V(P,T) = \frac{R}{P}dT - \frac{RT}{P^2}dP$$

102 **work heat energy**

102:1 *q and the cross-differentiation identity*

$$q = \left(\frac{\partial U}{\partial T}\right)_V dT + \left[\left(\frac{\partial U}{\partial V}\right)_T + P\right]dV$$

This form does not obey the cross-differentiation identy

$$\frac{\partial^2 U}{\partial V\,\partial T} \neq \frac{\partial^2 U}{\partial T\,\partial V} + \left(\frac{\partial P}{\partial T}\right)_V$$

as $(\partial P/\partial T)_V$ is not equal to zero.

102:2 *units and conversion*
   1 J = 1 N·m (newton x meter)
      = 1 kg·m·s^{-2}·m = 1 kg·m^2·s^{-2}
   1 Pa = 1 N·m^{-2} = 1 J·m^{-3}
   1 m^3 = 1 J·Pa^{-1}
   1 cm^3·atm = 10^{-6} m^3 x 101325 Pa = 0.101325 J

102:3 *reaction between zinc and sulphuric acid*

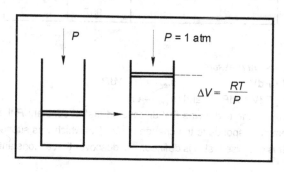

$\Delta W = -\int PdV$ at isobaric conditions $= -P\int dV = -P\Delta V = -RT$
$-RT = -8.3145$ J·K^{-1}·mol^{-1} x 298.15 K $= -2479$ J·mol^{-1}
$-P\Delta V = -1$ atm · RT/P x 10^3 $= -24.47$ atm x dm^3·mol^{-1} $= -24.47$ liter x atm·mol^{-1}

The work added to the system per mole generated hydrogen is -2479 J $= 24.47$ liter x atm

103   *heat capacity and enthalpy*

103:1   *a classroom calorimeter*
heat added 750 J·s^{-1} x 120 s $= 90000$ J
rise in temperature 9.05 K
heat capacity 90000 J / 9.05 K $= 9.94$ kJ·K^{-1}

103:2   *a drop calorimeter*
the effect is 64.80 x 333.5 $= 21.61$ kJ per 100 g
which is 13.73 kJ·mol^{-1}; therefore
$(H_T - H_{273})$ for $T = 800$ K is 13.7 kJ·mol^{-1}
the mean specific heat capacity of copper is 0.410 J·K^{-1}·g^{-1}.

103:3   *a cycle passed by a monatomic gas*

	P /atm	V /dm^3	T /K
A	1	24.45	298
B	1	48.91	596
C	2	24.45	596

change	nature	q_{rev}/J	w_{rev}/J	ΔU /J	ΔH /J
1	isobaric	6194	-2478	3717	6194
2	isothermal	-3435	3435	0	0
3	isochoric	-3717	0	-3717	-6194
cycle		-957	957	0	0

103:4   *heat at constant pressure*
$q_{rev} = dU + PdV$ $q_{rev} = dH - VdP$
$dH = dU + PdV + VdP$ and $(q_{rev})_P = dH$
By the way, in the text it was first discovered that at constant P the heat added to the system corresponds to the change of $(U + PV)$ which was subsequently defined as H; in this exercise first H is defined.... to descover that at constant P...

103:5   *choice of zero point*
$U = H - PV \Rightarrow U = -PV$
oxygen gas $U = -RT = -2479$ J·mol^{-1}
diamond $U = -1$ Pa x 3.417x10^{-6} m^{-3}·mol$^{-1} = -3.4$ x 10^{-6} J·mol^{-1}

103:6 *heat added to silver bromide*

$T = 313.71$ K.

50.763 J·K^{-1}·mol^{-1}

Robie et al. (1978) give 50.734 obtained with heat of melting of 9.163 kJ·mol^{-1} and the given C_P formula.

103:7 *an interpolation formula for C_P for diamond*

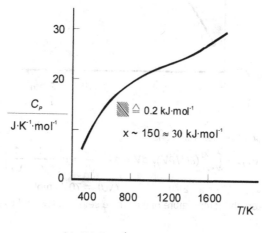

31.42 kJ·mol^{-1}

104 **the ideal gas – expansion and compression**

104:1 *a simulation of isothermal compression*
- From $pV = nRT$, n = 0.040 mol
- $\Delta W = nRT\ln 2 = 69.3$ J $= -\Delta Q$
- $\Delta T = -\Delta Q / (\text{heat capacity}) = 2$ mK

104:2 *adiabatic compression*
- $q_{rev} = 0$; $W_{rev} = -PdV = - (RT / V)\, dV$
  $dU = (C + C_V^{gas})\, dT$

  $dU = W_{rev} ; (C + C_V^{gas})\, dT/T = -R(dV/V)$

  $(C + C_V^{gas})\, \ln(T_2/T_1) = -R\ln(V_2/V_1)$

  for $C = 0$ $T_2 = 872$ K
- for $C = 500$ J · K^{-1} $T_2 = 306$ K

104:3 *adiabatic compression of helium*
- $\Delta S = C_P \ln (T_2 / T_1) - R \ln (P_2 / P_1)$, which in this case = 0. State B: $T = 749$ K

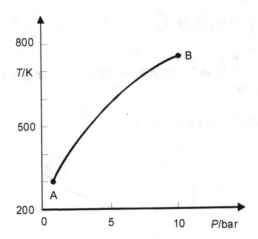

104:4 *expansion of a Van der waals gas*

$$(\Delta U)_T = \int_{V=V_1}^{V=V_2} (\partial U / \partial V)_T \, dV = a \int_{V_1}^{V_2} \frac{dV}{V^2} = a\left(-\frac{1}{V_2} + \frac{1}{V_1}\right)$$

with $V_2 = 2V_1$, $(\Delta U)_T = a / (2V_1)$ $(\Delta U)_T = 70$ J · mol^{-1}

70 J are taken: the temperature of the whole set-up is lowered by 0.005 K

104:5 *volume quotients for a cycle*

Adiabatic $dU = W_{rev}$, which for the ideal gas gives rise to

$$C_V \, dT = -\left(\frac{RT}{V}\right) dV; \qquad C_V \frac{dT}{T} = -R\frac{dV}{V}; \qquad C_V d\ln T = -Rd\ln V$$

$$C_V \ln\left(\frac{T_2}{T_1}\right) = -R\ln\left(\frac{V_2}{V_1}\right) = R\ln\left(\frac{V_D}{V_A}\right) = -R\ln\left(\frac{V_C}{V_B}\right)$$

so that $\dfrac{V_D}{V_A} = \dfrac{V_C}{V_B}$ or $\dfrac{V_B}{V_A} = \dfrac{V_C}{V_D}$

104:6 *an equation for an adiabatic*

- Ideal gas: $dU = C_V \, dT$; $C_P = C_V + R$

  adiabatic: $dU = W_{rev} = -P \, dV$

  taking P and V as the two independent variables, dT for the ideal gas can be given as $dT = (V/R) \, dP + (P/R) \, dV$; the adiabatic now satisfies the equation

  $C_V (V/R) \, dP + C_V (P/R) \, dV = -P \, dV$, from which it follows $(C_V/R) \, (dP/P) = -(C_P/R) \, (dV/V)$, or $d\ln P / d\ln V = -C_P/C_V$ Equation (1) upon integration it follows from Equation (1): $P \cdot V^{\,C_P/C_V} = constant$

- for each point in the PV plane $d\ln P/d\ln V$ has one single value: through the point there goes just one adiabatic; or, in other words, adiabatics do not intersect.

104:7 *ideal gas, expansivity and compressibility*

$$\alpha = \frac{1}{V}\left(\frac{\partial V}{\partial T}\right)_P \quad \text{ideal gas } \alpha = \frac{1}{T} \quad \kappa = -\frac{1}{V}\left(\frac{\partial V}{\partial P}\right)_T \quad \text{ideal gas } \kappa = \frac{1}{P}$$

$$dV = \left(\frac{\partial V}{\partial T}\right)_P dT + \left(\frac{\partial V}{\partial P}\right)_T dP = \alpha V dT - \kappa V dP \quad \text{or} \quad d\ln V = \alpha dT - \kappa dP$$

cross-differentiation identity: $\left(\dfrac{\partial \alpha}{\partial P}\right)_T = -\left(\dfrac{\partial \kappa}{\partial T}\right)_P$

$$\text{ideal gas} \qquad 0 \quad = \quad 0$$

104:8 *isothermal compression of Van der Waals gas*

$$\kappa = -\frac{1}{V}\left(\frac{\partial V}{\partial P}\right)_T \qquad \kappa^{-1} = -V\left(\frac{\partial P}{\partial V}\right)_T$$

$$\text{ideal gas} \qquad \kappa_{ID} = \frac{1}{P}; \qquad (1-\alpha)^{-1} \text{ for } \alpha \to 0 \text{ is } 1+\alpha$$

For the van der Waals gas, substitution $(\partial P/\partial V)_T$, see Exc 4.

$$\kappa_{VDW}^{-1} = \frac{RTV}{(V-b)^2} - \frac{2a}{V^2} = \frac{RT}{(V-b)} \cdot \frac{V}{(V-b)} - \frac{2a}{V^2} = \frac{RT}{(V-b)}\cdot\left(1+\frac{b}{V}\right) - \frac{2a}{V^2}$$

$$= \frac{RT}{(V-b)} - \frac{a}{V^2} + \frac{RT}{(V-b)}\cdot\frac{b}{V} - \frac{a}{V^2}$$

$$\approx P\left(1+\frac{b}{V}\right) - \frac{a}{V^2} = P\left(1+\frac{b}{V} - \frac{a}{PV^2}\right)$$

$$\kappa_{VDW} = \frac{1}{P}\left(1-\frac{b}{V}+\frac{a}{PV^2}\right)$$

correction term $= -\dfrac{b}{V} + \dfrac{a}{PV^2}$

105 chemical energy

105:1 *formation of liquid water*

- Strictly speaking, if one specifies the PT circumstances for the enthalpy effect, one should also specify P and T for the energy effect;
- the difference between the two effects is $\Delta_R H - \Delta_R U = P\cdot\Delta_R V$, where
  $$\Delta_R V = V_{H_2O} - V_{H_2} - 1/2\, V_{O_2};$$
  PV_{H_2} and PV_{O_2} are about equal to RT which is 2479 J·mol^{-1} at 298.15 K;
  PV_{H_2O} is about just 2 J·mol^{-1};

  as a result, the difference $P\cdot\Delta_R V$ is about $-1\,(1/2)RT$ and rounded to kJ it is -4.

105:2 *formation of gaseous water*

- -243 kJ·mol^{-1}
- for liquid H_2O we have at 1 atm (1 bar)
  $$\Delta_f H_{H_2O} = -284\times10^3 \text{ at } T = 373.15 \text{ K}$$
  $$\Delta_f H_{H_2O} = -286\times10^3 \text{ at } T = 298.15 \text{ K}$$

$$\Delta_f C_{P\,H_2O} \approx 2 \times 10^3 \,/\, 75 \approx 3 \times 10 \text{ J} \cdot \text{K}^{-1} \cdot \text{mol}^{-1}$$

by the way,

$$\Delta_f C_{P\,H_2O} = C_{P\,H_2O} - C_{P\,H_2} - 1/2\, C_{P\,O_2}$$

105:3 *reaction between graphite and carbon dioxide*
- At constant P, $\Delta Q = \Delta H = 172.8$ kJ·mol^{-1};
- the work performed by the system equals $+\, P\Delta V$, $\Delta V = 2V_{CO} - V_{CO_2} - V_C$, which is, in the ideal-gas approximation for CO and CO_2 and using the given density of graphite, $2RT/P - RT/P - 5.3 \times 10^{-6}$ m$^3 \cdot$mol^{-1}.
  $+\, P\Delta V = 2479 - 0.5(!) = 2478$ J·mol^{-1}
- ΔU = (heat taken by the system) – (work done by the system)
  $= 172.8 - 2.5 = 170.3$ kJ·mol^{-1}

(the influence of the volume of graphite is really negligible)

105:4 *enthalpy of formation*
Taking ethanol (C_2H_6O), the following two possibilities present themselves - with E for ethanol

a) $\Delta_f H^\circ_{400} = H_E(P = 1 \text{ unit}, T = 400 \text{ K}) - 2\, H_C(P = 1; T = 400) - 3\, H_{H_2}(P = 1; T = 400)$
  $- 0.5\, H_{O_2}(P = 1;\ T = 400)$

b) $\Delta_f H^\circ_{400} = H_E(P = 1,\ T = 400) \sum - 2\, H_C(P = 1; T = 298) - 3\, H_{H_2}(P = 1; T = 298)$
  $- 0.5\, H_{O_2}(P = 1;\ T = 298)$

The information needed, starting from $\Delta_f H^\circ_{298}$,

b) for E: C_P liquid from 298.15 K to boiling point at unit pressure (b.p); heat of vaporization at b.p; C_P gas from b.p. to 400 K

a) in addition: the C_P's of graphite, hydrogen and oxygen.

105:5 *reaction between hydrogen and chlorine*
In an isolated system the energy is constant: the chemical energy released by the reaction is stored in the vessel with its contents - rise in temperature, say from T_b to T_e. The easiest way to show that there is, under these conditions, no "conservation of enthalpy" is to consider the limiting case, in which the vessel has zero heat capacity (i.e. to consider its contents only). Assuming ideal-gas behaviour and taking 1/2 n mol H_2, then we have for $H_2 + Cl_2 \rightarrow \cdot 2HCl$

$H_{begin} = U_{begin} + nRT_b$
$H_{end} = U_{end} + nRT_e$
$\Delta H = \Delta U + nR(T_e - T_b) = 0 + nR(T_e - T_b) \neq 0$

105:6 *formation of magnesite*
$\Delta_f H_{MgCO_3} = -1113.28$ kJ·mol^{-1}

105:7 *standard formation energies*

	PV J·mol^{-1}	$\Delta_f U^o$ (298.15K) kJ·mol^{-1}
graphite	0.53	−0.001
silicon	1.21	−0.001
oxygen	2478.97	−2.479
carbon monoxide	2478.97	−111.769 ± 0.170
carbon dioxide	2478.97	−393.509 ± 0.130
quartz	2.27	−908.222 ± 1.000

$$CO: \Delta_f U^o_{CO} = U^o_{CO} - U^o_C - 1/2\, U^o_{O_2} = (H^o_{CO} - RT) - (H^o_C - PV_C) - 1/2(H^o_{O_2} - RT)$$
$$= 111769 \text{ J·mol}^{-1}$$

105:8 *acetic acid from methanol and carbon monoxide*

To start with, the heat effect at 1 atm and 298.15 K, i.e. ΔH^o_{298}, is calculated from the heats of combustion.

CH$_3$OH	+	3/2O$_2$	→	CO$_2$	+	2H$_2$O	−725.7
CO	+	1/2O$_2$	→	CO$_2$			−283.0
2CO$_2$	+	2H$_2$O	→	CH$_3$COOH			+874.4

$$\overline{\text{CH}_3\text{OH} \quad + \quad \text{CO} \quad\quad\quad \rightarrow \quad\quad \text{CH}_3\text{COOH} \quad\quad\quad -134.3 \text{ kJ·mol}^{-1}}$$

$$\Delta H^o_{298} = -134.3 \text{ kJ·mol}^{-1}$$

Next, for each of the three substances, the difference in H^o corresponding to the change in temperature from 298.15 K to 500 K, has to be calculated.

methanol

$$H^o_{500} - H^o_{298} = C^o_{P_{298}} (T_b - 298.15\text{K}) + \Delta H^o_V + C^{o\,vap}_{PT_b} (500\text{K} - T_b)$$
$$= 81.6\,(338 - 298.15) + 35300 + 44\,(500 - 338)$$
$$= 45.68 \text{ kJ·mol}^{-1}$$

carbon monoxide

$$H^o_{500} - H^o_{298} = 29.1\,(500 - 298.15)$$
$$= 5.87 \text{ kJ·mol}^{-1}$$

acetic acid

$$H^o_{500} - H^o_{298} = 123.4\,(391 - 298.15) + 44400 + 67\,(500 - 391)$$
$$= 63.16 \text{ kJ·mol}^{-1}$$

Finally

$$\Delta H^o_{500} - H^o_{298} = 63.16 - 45.68 - 5.87 = 11.6 \text{ kJ·mol}^{-1}$$

$$\Delta H^o_{298} = -134.3 + 11.6 = -122.7 \text{ kJ·mol}^{-1}$$

105:9 *combustion of benzoic acid*

- C_6H_5COOH + 7.5O_2 → 7CO_2 + 3H_2O; ΔH = −3227 kJ·mol^{-1}

  7CO_2 →   7C   +   7O_2 +2755

  3H_2O →   3H_2 +   1.5O_2 +858

  _____

  C_6H_5COOH →   7C   +   3H_2 + O_2 +386

  $\Delta_f H$ = −386 kJ·mol^{-1}

- C_6H_5COOH(sol) + 7.5O_2 → 7CO_2 +3H_2O(liq); ΔH = −3227 kJ·mol^{-1}

  C_6H_5COOH(liq)   → C_6H_5COOH(sol)   −17

  3H_2O(liq)   →   3H_2O(liq)   −123

  _____

  C_6H_5COOH(liq) +7.5O_2 → 7CO_2 + 3H_2O(vap); ΔH = −3121 kJ·mol^{-1}

105:10 *a reaction in a bomb calorimeter*

- Reaction $C_2H_4O_2 + 2O_2$ → 2CO_2 + 2H_2O

  equal amounts of gaseous species at both sides:

  $(\Delta Q)_V \sim (\Delta Q)_P = \Delta H = \Delta H°$ in ideal-gas approximation

  1.874 K x 15250 J·K^{-1} = 28578 J (per 1.96 g)

  molair mass: 60 g·mol^{-1}

  $\Delta_c H°$ = −875 kJ·mol^{-1} (heat of combustion)

  $$2\,CO_2 + 2H_2O \rightarrow C_2H_4O_2 + 2O_2 \qquad +875$$
  $$2\,O_2 + 2\,C \rightarrow 2\,CO_2 \qquad -787.0$$
  $$2\,H_2 + O_2 \rightarrow 2\,H_2O \qquad -571.7$$
  _____
  $$2\,H_2 + 2\,C + O_2 \rightarrow C_2H_4O_2 \qquad -484$$

  $\Delta_f H°$ = −484 kJ·mol^{-1}

106   entropy

106:1 *absolute entropy by graphical integration*

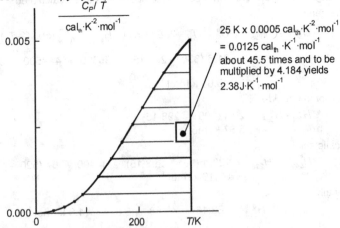

25 K x 0.0005 cal$_{th}$·K^{-2}·mol^{-1}

= 0.0125 cal$_{th}$·K^{-1}·mol^{-1}

about 45.5 times and to be

multiplied by 4.184 yields

2.38J·K^{-1}·mol^{-1}

See also Table 109:1a

106:2 *absolute entropy of liquid sodium chloride*
  The value given in Robie's table is 170.33 $J \cdot K^{-1} \cdot mol^{-1}$

106:3 *gaseous mercury and the Sackur-Tetrode equation*
  It is clear that the entropy values have been calculated by means of the S-T equation.
  The equation implies a constant C_P, having the value of 2.5 R.
  (from $q_{rev} = C_P \, dT$ it follows $dS = (q_{rev}/T =) (C_P/T) \, dT$ and $C_P = T \, (\partial S / \partial T)_P$)

106:4 *orientations up and down*
  $\Delta S = (R/N) \ln (W_b / W_a) = (R/N) \ln 2^N = R \ln 2$
  each possible orientation of one unit can be combined with two orientations of all the other units.

106:5 *substitutional disorder of red and blue molecules*
  See § 204 "the ideal mixture"

106:6 *cylinder with internal piston and entropy*

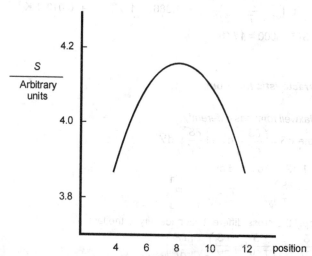

  NB If the piston can freely move, it will spontaneously move to position 8, i.e. the position of maximal entropy. See also Figure 001:2.

106:7 *uniform pressure at maximum entropy*
  $V_1 + V_2 = \text{const} = V$
  $S/R = n_1 \ln V_1 + n_2 \ln V_2 = \ln V_1^{n_1} \cdot V_2^{n_2} = \ln V_1^{n_1} \cdot (V - V_1)^{n_2}$
  condition $dS/dV_1 = 0$, or $\left(d \left(V_1^{n_1} \cdot (V - V_1)^{n_2} \right) / dV_1 \right) = 0$
  this leads to $n_1 \cdot V_2 = n_2 \cdot V_1$
  and with $V_2 = n_2 (RT / P_2)$ and $V_1 = n_1 (RT / P_1)$
  it follows $P_1 = P_2$!

106:8 *two vessels with helium in thermal contact*
Along with $T + |\Delta T|$ there is a $P + |\Delta P|$ and $P - |\Delta P|$ with $T - |\Delta T|$. From $PV = RT$,
$(\Delta P / \Delta T)_V = R/V = P/T$ and hence $\Delta P / P = \Delta T / T$.
Next

$$\Delta S = \frac{5}{2} R \ln\left\{ \frac{T^2}{(T + |\Delta T|)(T - |\Delta T|)} \right\} - R \ln\left\{ \frac{P^2}{(P + |\Delta P|)(P - |\Delta P|)} \right\}$$

$$= -\frac{5}{2} R \ln\left(1 - \left(\frac{\Delta T}{T}\right)^2\right) + R \ln\left(1 - \left(\frac{\Delta P}{P}\right)^2\right)$$

$$\approx \frac{5}{2} R \left(\frac{\Delta T}{T}\right)^2 - R \left(\frac{\Delta P}{P}\right)^2 = \frac{3}{2} R \left(\frac{\Delta T}{T}\right)^2$$

106:9 *supercooled water made to crystallize*
During a "reversible simulation" of the spontaneous change, first heat is added to
one mole of water to bring its temperature from −5°C to 0°C and next the same
amount of heat is withdrawn as a result of which part of the water crystallizes to ice.

- The amount of heat added is 5 x 75 = 375 J.
  The entropy change is
  $$75 \int_{263.15}^{273.15} \frac{dT}{T} - \frac{375}{273.15} = 1.386 - 1.373 = +0.013 \text{ J K}^{-1}$$
- 375 / 6000 = 1 / 16

107   **characteristic functions**

107:1 *the Maxwell relations differently*
Substitute $dS = \left(\dfrac{\partial S}{\partial T}\right)_V dT + \left(\dfrac{\partial S}{\partial V}\right)_T dV$

in $dU = T dS - P dV$ and obtain

$$dU = T \left(\frac{\partial S}{\partial T}\right)_V dT + \left[T \left(\frac{\partial S}{\partial V}\right)_T - P\right] dV$$

Next apply the cross-differentiation identity to the last dU

$$T \frac{\partial^2 S}{\partial V \partial T} = T \frac{\partial^2 S}{\partial T \partial V} + \left(\frac{\partial S}{\partial V}\right)_T - \left(\frac{\partial P}{\partial T}\right)_V$$

Hence $\left(\dfrac{\partial S}{\partial V}\right)_T = \left(\dfrac{\partial P}{\partial T}\right)_V$

107:2 *differential coefficients of energy*
From $dU = T dS - P dV$ it follows

$$\left(\frac{\partial U}{\partial P}\right)_T = T \left(\frac{\partial S}{\partial P}\right)_T - P \left(\frac{\partial V}{\partial P}\right)_T$$

$$= -T \left(\frac{\partial V}{\partial T}\right)_P - P \left(\frac{\partial V}{\partial P}\right)_T = -T \cdot \alpha \cdot V + P \cdot \kappa \cdot V$$

and $\left(\dfrac{\partial U}{\partial T}\right)_P = T \left(\dfrac{\partial S}{\partial T}\right)_P - P \left(\dfrac{\partial V}{\partial T}\right)_P = C_P - P \cdot \alpha \cdot V$

107:3 *total differential of enthalpy*

$$dH = TdS + VdP$$

$$= T\left\{\left(\frac{\partial S}{\partial T}\right)_P dT + \left(\frac{\partial S}{\partial P}\right)_T dP\right\} + VdP$$

$$= C_p dT - T\left(\frac{\partial V}{\partial T}\right)_P dP + VdP$$

$$= C_p dT - T\alpha VdP + VdP$$

So, if you want to "correct" H for changes in P and T you need to know the heat capacity; the volume; and the cubic expansion coefficient.

107:4 *model system with equation of state*

- $dH = TdS + VdP$

$$\left(\frac{\partial H}{\partial P}\right)_T = T\left(\frac{\partial S}{\partial P}\right)_T + V = -T\left(\frac{\partial V}{\partial T}\right)_P + V = b$$

$$\text{Maxwell} \quad \text{and} \quad V = \frac{RT}{P} + b$$

H_T does depend on P: $H_T(P_2) = H_T(P_1) + b(P_2 - P_1)$

- $dU = TdS - PdV$

$$\left(\frac{\partial U}{\partial P}\right)_T = T\left(\frac{\partial S}{\partial P}\right)_T - P\left(\frac{\partial V}{\partial P}\right)_T = -T\left(\frac{\partial V}{\partial T}\right)_P - P\left(\frac{\partial V}{\partial P}\right)_T$$

$$= -T\frac{R}{P} - P\left(-\frac{RT}{P^2}\right) = 0$$

At constant temperature the energy is independent of pressure.

107:5 *Van der Waals gas - change of enthalpy with volume*

From $dH = TdS + VdP$ it follows

$$\left(\frac{\partial H}{\partial V}\right)_T = T\left(\frac{\partial S}{\partial V}\right)_T + V\left(\frac{\partial P}{\partial V}\right)_T$$

and with one of the Maxwell relations

$$\left(\frac{\partial H}{\partial V}\right)_T = T\left(\frac{\partial P}{\partial T}\right)_V + V\left(\frac{\partial P}{\partial V}\right)_T$$

Rearranging the VDW equation, $P = RT/(V-b) - a/V^2$.

$$\left(\frac{\partial H}{\partial V}\right)_T = -RTb/(V-b)^2 + 2a/V^2$$

107:6 *change of heat capacity with pressure*

$$C_p = -T\left(\partial^2 G / \partial T^2\right)$$

$$\left(\frac{\partial C_p}{\partial P}\right)_T = -T\frac{\partial^3 G}{\partial P \partial T^2} = \text{(cross - differentation identity)}$$

$$= -T\frac{\partial^3 G}{\partial T^2 \partial P} = -T\frac{\partial^2 V}{\partial T^2}$$

107:7 *Gibbs energy of a hypothetical system*

$$S = -\beta - \beta \ln T - \gamma \ln P - \varepsilon - \varphi P;$$

$$V = \gamma(T/P) + \delta + \varphi T;$$

$$H = \alpha - \beta T + \delta P; \qquad C_p = -\beta$$

$$\left(\frac{\partial P}{\partial T}\right)_V = (\gamma P + \varphi P^2)/(\gamma T) \qquad \text{from Equation (107 : 25).}$$

$$U = \alpha - \beta T - \gamma T - \varphi P T;$$

$$dU = \left(\frac{\partial U}{\partial T}\right)_P dT + \left(\frac{\partial U}{\partial P}\right)_T dP;$$

$$C_V = \left(\frac{\partial U}{\partial T}\right)_V = \left(\frac{\partial U}{\partial T}\right)_P + \left(\frac{\partial U}{\partial P}\right)_T \left(\frac{\partial P}{\partial T}\right)_V = -\beta - \gamma - 2\varphi P - (\varphi^2/\gamma)P^2;$$

under isothermal ($T = T_a$) conditions

$$\Delta Q_{rev} = T_a \int (dS)_T = T_a \int_{P=P_1}^{P=P_2}\left(-\frac{\gamma}{P} - \varphi\right)dP$$

$$= -\gamma T_a \ln\frac{P_2}{P_1} - \varphi T_a(P_2 - P_1)$$

107:9 *Joule-Thomson coefficient*

From $\left(\frac{\partial H}{\partial T}\right)_P \left(\frac{\partial P}{\partial H}\right)_T \left(\frac{\partial T}{\partial P}\right)_H = -1$

$$\left(\frac{\partial H}{\partial T}\right)_P = C_p \quad \text{and} \quad \left(\frac{\partial P}{\partial H}\right)_T \quad \text{from}$$

$$dH = TdS + VdP, \text{ i.e, } \left(\frac{\partial H}{\partial P}\right)_T = T\left(\frac{\partial S}{\partial P}\right)_T + V$$

and one of the Maxwell relations.

107:10 *difference between C_P and C_V*

From $dU = \left(\frac{\partial U}{\partial T}\right)_P dT + \left(\frac{\partial U}{\partial P}\right)_T dP$ it follows

$$C_V = \left(\frac{\partial U}{\partial T}\right)_V = \left(\frac{\partial U}{\partial T}\right)_P + \left(\frac{\partial U}{\partial P}\right)_T \left(\frac{\partial P}{\partial T}\right)_V$$

See Exc 2 for $\left(\frac{\partial U}{\partial T}\right)_P$ and $\left(\frac{\partial U}{\partial P}\right)_T$ and Equation (107:25) for $\left(\frac{\partial P}{\partial T}\right)_V$.

107:11 *enthalpy as a characteristic function*

$$dH = TdS + VdP; \quad T = \left(\frac{\partial H}{\partial S}\right)_P \quad \text{and} \quad V = \left(\frac{\partial H}{\partial P}\right)_S$$

- $G = H - TS = H - S\left(\frac{\partial H}{\partial S}\right)_P$

- $U = H - PV = H - P\left(\frac{\partial H}{\partial P}\right)_S$

- $C_p = \left(\frac{\partial H}{\partial T}\right)_P = T\left(\frac{\partial S}{\partial T}\right)_P = T\left[\left(\frac{\partial T}{\partial S}\right)_P\right]^{-1} = \left(\frac{\partial H}{\partial S}\right)_P \left[\frac{\partial^2 H}{\partial S^2}\right]^{-1}$

- $C_V = \left(\dfrac{\partial U}{\partial T}\right)_V = T\left(\dfrac{\partial S}{\partial T}\right)_V$

use $dT = \left(\dfrac{\partial T}{\partial S}\right)_P dS + \left(\dfrac{\partial T}{\partial P}\right)_S dP$ from which

$$1 = \left(\dfrac{\partial T}{\partial S}\right)_P \left(\dfrac{\partial S}{\partial T}\right)_V + \left(\dfrac{\partial T}{\partial P}\right)_S \left(\dfrac{\partial P}{\partial T}\right)_V$$

and also $\left(\dfrac{\partial T}{\partial P}\right)_V = \left(\dfrac{\partial T}{\partial S}\right)_P \left(\dfrac{\partial S}{\partial P}\right)_V + \left(\dfrac{\partial T}{\partial P}\right)_S$;

$\left(\dfrac{\partial S}{\partial P}\right)_V$ from $\left(\dfrac{\partial S}{\partial V}\right)_P \left(\dfrac{\partial P}{\partial S}\right)_V \left(\dfrac{\partial V}{\partial P}\right)_S = -1$

107:12 *energy expressed in T and P*

If U would be characteristic for T and P then it would be possible to derive V from $U(T, P)$;

for the ideal gas U does not depend on V:

V cannot be derived just from $U(T)$;

Therefore

U impossibly can be characteristic for T and P

107:13 *a cosmetic imperfection?*

The expression in the thesis gives the dependence of a Gibbs energy property on T and P. Mathematically speaking, taking into account that the energy property is provided with (T,P), a more consequent expression would be

$\Delta G(T,P) = \Delta U(T,P) + P\Delta V - T\Delta S$.

If the latter is meant to state that energy is a function of T and P, and that the volume and entropy properties are constants, then the expression is incorrect from a thermodynamic point of view, as follows from the differential coefficients in Exc 2.

108   *Gibbs energy and equilibrium*

108:1   *entropy versus energy diagram*

$dU = TdS - PdV$

$$\left(\dfrac{\partial U}{\partial S}\right)_V = T$$

positive slope of U versus S

i.e. positive slope of S versus U

$$\dfrac{\partial^2 U}{\partial S^2} = \left(\dfrac{\partial T}{\partial S}\right)_V = \left(\dfrac{\partial S}{\partial T}\right)_V^{-1}$$

U versus S is convex (second derivative is positive)

i. e. S versus U is concave

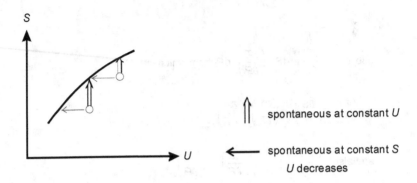

108:2 *a fancy device*

Take 1 mol water and suppose that n mol of it crystallizes. Then
$\Delta S = 75 \ln(273.15/268.15) - n\,6000 / 273.15$.
Constant entropy demands that $\Delta S = 0$.
As a result $n = 0.06308$.
$\Delta U = 5 \times 75 - n \cdot 6000 = 375 - 378.5 = -3.5$ J

108:3 *Gibbs energy of ideal gas*

$$S = -\left(\frac{\partial G}{\partial T}\right)_P = -\left(\frac{\partial G^\circ}{\partial T}\right)_P - R\ln P = S^\circ(T) - R\ln P$$

$$V = \left(\frac{\partial G}{\partial P}\right)_T = \frac{RT}{P}$$

$$H = G + TS = G^\circ + TS^\circ + RT\ln P - RT\ln P = H^\circ(T)$$

$$U = H - PV = H^\circ - P\frac{RT}{P} = H^\circ(T) - RT = U^\circ(T)$$

by the way, in H° and U° the superscript $^\circ$ has lost its meaning

108:4 *ammonia's Gibbs energy of formation*

$$\tfrac{1}{2}\,N_2 + \tfrac{3}{2}\,H_2 \rightarrow NH_3$$

for 25 °C and all gases at a certain pressure P

$$\begin{aligned}
\Delta_f G_{NH_3} &= G_{NH_3}^{obar} + RT\ln(P/\text{bar}) - \tfrac{3}{2}G_{H_2}^{obar} - \tfrac{3}{2}RT\ln(P/\text{bar}) - \tfrac{1}{2}G_{N_2}^{obar} - \tfrac{1}{2}RT\ln(P/\text{bar}) \\
&= \Delta_f G^{obar} - RT\ln(P/\text{bar}) \\
&= -16410 - RT\ln(P/10^5\text{Pa}) \\
&= -16410 + RT\ln 10^5 - RT\ln(P/\text{Pa}) \\
&= -16410 + 28540 \quad - RT\ln(P/\text{Pa}) \\
\Delta_f G_{NH_3}^{oPa} &= +12.13 \text{ kJ} \cdot \text{mol}^{-1}
\end{aligned}$$

108:5 *an equation of state for real gases*

On integration from 1 Pa at constant temperature
$G(P) = G(P{=}1\text{Pa}) + RT\ln P + B(P - 1\text{Pa})$;
on generalization, and neglecting Bx1Pa,
$G(T,P) = G^\circ(T) + RT\ln P + B(T)\,P$
$H(T,P) = H^\circ(T) + B(T) - T\,(\mathrm{d}B/\mathrm{d}T)$

108:6 *the concept of fugacity*

$$f(T,P) = P \cdot e^{B(T) \cdot P / RT}$$

for N_2 at 673 K and 101325 Pa $f = 101369$ Pa

108:7 *water + ice under a higher pressure*

The two lines have to be lifted up; the line for ice, with its greater molar volume, more than the line for water: the point of intersection is at a lower temperature.

108:8 *the calcium carbonate equilibrium under a higher pressure*

With $G_B^*(T = T) = G_B^*(T = T_o) - S_B^*(T - T_o)$

$$B = CO_2, CaO, CaCO_3$$

$$\ln P = \frac{G_{CaCO_3}^*(T_o) - G_{CaO}^*(T_o) - G_{CO_2}^o(T_o) - \left(S_{CaCO_3}^* - S_{CaO}^* - S_{CO_2}^o\right)(T - T_o)}{RT}$$

$$G_{CaCO_3}^*(T_o) - G_{CaO}^*(T_o) - G_{CO_2}^*(T_o) = 0 \text{ for } T_o = 1160 \text{ K}$$

$$S_{CaCO_3}^* - S_{CaO}^* - S_{CO_2}^o \qquad = -144 \text{ J} \cdot \text{mol}^{-1}$$

for $P = 1.25$ bar $T = 1175$ K

108:9 *the calcite and aragonite forms of calcium carbonate*

Calcite at 1160 K

$$G_{CaCO_3}^{CAL} - G_{CaO} - G_{CO_2} = 0$$

aragonite at 1160 K, as a result of $G^{ARA} > G^{CAL}$,

$$G_{CaCO_3}^{ARA} - G_{CaO} - G_{CO_2} > 0$$

$$G_{CaCO_3}^{ARA} > G_{CaO} + G_{CO_2}$$

aragonite has already passed the equilibrium curve: it changes spontaneously into CaO and CO_2

Conclusion: aragonite decomposes at a lower temperature

108:10 *liquid and gaseous water in equilibrium*

- 373.15 K; 1 atm.

  $\Delta G = \Delta H - T \Delta S = 0$ so that $\Delta S = 109.52$ J·K^{-1}·mol^{-1}

  $\Delta U = \Delta H - P \Delta V \approx \Delta H - P V^{vap} \approx \Delta H - R T = 37763$ J·mol^{-1}

- 372.15 K; 1 atm.

  $\Delta G = \Delta H - T \Delta S = 110$ J·mol^{-1}

- 374.15 K; 1 atm.

  $\Delta G = -110$ J·mol^{-1}

108:11 *lowest Helmholtz energy as a criterion for equilibrium*

Starting from Equation (6), and replacing Equation (8) by $A = U - TS$, the criterion expressed by Equation (11) becomes $(dA)_{T,V} \leq 0$.

109 *data and tables*

109:1 *partial derivatives of H-TS*

$$\left(\frac{\partial f}{\partial T}\right)_P = \left(\frac{\partial H}{\partial T}\right)_P - T\left(\frac{\partial S}{\partial T}\right)_P - S$$

$$= C_P - C_P - S = -S$$

$$\left(\frac{\partial f}{\partial P}\right)_T = \left(\frac{\partial H}{\partial P}\right)_T - T\left(\frac{\partial S}{\partial P}\right)_T$$

$$= \left\{T\left(\frac{\partial S}{\partial P}\right)_T + V\right\} - T\left(\frac{\partial S}{\partial P}\right)_T = V$$

109:2 *completion of a table*

5.74	0	0	0
33.15	0	0	0
205.15	0	0	0
197.67	−110.53	89.36	−137.17
213.79	−393.51	2.92	−394.38
42.62	−157.32	−93.11	−129.56

109:3 *the essential information of a table*

The information needed is:

the entropy of each of the five substances;

for the compounds the heat of formation:

for MgO and SiO_2 from the elements;

for Mg_2SiO_4 either from the elements or from the oxides.

109:4 *thermodynamic table for corundum*

- $\Delta_f H° = -1675.700$ kJ·mol^{-1} $\Delta_f S° = -313.505$ J·K^{-1}·mol^{-1}
  $\Delta_f G° = -1582.228$ kJ·mol^{-1}

- with the above values and $\Delta_f C_P° = -13.665$ J·K^{-1}·mol^{-1}

$$\left.\begin{array}{l}\Delta_f H° = -1677.092 \text{ kJ·mol}^{-1}\\ \Delta_f S° = -317.521 \text{ J·K}^{-1}\text{·mol}^{-1}\end{array}\right\}\quad \Delta_f G° = -1550.084 \text{ kJ·mol}^{-1}$$

109:5 *thermodynamic table for helium from the Sackur-Tetrode equation*

C_P, from $C_P = T\left(\frac{\partial S}{\partial T}\right)_P$, is constant and equal to $\frac{5}{2}R$.

$(H_T° - H_{298}°)/T = 14.589$ J·K^{-1}·mol^{-1}

$$S_T^o = 151.308 \ \text{J·K}^{-1}\text{·mol}^{-1}$$
$$-(G_T^o - H_{298}^o)/T = 136.72 \ \text{J·K}^{-1}\text{·mol}^{-1}$$

109:6 *extrapolation formula for Gibbs energy*

$$H_T - H_\theta = \int_\theta^T \left(C_{P_\theta} + C' \int_\theta^T dT \right) dT = \int_\theta^T \left(C_{P_\theta} + C'(T-\theta) \right) dT$$

$$= C_{P_\theta}(T-\theta) + C' \left[\left(\tfrac{1}{2}T^2 - \tfrac{1}{2}\theta^2 \right) - \theta(T-\theta) \right]$$

$$S_T - S_\theta = \int_\theta^T \frac{\left(C_{P_\theta} + \int_\theta^T C' dT \right)}{T} dT$$

$$= C_{P_\theta} \ln\left(\frac{T}{\theta}\right) + C' \left[(T-\theta)) - \theta \ln\left(\frac{T}{\theta}\right) \right]$$

$$G_T \qquad = H_T - T \, S_T$$

109:7 *silver oxide*

by least squares: $a = 49.397 \ \text{J·K}^{-1}\text{·mol}^{-1}$; $b = 5.5841 \times 10^{-2} \ \text{J·K}^{-2}\text{·mol}^{-1}$

T / K	$\left(H_T^o - H_{298}\right)$		S_T^o	
	calc	Table	calc	Table
400	7.016	7.042	141.502	141.570
500	14.469	14.509	158.108	158.212

109:8 *water versus steam*

		$(H_T^o - H_{298}^o)/T$	S_T^o	$-(G_T^o - H_{298}^o)/T$	C_P^o	$\Delta_f H^o$	$\Delta_f G^o$
298.15	water	0.000 [0]	69.95 [1] $=$	69.95	75.19 [1]	−285.830 [1]	−237.141 [1]
	steam	0.000 [0]	183.83 $=$	183.83	33.58	−241.814	−228.569
1800	water	59.236 [3]	259.37 [2]	200.13 [3]	49.68 [2]	−251.201 [2]	−147.035 [2]
	steam	34.783	259.37	224.59	49.68	−251.201	−147.035

[0] the selected zero points;
[1] independent of choice of zero point, but different owing to difference in state, i.e. liquid versus gaseous;
[2] independent of choice of zero point and equal, in view of equality of state;
[3] different values owing to difference in zero point.

109:9 *α-quartz - molar volume and compressibility*

Molar volume $\tfrac{1}{3} N_{Av} \cdot V_{unit\,cell} = 22.7 \times 10^{-6} \ \text{m}^3 \cdot \text{mol}^{-1}$

isothermal compressibility $\qquad 1.8 \times 10^{-11} \ \text{Pa}^{-1}$

348

109:10 *α-quartz - Gibbs energy at high pressure*

$\Delta G = V \Delta P \sim V P$ for high pressure and starting from low pressure

$= 113.5$ kJ·mol^{-1}

from $\quad V = V_0(1 - \kappa \cdot P)$

$\Delta G = V_0 \cdot P (1 - \frac{1}{2} \kappa \cdot P) = 108.4$ kJ·mol^{-1}

4 ½ % lower

109:11 *the change in Gibbs energy resulting from a change in pressure*

$$G(P) - G(\pi) = \int_\pi^P V(P)dP$$

$$= \int_\pi^P \left(V_\pi + \int_\pi^P \left(\frac{\partial V}{\partial P} \right)_T dP \right) dP$$

etc.

$$= V(\pi) \cdot (P - \pi) + \frac{1}{2}V'(\pi)(P - \pi)^2 + \frac{1}{6}V''(\pi)(P - \pi)^3$$

this result is just a *Taylor's series*

109:12 *bulk modulus*

Taking M_B as $M_B = -V \left(\frac{\partial P}{\partial V} \right)_T$, i. e. isothermal bulk modulus, we obtain

$$\kappa = M_B^{-1}$$

$$\left(\frac{\partial V}{\partial P} \right)_T = -\frac{V}{M_B}$$

$$\left(\frac{\partial^2 V}{\partial P^2} \right)_{TT} = \frac{V}{M_B^2} \left[1 + \left(\frac{\partial M_B}{\partial P} \right)_T \right]$$

109:13 *forsterite at high pressure*

$V(\pi) \cdot (P - \pi) = 235.8$ kJ·mol^{-1}

$\frac{1}{2}V'(\pi)(P - \pi)^2 = -6.70$

$\frac{1}{6}V''(\pi)(P - \pi)^3 = 0.13$

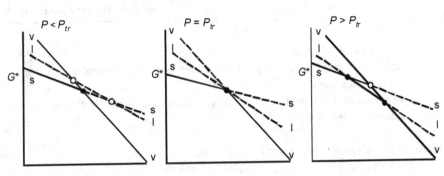

α = sol; β = liq; γ = vap

110 *pure substances*

110:1 *G*T diagrams around a triple point*
see bottom foregoing page

110:2 *benzophenon, a monotropic substance*

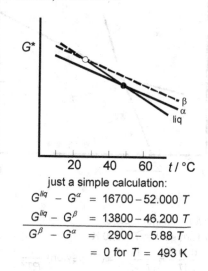

just a simple calculation:

$$G^{liq} - G^{\alpha} = 16700 - 52.000\,T$$
$$G^{liq} - G^{\beta} = 13800 - 46.200\,T$$
$$\overline{G^{\beta} - G^{\alpha} = 2900 - 5.88\,T}$$
$$= 0 \text{ for } T = 493\text{ K}$$

110:3 *a phase diagram analogue*

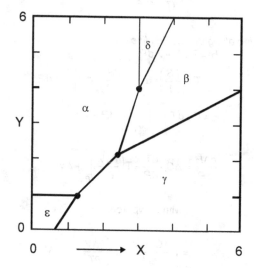

110:4 *a negative degree of freedom?*

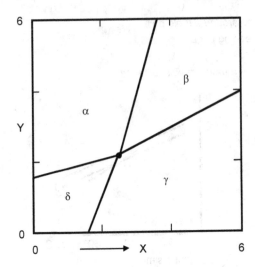

110:5 *Antoine equation for 1-aminopropane*
At n.b.p. $P = 760$ Torr, so that $T = 320$ K

$$\frac{d \ln P}{dT} = \frac{B}{(T-C)^2} = \text{(Clapeyron)} \frac{\Delta H}{RT^2}$$

$$\Delta H = \frac{B \cdot RT^2}{(T-C)^2} = 30.81\,\text{kJ}\cdot\text{mol}^{-1}$$

110:6 *heat of melting along the melting line*
$$dH = T\,dS + V\,dP$$

$$= T\left\{\left(\frac{\partial S}{\partial T}\right)_P dT + \left(\frac{\partial S}{\partial P}\right)_T dP\right\} + V\,dP$$

$$= C_P\,dT - T\left(\frac{\partial V}{\partial T}\right)_P dP + V\,dP$$

$$\frac{d\Delta H}{dP} = \Delta C_P \frac{dT}{dP} - T\left(\frac{\partial \Delta V}{\partial T}\right)_P + \Delta V$$

with Clayperon

$$= T\frac{\Delta C_P \cdot \Delta V}{\Delta H} - T\left(\frac{\partial \Delta V}{\partial T}\right)_P + \Delta V$$

110:7 *vapour pressure over 1,4-dibromobenzene*

temperature 25 °C	H° (kJ·mol^{-1})	S° (J·K^{-1}·mol^{-1})	G° (kJ·mol^{-1})
standard pressure 1 Pa	74.00	458.57	−62.72
standard pressure 1 atm	74.00	362.73	−34.15

NB   the values shown are of limited significance as C_P influences and the dependence of G^{*sol} on pressure could not be taken into account

110:8 *monoclinic and orthorhombic sulphur*

$$H_{SO_2} - H_{O_2} - H_{S^\circ} = -296810$$

$$\underline{H_{SO_2} - H_{O_2} - H_{S_m} = -297210}$$

$$\Delta H = H_{S_m} - H_{S^\circ} = 400 \, \text{Jmol}^{-1}$$

$$\Delta G = \Delta H - T\,\Delta S = 75 \, \text{Jmol}^{-1}$$

- From $\Delta G = \Delta H - T\,\Delta S = 0$ it follows $T = 367$ K.
- It follows that o has the lowest Gibbs energy at 25 °C. In other words, formation properties of sulphur compounds refer to elementary sulphur in the orthorhombic form.
  The answer is −296.81 kJ · mol^{-1}

110:9 *caesium chloride*

by subtraction: $\Delta_\alpha^\beta G \equiv G^\beta - G^\alpha = -2900 + 3.91 \, T$

transition	$\alpha \to$ liq	$\beta \to$ liq	$\alpha \to \beta$
temperature (K)	913	887**	742
heat effects (kJ·mol^{-1})	20.116	23.016	−2900*

* minus sign: α is the high-, and β the low-temperature form
** metastable melting point
stability order: β (742 K) α (913 K) liquid

110:10 *diamond out of graphite*

The equilibrium pressure (P_e) at 1700 K follows from
ΔG (1700 K, P_e) = ΔG (1700 K, 1 bar) + ΔV (P_e − 1 bar) = 0.
$P_e = 4.8 \cdot 10^9$ Pa
Let's say, order of magnitude needed, at least 10^{10} Pa. By the way, the value of P_e for 1700 K, given by Whittaker (1978) is $6 \cdot 10^9$ Pa.

110:11 *heat capacity change from the shape of the arc*
Taylor: $F = F_o + (X-X_o)(dF/dX) + \frac{1}{2}(X-X_o)^2(d^2F/dX^2) +$ etc.
For $F = \ln f$
$F - F_o = -h$; at the top $dF/dX = 0$; $(X-X_o) = \frac{1}{2}b$; $d^2F/dX^2 = (\Delta C_P/R)(T_{max})^2$

110:12 *the water arc*
Read from the arc, $T_{max} = 297.6$ K; $h = 0.00605$; $b = 0.000338$ K^{-1}.
With these numbers, Equation (30) gives $\Delta C_P = -39.8$ J$\cdot$K^{-1}mol^{-1}

110:13 *a different arc*
See Exc 11; this time the second derivative of $\ln f$ is given by $[2\,\beta T + (\Delta C_P/R)]/T^2$
Read from the arc, $\Theta = 297.85$ K; $h = 0.0512$; $b = 30$ K.
With these numbers, $\Delta C_P = -39.7$ J$\cdot$K^{-1}mol^{-1}.

110:14 *naphthalene: the assessment of a data set*
Among the data for vapour pressures over solids, the set of given data are of a
superior quality (see van der Linde et al. 1998); yet the construction of the arc leaves
some space for a subjective interpretation.
For two extreme arcs (having $h = 0.065$; $b = 0.00075$ K^{-1}; and the other $h = 0.040$;
$b = 0.00100$ K^{-1}) the calculated properties are (SI units):

	ΔG^o	ΔH^o	ΔC_P^o	$T (P = 800$ Pa$)$
$h = 0.065$	−6019	72500	−86.5	350.76
$h = 0.040$	−5957	72500	−30.0	349.90

110:15 *second-order transition according to Ehrenfest*
From $d(\Delta S^*) = 0$, the relationship $dP/dT = (\Delta C_P/T)/(\partial \Delta V^*/\partial T)_P$;
From $d(\Delta V^*) = 0$, the relationship $dP/dT = -(\partial \Delta V^*/\partial T)_P/(\partial \Delta V^*/\partial P)_T$

111   **chemical reactions and equilibrium**

111:1 *the strontium oxides*
- lowest G for SrO_2
- with 1 mol O_2: either 2 mol SrO or 1 mol SrO_2;
  lowest G for 2 mol of SrO
- now SrO has a lower G then SrO_2;
  with 1 mol O_2: 1 mol SrO and $\frac{1}{2}$ mol O_2

111:2 *Alexander von Humboldt's discovery*
- at equilibrium $\Delta G = \Delta G^o + RT \ln P$ is equal to zero;
  $\Delta H^o = 154$ kJ$\cdot$mol^{-1}
  $\Delta S^o = 139$ J$\cdot$K$^{-1}\cdot$mol^{-1}
- From Equation (007:7):
  $\Delta G = \Delta G^o + RT \ln(X_{O_2} \cdot P) = \Delta H^o - T\Delta S^o + RT \ln(X_{O_2} \cdot P)$
  $= 154000 - 139\,T + RT \ln 0.2$

ΔG is zero for $T = 1014$ K: up to 1014 K BaO will absorb O_2 from the air.

111:3 *magnesium carbonate*
- straight line, passing zero at about 682 K;
- during the whole experiment the gas in the cylinder is pure CO_2 at 1 bar : isothermal decomposition at ~ 682 K.

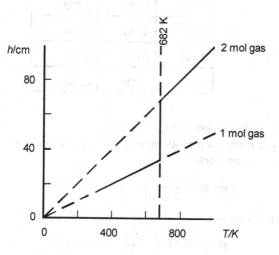

111:4 *air as a CO_2 buffer*

$\Delta_{fox}H° = -115.4$ kJ·mol^{-1}

$\Delta_{fox}S° = -169.2$ J·K^{-1}·mol^{-1}

for the reaction $MgCO_3 \rightarrow MgO + CO_2$ the ΔG is given by (use Equation (007:7))

$$\Delta G = -\Delta_{fox}G^°_{M\,gCO_3} + RT \ln(X_{CO_2} \cdot P)$$

the asked temp is the one for which ΔG passes zero; the answer is 488 K

111:5 *the ammonia equilibrium - the role of pressure*
- extrapolated value $6.50 \cdot 10^{-3}$ atm^{-1} = $6.42 \cdot 10^{-3}$ bar^{-1}
- $RT \ln K = -\left\{ G^°_{NH_3} - \frac{1}{2} G^°_{N_2} - \frac{3}{2} G^°_{H_2} \right\} = -\Delta_f G^°_{NH_3}$

  result $\Delta_f G^°_{NH_3}$ (723.15 K) $= 30.35$ kJ·mol^{-1}
- data 700: $\Delta_f S° = -114$ J·K^{-1}·mol^{-1}
- taking, for the extrapolation to 723.15 K, $\Delta_f H°$ and $\Delta_f S°$ constant, then

  $\Delta_f G° = 29.76$ kJ·mol^{-1}

111:6 *ammonia's degree of dissociation*

In ideal gas mixture for substance B, Equation (007:7),

$$\mu_B = G^°_B + RT \ln(X_B \cdot P)$$

$X_{NH_3} = (1-\alpha)/(1+\alpha); \quad X_{N_2} = (\frac{1}{2}\,\alpha)/(1+\alpha); \quad X_{H_2} = (\frac{3}{2}\alpha)/(1+\alpha)$

$$RT \ln \left\{ \frac{3\alpha^2 \sqrt{3}}{4\,(1-\alpha^2)} \cdot P \right\} = \Delta_f G^°_{NH_3}$$

$$\alpha = \left\{ \frac{e^{\Delta_f G^\circ/RT}}{e^{\Delta_f G^\circ/RT} + (0.75\sqrt{3})\cdot P} \right\}^{\!\!\frac{1}{2}}$$

P /bar →	1	10	100
	0.032	0.010	0.003
	0.336	0.112	0.036
	0.841	0.441	0.154
	0.974	0.805	0.394

111:7 *vapour in equilibrium with solid salammoniac*

There are two equations (equilibrium conditions) to find P and α
dissociation of salammoniac

$$\ln P + \ln\left\{ \frac{(1-\alpha)^{\frac{1}{2}}}{(2+\alpha)} \right\} = -1.07393$$

dissociation of ammonia

$$\ln P + \ln\left\{ \frac{(\tfrac{1}{2}\alpha)^{\frac{1}{2}} \cdot (\tfrac{3}{2}\alpha)^{\frac{3}{2}}}{(1-\alpha)\,(2+\alpha)} \right\} = 3.17537$$

solution $\alpha = 0.936$; $P = 3.96$ bar;
mole fractions in the vapour phase:

HCl (0.341)   NH$_3$ (0.022)   N$_2$ (0.159)   H$_2$ (0.478)

111:8 *dissociation of water at 1800 K*

	H$_2$O	⇌	H$_2$	+	1/2 O$_2$
relative amounts	$1-\alpha$		α		$1/2\,\alpha$
mole fractions	$\dfrac{(1-\alpha)}{(1+1/2\alpha)}$		$\dfrac{\alpha}{(1+1/2\alpha)}$		$\dfrac{1/2\,\alpha}{(1+1/2\alpha)}$

$$\tfrac{1}{2} RT \ln P + RT \ln\left\{ \frac{\alpha\,(\tfrac{1}{2}\,\alpha)^{\frac{1}{2}}}{(1-\alpha)\,(1+\tfrac{1}{2}\,\alpha)^{\frac{1}{2}}} \right\} = \Delta_f G^\circ_{H_2O}$$

this equation most easily can be solved numerically; solution: $\alpha = 0.0023$

111:9 *the simple model - interdependent reactions*

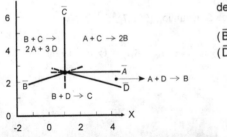

dependent reactions

($\bar{B}$)   A + 2 D → C
($\bar{D}$)   A + C → 2 B

111:10 *virtual experiments related to Figure 3*

	$\overline{C}\,\overline{D}$	$\overline{D}\,\overline{A}$	$\overline{A}\,\overline{B}$	$\overline{B}\,\overline{C}$
A + B	A + B	C	C	1/2 A + 1/2 D
A + C	2 A + B	A + C	A + C	3/2 A + 1/2 D
A + D	2 A + 2 B	2 C	2 C	A + D
B + C	A + 2 B	B + C	D	D
B + D	A + 3 B	2 B + C	B + D	B + D
C + D	2 A + 3 B	B + 2 C	C + D	C + D

111:11 *one of Professor Schuiling's favourites*

($\overline{A}$) $Al_2Si_2O_5(OH)_4 + 2\ SiO_2\ \rightarrow\ Al_2Si_4O_{10}(OH)_2 + H_2O$

($\overline{Q}$) $Al_2Si_4O_{10}(OH)_2 + 2\ Al_2SiO_5 + 5\ H_2O\ \rightarrow\ 3\ Al_2Si_2O_5(OH)_4$

($\overline{W}$) $2\ Al_2Si_4O_{10}(OH)_2\ \rightarrow\ Al_2Si_2O_5(OH)_4 + Al_2SiO_5\ + 5\ Si_2O$

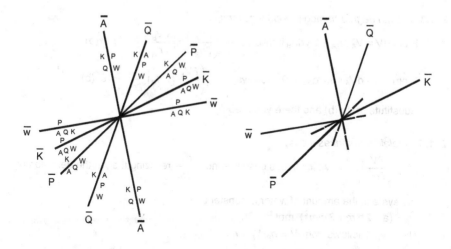

201   *mixtures and partial quantities*

201:1   *partial volumes from densities*

Referred to the same amount = 1 mol of water in each of the two mixtures:

wt % alc	n_{alc}/mol	mass /g	V /cm^3
72	1.00621	64.2857	74.1387
73	1.05797	66.6666	77.0980

$\Delta n_{alc} = 0.05176$ mol $\Delta V = 2.9593$ cm^3

$$V_{alc} \approx \left(\frac{\Delta V}{\Delta n_{alc}} \right)_{n_w} = 57.17 \text{ cm}^3 \cdot \text{mol}^{-1}$$

201:2   *from integral volume to partial volumes*

$$V_A = V_A^* + C \frac{n_B^2}{(n_A + n_B)^2}$$

$$V_B = V_B^* + C \frac{n_A^2}{(n_A + n_B)^2}$$

201:3   *zeroth degree homogeneous functions*

From $dV = V_A \, dn_A + V_B \, dn_B$ it follows $\left(\dfrac{\partial V_A}{\partial n_B} \right) = \left(\dfrac{\partial V_B}{\partial n_A} \right)$ (a)

From $n_A \, dV_A + n_B \, dV_B = 0$ it follows $n_A \left(\dfrac{\partial V_A}{\partial n_A} \right) + n_B \left(\dfrac{\partial V_B}{\partial n_A} \right) = 0$ (b)

Substitute (a) in (b) and there you are!

201:5   *molality(m) makes it easy*

$$V_B = \left(\frac{\partial V}{\partial n_B} \right)_{T,P,n_{H_2O}}$$ which in this case is simply $\dfrac{dV}{dm}$ realizing that, in the definition of

the system, the amount of water is constant

$V_B = (a + 2\,b \cdot m + 3\,c \cdot m^2) \cdot \text{mol}^{-1}$

Next, V_{H_2O} follows from $V = n_B \cdot V_B + n_{H_2O} \cdot V_{H_2O}$

$V_{H_2O} = 18.015 \times 10^{-3} \cdot (V^{\circ} - b \cdot m^2 - 2\,c \cdot m^3) \cdot \text{mol}^{-1}$

201:6   *partial volumes of sodium chloride and water in their liquid mixture*

m	V_{NaCl} cm$^3 \cdot$ mol^{-1}	V_{H_2O} cm$^3 \cdot$ mol^{-1}
0	17.4	18.05
3.0	21.6	17.94
6.01	23.0	17.84

202 the open system, chemical potentials

202:1 the ammonia equilibrium from different angles

	NH₃	=	½ N₂	+	³⁄₂ H₂
n_B	$1-\alpha$		$\tfrac{1}{2}\alpha$		$\tfrac{3}{2}\alpha$
X_B	$\dfrac{1-\alpha}{1+\alpha}$		$\dfrac{\tfrac{1}{2}\alpha}{1+\alpha}$		$\dfrac{\tfrac{3}{2}\alpha}{1+\alpha}$

$$\mu_B = G_B^o + RT \ln(X_B \cdot P)$$

$G = \sum_B n_B \cdot \mu_B$, which gives rise to (P being 1 bar)

- (a)

$$= (1-\alpha)\Delta_f G_{NH_3}^o + \tfrac{1}{2}G_{N_2}^o + \tfrac{3}{2}\,G_{H_2O}^o + RT\{(1-\alpha)\ln(1-\alpha) + \tfrac{1}{2}\alpha\ln(\tfrac{1}{2}\alpha) + \tfrac{3}{2}\alpha\ln(\tfrac{3}{2}\alpha)$$
$$- (1+\alpha)\ln(1+\alpha)\}$$

where $G_{N_2}^o$ and $G_{H_2}^o$ are arbitrary constants, which in the following are given the value zero.

- (b) $\sum_B v_B \cdot \mu_B = \tfrac{1}{2}\mu_{N_2} + \tfrac{3}{2}\mu_{H_2} - \mu_{NH_3}$
$$= -\Delta_f G_{NH_3}^o + RT\{\tfrac{1}{2}\ln(\tfrac{1}{2}\alpha) + \tfrac{3}{2}\ln(\tfrac{3}{2}\alpha) - \ln(1-\alpha) - \ln(1+\alpha)\}$$

NB function (b) is obtained when (a) is differentiated with respect to α.

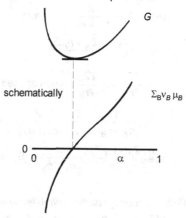

All mixtures of $(1 - \alpha)$ mole of NH₃, ½ α mole of N₂ and ³⁄₂ α mole of H₂ can be prepared from 1 mole of NH₃. Therefore, α is identical with the "classical" degree of dissociation. Its equilibrium value corresponds to the minimum in the plot of G versus α. For $T = 400$ K and $P = 1$ bar this values is 0.33.

202:2 the electrochemical cell

$\Delta_r S^o = -163.30$ J·K⁻¹·mol⁻¹

$\Delta_r G^o = -285830 + 163.30\ (T/K)$

E(taken positive) = 1.23 V

At increasing temperature the Gibbs energy change becomes less negative; the temperature coefficient, related to the entropy change, is −0.00085 V·K⁻¹.

At increasing pressure the Gibbs energies of the reactants increase substantially more than the Gibbs energy of the product: the Gibbs energy change becomes more negative and, as a result, the emf increases.

The reversible heat is the product of temperature and entropy change = −48689 J·mol⁻¹; heat is given off.

202:3   *the influence of gravity*

From the cross-differentiation identity, the change is given by the change of Mg with n_B, the amount of B. This is $M_B g$, where M_B is B's molar mass.

The function recipe is $\mu_B = G_B^\circ + RT \ln P + M_B \cdot g \cdot h$

From one layer to the other:

$$d\mu_B = RT \cdot d(\ln P) + M_B \cdot g \cdot h = 0;$$

with pressure P at altitude h, and pressure P° at $h = 0$, upon integration and after rearranging the expression, Equation (002:7) is obtained.

NB:

In atmospheric reality, not only the pressure of the air, but also its composition changes with altitude, and not to forget temperature.

A better reflection of the atmosphere is a two-component system, for which the gravitational term has to be added to Equation (5)

Neglecting the role of temperature, you could calculate the change in altitude as a result of which the percentage of oxygen in a mixture with nitrogen falls from 20% to 19%. Realize that there are two conditions with terms in dP, dX and dh; you can solve the two 'unknowns' dX/dh and dP/dh.

203   *change to molar quantities. Molar Gibbs energy*

203:1   *integral molar quantities are zeroth degree homogeneous functions*

Make use of the properties

(A) $Z_m = \dfrac{Z}{n_A + n_B}$ and

(B) $Z_m = \dfrac{n_A Z_A + n_B Z_B}{n_A + n_B}$

$n_A \dfrac{\partial Z_m}{\partial n_A} + n_B \dfrac{\partial Z_m}{\partial n_B}$ when applied to (A) gives

$$= n_A \left\{ \frac{1}{n_A + n_B} \frac{\partial Z}{\partial n_A} - \frac{Z}{(n_A + n_B)^2} \right\} + n_B \left\{ \frac{1}{n_A + n_B} \frac{\partial Z}{\partial n_B} - \frac{Z}{(n_A + n_B)^2} \right\}$$

$$= \frac{n_A \cdot Z_A + n_B \cdot Z_B}{n_A + n_B} - \frac{Z}{n_A + n_B} = (B) - (A) = Z_m - Z_m = 0$$

203:2   *recipes for partial quantities in a ternary system*

$(dZ)_{T,P} = Z_A\, dn_A + Z_B\, dn_B + Z_C\, dn_C$ (I)

$Z \quad = n_A \cdot Z_A + n_B \cdot Z_B + n_C \cdot Z_C$ (II)

change into $(dZ_m)_{T,P} = (Z_B - Z_A)\, dX + (Z_C - Z_A)\, dY$ (I')

$Z_m \quad = (1 - X - Y)\, Z_A + X\, Z_B + Y\, Z_C$ (II')

from (I'): $\dfrac{\partial Z_m}{\partial X} = Z_B - Z_A$ (III')

$\dfrac{\partial Z_m}{\partial Y} = Z_C - Z_A$ (IV')

Z_A, Z_B and Z_C can be solved from (II'), (III') and (IV').

203:3 *a Gibbs-Duhem exercise*

Gibbs-Duhem: $(1-X)\dfrac{\partial Z_A}{\partial X} + X\dfrac{\partial Z_B}{\partial X} = 0;$

from given $Z_A(X)$: $\dfrac{\partial Z_A}{\partial X} = -20\,X;$

so that $\dfrac{\partial Z_B}{\partial X} = 20\,(1-X)$

and $Z_B(X) = Z_B(X=0) + \displaystyle\int_{X=0}^{X=X} \dfrac{\partial Z_B}{\partial X}\,dX$

$= 0 \qquad + 20\,X - 10\,X^2;$

$Z_m = (1-X)\,Z_A + X\,Z_B \quad = 8 - 8\,X + 10\,X^2.$

203:4 *water + methanol: volumes, integral and partial*

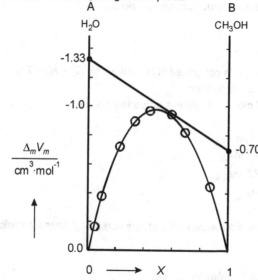

for $X = 0.591$:

$V_A = 18.047 - 1.33 = 16.72\ \text{cm}^3\cdot\text{mol}^{-1}$

$V_B = 40.46 - 0.70 = 39.76$

check: $0.409 \times 16.72 + 0.591 \times 39.76 = 30.34$

in agreement with the value given in the table

203:5 *the ideal mixture*

$G = -4808\ \text{J}$

204 *the ideal mixture*

204:1 *a Gibbs-Duhem exercise*

Yes it is:

from $\mu_A = G_A^* + RT\ln X_A = G_A^* + RT\ln(1-X)$

it follows - Gibbs-Duhem - that

$$\left(\frac{\partial \mu_B}{\partial X}\right)_{T,P} = \frac{RT}{X}$$

so that ($T\,P$ constant)

$$\mu_B(X = X) = \mu_B(X = 1) + \int_{X=1}^{X=X} \frac{RT}{X}dX = G_B^* + RT \ln X$$

204:2 *mathematical analysis of a function*

$$f' = \ln\left(\frac{X}{1-X}\right)$$

$$f'' = \frac{1}{X} + \frac{1}{1-X} = \frac{1}{X\,(1-X)}$$

205 non ideal behaviour. Excess functions

205:1 *ideal or not?*

Not necessarily: G^E is composed of H^E and S^E; (in $G^E = H^E - T \cdot S^E$) H^E and $T \cdot S^E$ may compensate each other;

in a really ideal mixture all excess properties are zero

205:2 *Gibbs-Duhem and activity*

$$\mu_A = \mu_A^{st\,st\,A} + RT \ln a_A$$

$$\mu_B = \mu_B^{st\,st\,B} + RT \ln a_B$$

$$\frac{da_A}{da_B} = -\frac{X \cdot a_A}{(1-X) \cdot a_B}$$

a_A increases when a_B decreases and vice versa no matter the choice of standard states

205:3 *NaCl's activity in saturated solution*

Saturated solution corresponds to

$$\mu_{NaCl}^{liq} = \mu_{NaCl}^{sol} ;$$

substitution of μ - recipes:

$$G_{NaCl}^{*\,sol} + RT \ln a_{NaCl}^{liq} = G_{NaCl}^{*\,sol} ;$$

as a result $a_{NaCl}^{liq} = 1$.

205:4 *equality of activities*

Equilibrium between phases α and β containing B; condition $\mu_B^\alpha = \mu_B^\beta$;

function recipes in terms of activities in terms of activities:

$$\mu_B^\alpha = \mu_B^{st\,st} + RT \ln a_B^\alpha$$

$$\mu_B^\beta = \mu_B^{st\,st} + RT \ln a_B^\beta ;$$

it follows $a_B^\alpha = a_B^\beta$.

NB this is what is done with fugacities; see Prausnitz et al., 1986.

205:5 *heat of mixing and activity coefficients*

The molar Gibbs energy of mixing $\{(1 - X) A + X B\}$ is given by

$$\Delta_m G = RT \{(1 - X) \ln f_A (1 - X) + X \ln f_B X\}$$

The reaction for $\Delta_m H$ follows from the general property

$$\left\{ \frac{\partial(G/T)}{\partial T} \right\}_{P,X} = -\frac{H}{T^2}$$

205:6 *a convenient formula*

$$S^E = \frac{A}{\Theta} X(1 - X) [1 + B(1 - 2X)] = \frac{H^E}{\Theta}$$

$$C_P^E = 0$$

$$\mu_1^E = A \left(1 - \frac{T}{\Theta}\right) X^2 [1 + B(3 - 4X)] = RT \ln f_1$$

$$\mu_2^E = A \left(1 - \frac{T}{\Theta}\right) (1 - X)^2 [1 + B(1 - 4X)] = RT \ln f_2$$

205:7 *alcohol is mixed with water - a classroom experiment*

The cylinder contains 2.85 mol of water and 0.835 mol of alcohol, the mole fraction of the latter being 0.23.

The enthalpy effect, the heat of mixing is negative: after mixing the system gives off heat to the surroundings, to go back to room temperature. The volume change on mixing is also negative.

The temperature t_0 the mixture would have if the heat of mixing were zero follows from

$$2.85 \times (t_0 - 19.2) \times 75.3 = 0.835 \times (21.9 - t_0) \times 111.5 \rightarrow t_0 = 20.0 \,^\circ C$$

The heat capacity of the homogeneous mixture, assuming $C_P^E = 0$, is

$$(2.85 \times 75.3 + 0.835 \times 111.5 = 308) \, J \cdot K^{-1}.$$

The enthalpy effect is

$-(27.3 - 20.0) \times 308 \, J$; molar excess enthalpy $H^E = -0.61 \, kJ \cdot mol^{-1}$.
$V^E = -1.1 \, cm^3 \cdot mol^{-1}$

206   *magic formulae*

206:1 *volume properties of supercritical carbon dioxide*

$V = 0.20287 \, dm^3 \cdot mol^{-1}$
$\kappa = 0.0275 \, bar^{-1}$
$\alpha = 0.0143 \, K^{-1}$

206:2 *Van der Waals and Helmholtz*

From $dA = -S \, dT - P \, dV$: $(\partial A/\partial V)_T = -P$, which means that the curves given correspond to the derivatives of the curves asked for, with opposite sign.

In all cases A decreases with increasing V. The PV curve for $T < T_C$ gives rise to an AV curve with two points of inflexion.

These two points coincide for $T = T_C$ and are absent for $T > T_C$.

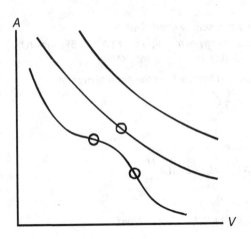

206:3 *critical coordinates from Van der waals constants*

T_C, P_C and V_C can be solved from the three equations

(1) $$P = \frac{RT}{V-b} - \frac{a}{V^2}$$

(2) $$\left(\frac{\partial P}{\partial V}\right)_T = -\frac{RT}{(V-b)^2} + \frac{2a}{V^3} = 0$$

(3) $$\left(\frac{\partial^2 P}{\partial V^2}\right) = \frac{2RT}{(V-b)^3} - \frac{6a}{V^4} = 0$$

206:4 *minimal Helmholtz energy*

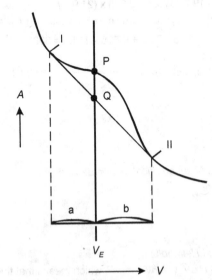

Q is the intersection of the line $V = V_E$ with the double tangent line; the lowest possible A the system can reach is represented by Q and corresponds to equilibrium between two phases (I) and (II) of which the molar volumes are the abscissae of the points of contact;

the amounts of the phases (I) and (II) are b / (a + b) mole and a / (a + b) mole, respectively.

206:5 *critical temperature for a model system*

$$T_c = \frac{A \cdot \Theta}{A + 2R\Theta}$$

yes, if the condition $2R\Theta < |A|$ is satisfied

206:6 *spinodal and critical point for a given function*

$$R\,T_{spin}(X) = 2\,A\,X\,(1 - X)\,[1 + 3\,B\,(1 - 2X)]$$

$$X_c = \frac{(9\,B + 1) - \sqrt{27\,B^2 + 1}}{18\,B}$$

207   dilute solutions

207:1 *the ideal isothermal vaporus*

from $(P_B =)\ X^{vap} \cdot P = X^{liq} \cdot P_B^o$ Equ (17)

and $(P_A =)\ (1 - X^{vap}) \cdot P = (1 - X^{liq}) \cdot P_A^o$

it follows, on elimination of X^{liq},

$$P(X^{vap}) = \frac{P_A^o \cdot P_B^o}{X^{vap} \cdot P_A^o + (1 - X^{vap}) \cdot P_B^o}$$

this is a part of a rectangular hyperbola

207:2 *Raoult, Henry, and Gibbs-Duhem*

In the vicinity of $X = 0$, say for $0 < X < a$ with a <<1, μ_B is given by

$R\,T \ln X$ + constant.

From Gibbs-Duhem, i.e. from

$$(1 - X)\frac{\partial \mu_A}{\partial X} + X\frac{\partial \mu_B}{\partial X} = 0\ ,$$

it follows $(1 - X)\dfrac{\partial \mu_A}{\partial X} + RT = 0$.

So that $\dfrac{\partial \mu_A}{\partial X} = -\dfrac{RT}{(1 - X)} = -\dfrac{RT}{X_A} = -\dfrac{\partial \mu_A}{\partial X_A}$.

Let $X_A' < (1 - a)$, then

$$\mu_A = \mu_A(X_A = 1) + \int_{X_A=1}^{X_A = X_A'} \frac{RT}{X_A} dX_A = G_A^* + RT \ln X_A$$

$$= G_A^* + RT \ln(1 - X)$$

207:3   *the activity coefficient*

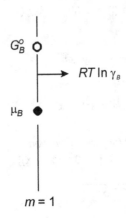

207:4   *the solute's chemical potential*

We know that the chemical potential of B (μ_B^{liq}) in the "solution phase" has to be equal to the chemical potential of B in the gas phase (μ_B^{vap}); the latter is

$$\mu_B^{vap} = G_B^{o\ vap} + RT \ln P \; ;$$

as a result $\mu_B^{liq} = G_B^{o\ vap} + RT \ln P \; ;$

substitution of Henry's Law:

$$\mu_B^{liq} = G_B^{o\ vap} + RT \ln(m_B/\kappa)$$
$$= \text{cons tan} t + RT \ln m_B$$
$$= \mu_B^{liq}(m = 1) + RT \ln m_B$$
$$= G_B^{o\ liq} + RT \ln m_B$$

207:5   *a strange question (?)*

This is almost a trick question: the differential coefficient ($\partial U / \partial n_1$) under the conditions given is the chemical potential potential μ_1.

The difference, taking the ideal dilute solution, is $- RT \Delta X = -24.8 \; \text{J·mol}^{-1}$

207:6   *a trigonometric excess function*

$$\left(\frac{\partial \mu_A^E}{\partial X} \right) = a_n n^2 \pi^2 X \sin n\pi X$$

which is zero for $X = 0$

207:7   *the ideal solution's quantities*

$$G_A = \mu_A = G_A^* + RT \ln(1 - X) \qquad G_B = \mu_B = G_B^o + RT \ln X$$
$$S_A = S_A^* - R\ln(1 - X) \qquad\qquad S_B = S_B^o - R\ln X$$
$$H_A = H_A^* \qquad\qquad\qquad\qquad\quad H_B = H_B^o$$
$$V_A = V_A^* \qquad\qquad\qquad\qquad\quad V_B = V_B^o$$

$$G_m = (1 - X)G_A^* + XG_B^o + RT\{(1 - X)\ln(1 - X) + X \ln X\}$$
$$S_m = (1 - X)S_A^* + XS_B^o - R\{(1 - X)\ln(1 - X) + X \ln X\}$$

$$H_m = (1-X)H_A^* + XH_B^o$$
$$V_m = (1-X)V_A^* + XV_B^o$$

208   *the solvent laws*

208:1   *the solute is present in both phases*

Equations of the type $G_A^{*\alpha} = G_A^{*\beta} - RTX$ will change into $G_A^{*\alpha} - RTY = G_A^{*\beta} - RTX$, as a result of which the effect, such as $(\Delta P/P_A^o)$ will stand in relation to $(X - Y)$ instead of X; and eventually may change sign (when $Y > X$).

208:2   *chemical potentials of A in liquid and vapour versus pressure*

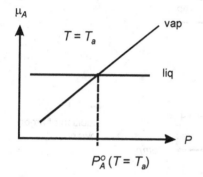

substance A
In reality for straight
lines: μ_A preferably
against ln P

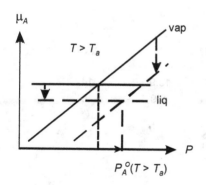

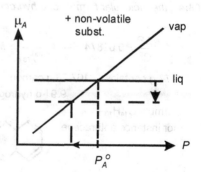

changes with dashed lines

208:3 *chemical potentials of H_2O in liquid and vapour versus temperature*

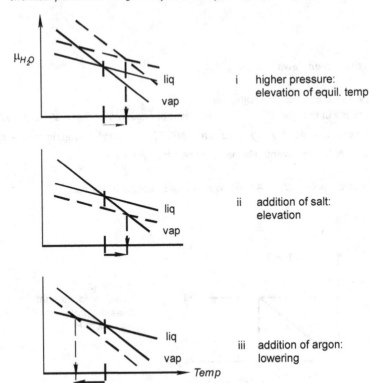

i higher pressure:
   elevation of equil. temp

ii addition of salt:
    elevation

iii addition of argon:
     lowering

100°C

208:4 *the molecular formula of a hydrocarbon*

$$X = \frac{\Delta P}{P^o} = 0.00874 = \frac{^2/_{M_2}}{^2/_{M_2} + ^{100}/_{78}} \Rightarrow M_2 = 177$$

| 177 g contains | 167.08 g carbon | → | 13.9 | → | 14 |
| | 9.91 g hydrogen | → | 9.9 | → | 10 |

formula $C_{14}H_{10}$
for instance anthracene

208:5 *the molar mass of the solvent*

$$X = -\Delta T \frac{\Delta H_A^*}{RT_A^{o2}} = 0.00868 \text{ with } M_1 = 78$$

from which M_2 is calculated as 178

$$X = \qquad = 0.01112 \text{ with } M_1 = 100$$

M_2 is calculated as 178

• the answer is "yes" !

208:6 *initial slope of solidus*
The result, indeed, is in full agreement with the calculated solidus in Figure 213:4

208:7 *the intervention of a foreign gas*
The liquid is put under pressure: its μ_A increases, because of $(\partial \mu_A / \partial P)_T = V_A$.

The μ_A of gaseous A has to increase; this can be done by "sending" more A into the gas phase: the partial pressure of A increases.

208:8 *an equation for the relative change in partial pressure*
increase in μ_A^{liq} : $\quad V_A^{*\,liq} \cdot P \qquad (= a)$

change in μ_A^{vap} : $\quad RT \ln \dfrac{P_A^\circ + \Delta P_A}{P_A^\circ} = RT \dfrac{\Delta P_A}{P_A^\circ} \qquad (= b)$

from   a = b
$$\frac{\Delta P_A}{P_A^\circ} = V_A^{*\,liq} \cdot \frac{P}{RT}$$

if V^{gas} is the volume of the gas phase, then $V^{gas} = n^{gas} \dfrac{RT}{P}$ as a result

$$\frac{\Delta P_A}{P_A^\circ} = n^{gas} \frac{V_A^{*\,liq}}{V^{gas}}$$

209   the solute laws

209:1 *the Kritchevsky-Kasarnovsky equation*
- m_B^P : molality of B in solution in equilibrium whit gaseous B whose pressure is P;
  m_B° : likewise, gaseous pressure P°;
  V_B° : partial volume of B in solution phase.
- recipes of μ's: $\mu_B^{vap} = G_B^{\circ\,vap} + RT \ln P$
  $$\mu_B^{liq} = G_B^{\circ\,liq} + RT \ln m_B + V_B^\circ (P - P^\circ)$$
  $$\sim V_B^\circ \cdot P \quad \text{for } P \gg P^\circ$$

substitution of recipes in equil. condition $(\mu_B^{liq} = \mu_B^{vap})$:

under high pressure P: $G_B^{\circ\,liq} + RT \ln m_B^P + V_B^\circ \cdot P = G_B^{\circ\,vap} + RT \ln P^{*)}$
at low pressure P° : $\dfrac{G_B^{\circ\,liq} + RT \ln m_B^\circ \qquad = G_B^{\circ\,vap} + RT \ln P^\circ}{}$

$$RT \ln \frac{m_B^P}{m_B^\circ} \quad = \quad RT \ln \frac{P}{P^\circ} - V_B^\circ \cdot P$$

$$\frac{m_B^P}{m_B^\circ} \quad = \quad \frac{P^{*)}}{P^\circ}\, e^{-(V_B^\circ / RT) \cdot P}$$

- *) only this P can be replaced by fugacity.

209:2 *helium in deep ocean water*

NB helium's molar mass is unnecessarily given.

- 10 ml of ideal gas at 25 °C, 1 atm is an amount of 0.0004 mol; this amount dissolves in 1 kg water:

$$m^o_{He} = 0.0004 \, \text{mol} \cdot \text{kg}^{-1}$$

- P is about 800 times P^o

$$m^P_{He} = m^o_{He} \cdot \frac{P}{P^o} \, e^{-(V^o_{He}/RT) \cdot P} = 0.12 \, \text{mol} \cdot \text{kg}^{-1}$$

$$m^P_{He} \sim 300 \, m^o_{He}$$

209:3 *calculation of molality and Gibbs energy*

a) $G^{*sol}_B = G^o_B + RT \ln m^{sat} \Rightarrow m^{sat} = 0.100 \, \text{mol} \cdot \text{kg}^{-1}$

$$X = 0.0099$$

b) $\mu^{liq}_A = G^{*liq}_A - RTX \quad = -1025$

$\mu^{liq}_B = G^o_B + RT \ln m^{sat} \quad = -2500 \quad (= \text{still } G^{*sol}_B)$

$G^{liq}_m = (1-X)\mu^{liq}_A + X\mu^{liq}_B = -1040 \, \text{J} \cdot \text{mol}^{-1}$

209:4 *simultaneous saturation?*

The answer is yes.

The partition of B over the two liquid phases I (mainly A) and II (mainly C) is "controlled" by

$\mu^I_B = \mu^{II}_B$ (1)

The equilibrium between solid B and (saturated) liquid I implies

$\mu^{sol}_B = \mu^I_B$ (2)

From the combination of (1) and (2) it follows that I and II will be saturated at the same time; see also Exc 006:12.

209:5 *a system for storage of thermal energy*

$$f = M \, [T, m] - N \left[\mu_{decah.} = \mu_{anh.} + 10 \, \mu_{H_2O}; \, \mu^{sol}_{Na_2SO_4} = \mu^{in \, soln}_{Na_2SO_4} \right] = 2 - 2 = 0$$

Make a plot of $\ln m$ versus $1/T$: two straight lines can be distinguished; from the difference between the slopes the heat effect is calculated as about 58 kJ·mol⁻¹. The point of intersection of the two lines is at about 32.5°C

209:6 *mixing of salt and water*

The source of the effect is the heat of solution. Apparently, the compiler of the table in question did not give attention to the solubility of the salts. If the solubility is less than 50 g per 50 g of water, it may happen that the fall in temperature with less than 50 g of salt is greater than the figure in the table - owing to the fact that the heat capacity of the combination is lower than in the 50/50 case.

210 *ideal equilibria*

210:1 *deviation from ideal vapour pressure*

In Scheme 2 insert the excess part of A's chemical potential in the liquid phase. Next, with the modified chemical potential(s), repeat the route from Equation (1) to Equation (3). The diagnostic formula is $\Omega = 4\,R\,T\ln(P/P^{id.})$

210:2 *like and unlike*

Methanol (A) + 1-propanol (B) $\Omega = 161\ \text{J·mol}^{-1}$
Isobutylalcohol (A) + toluene (B) $\Omega = 2948\ \text{J·mol}^{-1}$

210:3 *entropies of vaporization from phase diagram data*

Equations

$$(1 - X^{liq}) = (1 - X^{vap})\exp\left\{\Delta S_A^*(T_A^o - T)/RT\right\}$$

$$X^{liq} = X^{vap}\exp\left\{\Delta S_B^*(T_B^o - T)/RT\right\}$$

mean values found $\Delta S_A^* = 9.54\ R$

$\Delta S_B^* = 9.80\ R$

210:4 *point of inflexion in ideal liquidus*

$$\frac{d^2 X}{dT^2} = X\left(\frac{\Delta S_A^* \cdot T_A^o}{R}\right)\frac{1}{T^3}\left(\frac{\Delta S_A^* \cdot T_A^o}{RT} - 2\right)$$

point of inflexion if $\left(\dfrac{\Delta S_A^* \cdot T_A^o}{RT} - 2\right) = 0$

i.e. for $T = \dfrac{\Delta S_A^* \cdot T_A^o}{2R}$

and this T is $\leq T_A^o$ if $(\Delta S_A^*/2R) \leq 1$

210:5 *solubility of anthracene in benzene*

Anthracene (A) + benzene (B)
condition $\mu_A^{sol} = \mu_A^{liq}$

substitution recipes, assuming ideal liquid mixing:

$$G_A^{*sol}(T) = G_A^{*liq}(T) + RT\ln X_A$$

$$RT\ln X_A = -\Delta G_A^*(T) = \Delta S_A^*(T - T_A^o)$$

$X = 0.0106$; 1 mole per cent.

NB This is equivalent to calculating the anthracene liquidus in the system anthracene + benzene.

Observe that the same result will be obtained for any B that mixes ideally with A.

210:6   *from TX to PX*

A = n-pentane   B = n-hexane

normal boiling points T_A^o = 308.75 K T_B^o = 341.15 K

heat of vaporization $\Delta H_A^{\cdot}$ = 2945 R · K $\Delta H_B^{\cdot}$ = 3343 R · K

equil. vap pressures at 60°C with Clapeyron's equation:

$$P_A^o = 2.01\,\text{bar} \qquad\qquad P_B^o = 0.79\,\text{bar}$$

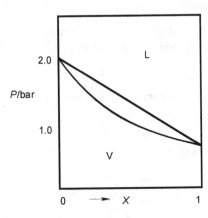

210:7   *narrow two-phase regions?*

Assuming ideal-mixing behaviour in each of the three forms one can write, e.g.,

$$G_{Er}^{*hcp} + RT \ln X^{hcp} = G_{Er}^{*liq} + RT \ln X^{liq}$$

which gives rise to $\ln \dfrac{X^{hcp}}{X^{liq}} = \dfrac{\Delta G_{Er}^{\cdot}}{RT} = \dfrac{\Delta S_{Er}^{\cdot}}{RT}(T - T_{Er}^{o})$.

At the three-phase equil. temp (1456°C) X^{liq} is about 0.45, as a result X^{hcp} is calculated as 0.476.

This is really a borderline case of drawing a phase diagram with open two-phase fields.

210:8   *heteroazeotrope*

Thermodynamically, the case is analogous to the simple eutectic phase diagram, Figure 3. The answer is found by means of Equations (18) and (19). The temperature of the heteroazeotrope given by the Handbook is 84.1 °C.

210:9   *the ortho and para forms of H_2*

- Boiling points: para 20.28 K; ortho (by extrapolation) 20.44 K.
  Mole fractions at 20.36 K: vapour 0.4947; liquid 0.5053.
- The introduction of the catalyst has no influence: equality of chemical potentials was already realized by the liquid+vapour equilibrium.
- The set M contains five variables: pressure, temperature, and, for each of the three phases, the mole fraction of ortho.
  The set N shows five signs of equality between the six chemical potentials:
  (ortho sol =para sol) = (ortho liq = para liq) = (ortho vap = para vap).
  The system is invariant: $f = M - N = 5 - 5 = 0$.

211   *non-ideal systems - geometrically*

211:1   *validation of linear contributions*

$$\frac{\Delta_e G\,'}{\Delta_e X} = \frac{G^\beta\,(X^\beta = X_e^\beta) + C\cdot X_e^\beta - G^\alpha\,(X^\alpha = X_e^\alpha) - C\cdot X_e^\beta}{X_e^\beta - X_e^\alpha} = \frac{\Delta_e G}{\Delta_e X} + C$$

$$\left(\frac{\partial G'^\alpha}{\partial X^\alpha}\right)_{X^\alpha = X_e^\alpha} = \left(\frac{\partial G^\alpha}{\partial X^\alpha}\right)_{X^\alpha = X_e^\alpha} + C$$

$$\left(\frac{\partial G'^\beta}{\partial X^\beta}\right)_{X^\beta = X_e^\beta} = \left(\frac{\partial G^\beta}{\partial X^\beta}\right)_{X^\beta = X_e^\beta} + C$$

if the original functions G^α and G^β satisfy the conditions for X_e^α and X_e^β, then the functions G'^α and G'^β will also satisfy the conditions for X_e^α and X_e^β

211:2   *from G-curves to phase diagram*

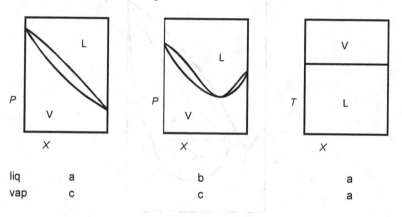

| liq | a | b | a |
| vap | c | c | a |

211:3   *metastable extensions*
For instance for the situation below the three-phase equilibrium temperature

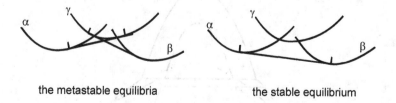

the metastable equilibria   the stable equilibrium

The points of contact, on the G curves of α and β, of the "metastable common tangent lines" have inward positions with respect to the points of contact of the 'stable common tangent line'.

211:4 *isothermal solid+vapour equilibrium*

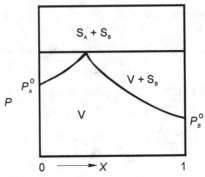

NB See also Figure 006:2.

211:5 *phase diagram for given ROD and EGC*

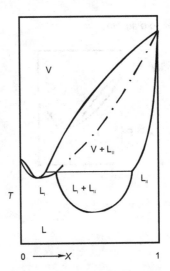

211:7 *phase diagram for EGC and two ROD's*

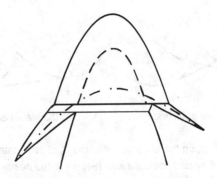

211:8 *overlapping two-phase regions*

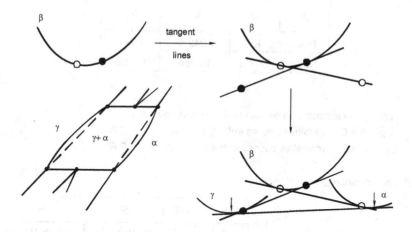

See also Exc.005:6

211:9 *the system formulation for a symmetrical binary system*

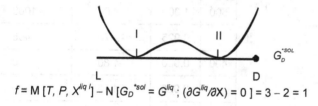

$$f = M\,[T, P, X^{liq\,I}] - N\,[G_D^{*sol} = G^{liq}\,;\,(\partial G^{liq}/\partial X) = 0\,] = 3 - 2 = 1$$

211:10 *azeotropy and Gibbs-Duhem*

After subtraction:

$$(1 - X)\Big[d\mu_A^{vap} - d\mu_A^{liq}\Big] + X\Big[d\mu_B^{vap} - d\mu_B^{liq}\Big] + (S^{vap} - S^{liq})dT = 0$$

Along the liquidus and the vaporus A's chemical potentials change, but invariably have the same value: the difference between the changes in vap and liq, as a result, is zero (same for B's potentials). For the special case of equality of composition, therefore, $dT = 0$.

If the compositions are not equal, one has, after subtraction, for A's potentials

$$(1 - X^{vap})d\mu_A^{vap} - (1 - X^{lliq})d\mu_A^{liq}, \text{ which is not equal to zero.}$$

212 *non-ideal systems - analytically*

212:1 *spinodal and binodal for a trigonometric excess function*

$RT_{spin}(X) = 4 \cdot C\,\pi^2\,X\,(1 - X)\,\sin 2\,\pi\,X$

binodal points at zero Kelvin: $X^I = 0$; $X^{II} = 0.715$

$T_C = 948.5$ K; $0.75\,T_C = 711$ K: $X^I = 0.132$; $X^{II} = 0.469$; $X_C = 0.298$

212:2 *small solid-state solubility*

A	B	X_B^I	X_A^{II}
NaCl + KCl		0.042 %	0.56 %
NaCl + RbCl		16 ppb	6 ppm

212:3 *conditions imposed on excess enthalpy and entropy*

u c p $A > 0$ (positive excess enthalpy); $A/\theta > -2R$
l c p $A < 0$ (negative excess enthalpy); $A/\theta < -2R$

212:4 *'re-entrant region of demixing'*

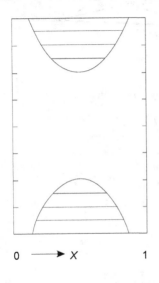

0 ——→ X   1

T /K	$\omega/2RT$	S^E	H^E
		$R \cdot X(1-X)$	$RK \cdot X(1-X)$
600	1.300	−7.40	−2880
	1.125	−5.55	−1815
500	1.000	−4.00	−1000
	0.925	−2.75	−405
400	0.900	−1.80	0
	0.925	−1.15	245
300	1.000	−0.80	360
	1.125	−0.75	375
200	1.300	−1.00	320

$\omega = (5 - 0.016\,T + 0.00002\,T^2)\,RT$
The excess heat capacity is given by $-X(1-X)\cdot RT\cdot(2b + 6cT)$

212:5 *indirect evaluation of heat of mixing*

$\ln P$ vs. X diagram : $G^E(X = 0.4;\ T = 308.35\,K) = -577\ J\cdot mol^{-1}$
T vs. X diagram : $G^E(X = 0.4;\ T = 337.5\,K) = -463\ J\cdot mol^{-1}$

$H^E(X = 0.4;\ T \approx 323\,K) = -1783\ J\cdot mol^{-1}$

212:6 *isothermal EGC for change from solid to liquid*

$$P_{EGC}(X) = \frac{(1-X)\cdot P_A^o\cdot \Delta V_A^* + X\cdot P_B^o\cdot \Delta V_B^*}{(1-X)\Delta V_A^* + X\Delta V_B^*} + \frac{G^{E\,sol}(X)}{(1-X)\Delta V_A^* + X\Delta V_B^*}$$

212:7 *types of phase diagram*

type of EGC	absence of ROD	presence of ROD
0	[0]	[p]
+	[+]	[p +]
−	[−]	[e] [−p]
− +	[− +]	[e +] [− p +] [p − +]

For details, and a further extension, see Oonk and Sprenkels (1969)

212:10 *the pressure above which mixed crystals are stable*
ΔA's value is −13 kJ·mol^{-1}.
The EGC in the PX plane runs from −1.07 kbar at $X = 0$, through its maximum at $X = 0.535$; $P = 3.87$ kbar to $P = 0.00$ kbar at $X = 1$.
The stable part of the phase diagram is above $P = 0$; at the NaCl side, the EGC intersects $P = 0$ at $X = 0.062$, the compositions of the coexisting solid and liquid phases being about 0.03 and 0.11, respectively (see Exc 213:6).

212:11 *heat effects derived from phase diagrams*
D, and L, heat of melting about 19 kJ·mol^{-1};
DL (composed of 0.5 D and 0.5 L), heat of melting about 28 kJ·mol^{-1};
Heat effect of racemization : (−56 +38 = −18) kJ·mol^{-1}.

212:12 *a rule of thumb for minimum azeotropes*
The absence of a minimum is equivalent to the property that the EGC emanates from the lower boiling point, which is the boiling point of the first component, with a positive slope. It implies that Ω may not exceed a certain value. The non-interference of the region of demixing comes down to a second condition, which is $\Omega < 2RT_A^{\circ}$.

213   *non-ideal systems – numerically*

213:1 *Chanh's data set*
$g_1 = (19729 - 9.13853\ T/K)$ J·mol^{-1}; and $g_2/g_1 = 0.2$.
This result is quite close to the real properties.
The last sentence could be rephrased to "Generally, and in general terms, TX phase diagrams are capable of providing the true Gibbs energy function for their mean temperature, but not - or only exceptionally, and probably depending on the structure of the data set and the mathematical model used - the precise change of the Gibbs energy with temperature".

NB. For the data set by Bunk and Tichelaar, Table 2, a physically realistic solution is obtained in terms of the model with three h and two s parameters (Oonk 1981).

213:2 *excess function constants from critical coordinates*

$a_1 = (-1+6X-6X^2)/\left[4X^2(1-X)^2\right]$; $a_2 = (1-2X)/\left[12X^2(1-X)^2\right]$, where X is short for X_C.

The uncertainties in g_1 and g_2 reach values of 1 kJ·mol^{-1}.

213:3 *analytical check*

$$RT \ln X^{liq} = RT \ln X^{sol} - \Delta H_B^* + T\Delta S_B^* + (1-X^{sol})^2\left[2135+429(1-X^{sol})\right] J\cdot mol^{-1}.$$

The mean absolute difference between the calculated mole fractions of the liquid phase and those in the first column of Table 4 is just 0.0012.

213:5 *elementary analysis of isothermal liquid+vapour data*

In rounded values: $g_1/RT = 0.36$; $g_2/RT = -0.10$; $\Delta_P = 1$ Pa; $\Delta_X = 0.006$

213:6 *an application of LOOP*

0.0273 (solid) and 0.1128 (liquid).

213:7 *an application of ROD*

Starting values 0.005 and 0.968; final values 0.00505 and 0.96795.

213:8 *unilateral region of demixing*

$RT_{spin}(X) = BX(1-X)^{n-1}(2n-nX-n^2X)$; the critical mole fraction follows from $dT_{spin}/dX=0$, which gives rise to an equation which is quadratic in X.

213:9 *unilateral ROD in SiO_2 + MgO*

$n = 3.5$; $B(2240 K) = 40.2$ kJ·mol^{-1}; $B(1987 K) = 52.6$ kJ·mol^{-1}; points of contact at double tangent $X' = 0.021$; $X'' = 0.402$; $h = 150$ kJ·mol^{-1}; $s = 49$ J·K^{-1}·mol^{-1}.

213:10 *an amusing side-effect*

Obviously, in the vicinity of the two maxima there are two regions of demixing, separated from one another. The inner boundaries of these two ROD's, the inner parts of the two binodals, intersect at a temperature above the spinodal's intermediate minimum. That temperature is the one at which three phases are in equilibrium. Below the three-phase temperature the stable equilibrium is between the two outer phases.

Instructive GX sections, all having four points of inflexion, are for the three-phase temperature, and for temperatures just above and below it.

REFERENCES

1824   Carnot S (1824) Réflexions sur la puissance motrice du feu et sur les machines propres à developer cette puissance.

1833   Clapeyron E (1833) Mémoire sur la puissance motrice de la chaleur. Journal de l'École Polytechnique 14:153-190

1854   Clausius R (1854) Ueber eine veränderte Form des zweiten Hauptsatzes der mechanische Wärmetheorie. Annalen der Physik und Chemie 93: 481-506

1865   Clausius R (1865) Ueber verschiedene für die Anwendung bequeme Formen der Hauptgleichungen der mechanischen Wärmetheorie. Annalen der Physik und Chemie 125: 353-400

1867   Debray H (1867) Recherches sur la dissociation. Comptes Rend. 64: 603-606

1869   Horstmann A (1869) Dampfspannung und Verdampfungswärme des Salmiaks. Ber. D. Chem. Geselsch. 2:137-140

1876   Gibbs JW (1876) On the equilibrium of heterogeneous Substances. Trans. Connect. Acad. III:108-248; (1878)343-524

1881   Konowalow D (1881) Ueber die Dampfspannungen der Flüssigkeitsgemische. Annalen der Physik und Chemie 250:34-52; 219-226

1888   Antoine Ch (1888) Tensions des vapeurs: nouvelle rélation entre les tensions et les températures. Comptes Rend. 107:681-684; 836-837

1890   van der Waals JD (1890) Molekular-theorie eines Körpers, der aus zwei verschiedenen Stoffen besteht. Z. physik. Chem. 5:133-173

1891   Nernst W (1891) Verteilung eines Stoffes zwischen zwei Lösungsmitteln und zwischen Lösungsmittel und Dampfraum. Z. physik. Chem. 8:110-139

1893   van Rijn van Alkemade AC (1893) Graphische Behandlung einiger thermodynamischer Probleme über Gleichgewichtszustände von Salzlösungen mit festen Phasen. Z. physik. Chem. 11:289-327

1899   Bakhuis Roozeboom HW (1899 Erstarrungspunkte der Mischkristalle zweier Stoffe. Z. physik. Chem. 30:385-412

      Wind CH (1899) Zur Gibbsschen Phasenregel. Z. physik. Chem. 31: 390-397

1901   Schreinemakers FAH (1901) Dampfdrucke ternärer Gemische. Z. physik. Chem. 36:257-289; 36:413-449; 36:710-740; 37:129-156; (1903)43:671

Bakhuis Roozeboom HW (1901) Die heterogenen Gleichgewichte vom Standpunkt derPhasenlehre, I. Die Phasenlehre – Systeme aus einer Komponente. Vieweg, Braunschweig

1904   Bakhuis Roozeboom HW (1904) Die heterogenen Gleichgewichte, II, part 1, Systeme aus zwei Komponente. Vieweg, Braunschweig

Jaeger FM (1904) Ueber molekulare und krystallographische Symmetrie von stellungsisomeren Benzolabkömmlingen. Z. physik. Chem. 38: 555-601

Geer WC (1904) Crystallization in three-component systems. J. Phys. Chem. 8:257-287

1905   Schreinemakers FAH (1905) Mischkristalle in Systemen dreier Stoffe. Z. physik. Chem. 50:169-199; 51:547-576; 52:513-550

1907   Kurnakov NS, Zhemchuzhnii SF (1907) Isomorphismus der Natrium- und Kaliumverbindungen. Z. anorg. Chem. 52:187-201

1908   van Laar JJ (1908) Die Schmelz- oder Erstarrungskurven bei binären Systemen, wenn die feste Phase ein Gemisch (amorphe feste Lösung oder Mischkristalle) der beiden Komponenten ist. Z. physik. Chem. 63:216-253; 64:257-297

1909   van Laar JJ (1909) Die Schmelz- oder Erstarrungskurven bei binären. Systemen, wenn die feste Phase ein Gemisch der beiden Komponenten ist, und eine Verbindung auftritt. Z. physik. Chem. 66:197-237

1910   van Laar JJ (1910) Ueber Dampfspannungen von binären Gemischen. Z. physik. Chem. 72:723-751

1911   Schreinemakers FAH (1911) H.W. Bakhuis Roozeboom. Die heterogenen Gleichgewichte. III, part 1, Die ternären Gleichgewichte. Vieweg, Braunschweig

1912   Hildebrand JH (1912) The thermal dissociation of barium peroxide. J. Am. Chem. Soc. 34:246:258

Tetrode H (1912) Die chemische Konstante der Gase und das elementare Wirkungsquantum. Annalen der Physik, vierte Folge 38:434-442; 39: 255-256

1913 Sackur O (1913) Die universelle Bedeutung des sog. Elementaren Wirkungsquantums. Annalen der Physik, vierte Folge 40:67-86

Schreinemakers FAH (1913) H.W. Bakhuis Roozeboom. Die heterogenen Gleichgewichte, III, part 2, Die ternären Gleichgewichte. Vieweg, Braunschweig

Hasselblatt M (1913) Linear velocity of crystallization of isomorphous mixtures. Z. physik. Chem. 83:1-40

1914 Bowen NL, Anderson O (1914) The binary system: MgO-SiO_2. Am. J. Sci. 37:487-500

1918 Nacken R (1918) Über die Grenzen der Mischkristallbildung zwischen Kaliumchlorid und Natriumchlorid. Sitzungsber. Preuss. Akad. Wiss. Phys. Math. Kl 192

1923 Larson AT, Dodge RL (1923) The ammonia equilibrium. J. Am. Chem. Soc. 45:2918-2930

1924 Larson AT (1924) The ammonia equilibrium at high pressure. J. Am. Chem. Soc. 46:367-372

Tammann G (1924) Lehrbuch der heterogenen Gleichgewichte. Vieweg, Braunschweig

1927 Greig JW (1927) Immiscibility in silicate melts. Am. J. Sci. 13:1-44, 113-153

1933 Ehrenfest P (1933) Phase changes in the ordinary and extended sense classified according to the corresponding singularities of the thermodynamic potential. Leiden Comm. Suppl.75b

1935 van Laar JJ (1935) Die Thermodynamik einheitlicher Stoffe und binärer Gemische, mit Anwendungen auf verschiedene physikalisch-chemische Probleme. Noordhoff, Groningen

1937 Borelius G (1937) The theory of the transformation of metallic mixed phases. IV. The separation of disordered mixed phases. Ann. Physik, 5 Folge 28:507-519

1938 Scatchard G, Raymond CL, Gilmann HH (1938) Vapor-liquid equilibrium.I. Apparatus for the study of systems with volatile components. J. Am. Chem. Soc. 60:1275-1278

Scatchard G, Raymond CL (1938) Vapor-liquid equilibrium.II. Chloroform-ethanol mixtures at 35, 45 and 55 °C. J. Am. Chem. Soc. 60:1278-1287

1940   Lipson H, Wilson AJC (1940) Some properties of alloy equilibrium diagrams derived from the principle of lowest free energy. J. Iron Steel Inst. 142:107-122

       Reinders W, de Minjer CH (1940) Vapour-liquid equilibria in ternary systems.II. The system acetone-chloroform-benzene. Rec. Trav. Chim. Pays-Bas 59:369-391

1944   Wilson AJC (1944) Binary equilibrium. J. Inst. Metals 70:543-556

1945   Temkin M (1945) Mixtures of fused salts as ionic solutions. Acta Physicochem. URSS 20:411-420

1947   Fredga A , Miettinen JK (1947) Steric relationships of optically active α-isopropylglutaric acid, fenchone, and camphor. Acta Chem. Scand. 1: 371-378

1948   Campbell AN, Prodan LA (1948) An apparatus for refined thermal analysis exemplified by a study of the system p-dichlorobenzene – p-dibromobenzene – p-chlorobromobenzene. J. Am. Chem. Soc. 70: 553-561

       Redlich O, Kister AT (1948) Thermodynamics of nonelectrolytic solutions. Algebraic representation of thermodynamic properties and the classification of solutions. J. Ind. Eng. Chem. 40:845-848

1949   Haase R (1949) Zur Thermodynamik flüssiger Dreistoffgemische. Z.Naturforsch. A 4:342-352

1950   Haase R (1950) Verdampfungsgleichgewichte von Mehrstoffgemischen VII: Ternäre azeotrope Punkte. Z. physik. Chem. 195:362-385

       Guggenheim EA (1950) Thermodynamics: an advanced treatment for chemists and physicists. 2nd Ed. North-Holland Publ. Comp., Amsterdam

       Meijering JL (1950) Segregation in regular ternary solutions, part 1. Philips Res. Rep. 5:333-356

1952   Brown I (1952) Liquid-vapor equilibria.III. The systems benzene-heptane, hexane-chlorobenzene, and cyclohexane-nitrobenzene. Austral. J. Sci. Res. Ser. A 5:530-540

1953   Barker JA (1953) Determination of activity coefficients from total-pressure measurements. Austral. J. Chem. 6:207-210

1953   Bunk AJH, Tichelaar GW (1953) Investigations in the system NaCl-KCl. Proc. K. Nedl. Akad. Wet. Ser. B. 56:375-384

       De Sorbo W (1953) Specific heat of diamond at low temperatures. J. Chem. Phys. 21:876-880

1954   Barrett WT, Wallace WE (1954) Sodium chloride-potassium chloride solid solutions. I. Heats of formation, lattice spacings, densities, Schottky defects, and mutual solubilities. J. Am. Chem. Soc. 76:366-369

1955   Gromakov SD, Gromakova LM (1955) The liquidus curve (mathematical) application in binary systems. Zh. Fiz. Khim. 29:745-749

       Mellor JW (1955) Higher mathematics for students of chemistry and physics. Dover Publications

1958   Rose A, Papahronis BT, Williams ET (1958) Experimental measurement of vapor-liquid equilibrium for octanol-decanol and decanol-dodecanol binaries. Chem. & Eng. Data Ser. 3:216-219

1959   Korvezee AE, Meijering JL (1959) Validity and consequences of Schreinemakers' theorem on ternary distillation lines. J. Chem. Phys. 31:308-313

1960   Hill TL (1960) An Introduction to Statistical Thermodynamics. Addison-Wesley, Reading, Massachusetts

1963   Tenn FG, Missen RW (1963) Study of the condensation of binary vapors of miscible liquids.I.The equilibrium relations. Canad. J. Chem. Eng. 41:12-14

1964   Nguyen-Ba-Chanh   (1964)   Équilibres   des   systèmes   binaires d'halogénures de sodium et potassium à l'état solide. J. Chim. Phys. 61:1428-1433

       Short JM, Roy R (1964) Use of interdiffusion to determine crystalline solubility in alkali halide systems. J. Am. Ceram. Soc. 47:149-151

1965   Holm JL (1965) Phase relations in the systems NaF-LiF, NaF-KF, and NaF-RbF. Acta Chem. Scand. 19:638-644

       Kleppa OJ, Meschel SV (1965) Heats of formation of solid solutions in the systems (Na-Ag)Cl and (Na-Ag)Br. J. Phys. Chem. 69:3531-3537

       Morcom KW, Travers DN (1965) Heat of mixing of the system acetone+chloroform; temperature dependence and deuterium isotope effect. Trans. Faraday Soc. 61:230-234

1966   Clarke ECW, Glew DN (1966) Evaluation of thermodynamic functions from equilibrium constants. Trans. Faraday Soc. 62:539-5

       E-an Zen (1966) Construction of pressure-temperature diagrams for multicomponent systems after the method of Schreinemakers – geometric approach. U.S. Geological Survey, Bulletin 1225. U.S. Government Printing Office, Washington

       Luova P, Tannila O (1966) Miscibility gaps in the systems KBr-KI and NaCl-KCl. Suom. Kemistil. B39:220-224

       Fischer WA, Lorenz K, Fabritius H, Hoffmann A, Kalwa G (1966) Phase transformations in pure iron alloys with a magnetic balance. Arch. Eisenhüttenwes. 37:79-86

1967   Lupis CHP, Elliot JF (1967) Prediction of enthalpy and entropy interaction coefficients by the central atoms theory. Acta Metall. 15:265-276

1968   Osborn AG, Douslin DR (1968) Vapor pressure relations of 13 nitrogen compounds related to petroleum. J. Chem. Eng. Data 13:534-537

       Ahtee M, Koski H (1968) Solubility gap in the system rubidium bromide – rubidium iodide. Ann. Acad. Sci. Fenn. Ser.A6 297:8pp

1969   Althaus E (1969) Evidence that the reaction of kyanite to form sillimanite is at least bivariant. Amer. J. Sci. 267:273-277

       Oonk HAJ, Sprenkels A (1969) Types of TX phase diagrams derived from Gibbs functions with the help of the equal-G curve. Rec. Trav. Chim. Pays-Bas 88:1313-1331

       Thompson JB, Waldbaum DR (1969) Analysis of the two-phase region halite-sylvite in the sodium chloride-potassium chloride system. Geochim. Cosmochim. Acta 33:671-690

1970   Fletcher NH (1970) The chemical physics of ice. Cambridge

       Rant D, Nietzschmann B, Werner G, Schuberth H (1970) Thermal and caloric properties of the ternary system n-hexane-methylcyclohexane-methanol at 30 °C. Z. physik. Chem. 244:387-400

1971   Oonk HAJ, Sprenkels A (1971) Isothermal binary liquid-vapour equilibria and the EGC method. I. General aspects. Z. physik. Chem. Neue Folge 75:225-233

1971 Oonk HAJ, Brouwer N, de Kruif CG, Sprenkels A (1971) Isothermal binary liquid-vapour equilibria and the EGC method. II. The interpretation of experimental phase diagrams. Z. physik. Chem. Neue Folge 75:234-241.

1972 Sinistri C, Riccardi R, Margheritis C, Tittarelli P (1972) Thermodynamic properties of solid systems AgCl+NaCl and AgBr+NaBr from miscibility gap measurements. Z. Naturforsch. 27a:149-154

1973 Kirchner G, Nishizawa T, Uhrenius B (1973) Distribution of chromium between ferrite and austenite and the thermodynamics of the α/γ equilibrium in the iron-chromium and iron-manganese systems. Metall. Trans. 4:167-174

Würflinger A, Schneider GM (1973) Differential Thermal Analysis under high Pressures II: Investigation of the rotational transition of several n-alkanes. Ber. Bundesges. Phys. Chem. 77:121-128

1974 Il'yasov I I, Volchanskaya VV (1974) Liquidus surface of the rubidium chloride, bromide, iodide system. Ukr. Khim. Zh. 40:732-734

1975 Kuznetsov GM, Leonov MP, Luk'yanov AS, Kovaleva VA, Shapovalov MP (1975) Description and liquidus curves of AmBn chemical compounds. Dokl. Akad. Nauk SSSR 223:124-126

Ambrose D, Lawrenson IJ, Sprake CHS (1975) Vapour pressure of naphthalene. J. Chem. Thermodynamics 7:1173-1176

1976 Shannon RD (1976) Revised effective ionic radii and systematic studies of interatomic distances in halides and chalcogenides. Acta Cryst. A32: 751-767

1977 Nývelt J (1977) Solid-liquid phase equilibria. Elsevier, Amsterdam

1978 Powell R (1978) Equilibrium thermodynamics in petrology. Harper & Row, London

Robie RA, Hemingway BS, Fisher JR (1978) Thermodynamic properties of minerals and related substances at 298.15 K and 1 bar (10^5 pascals) pressure and at higher temperatures. U.S. Geological Survey, Bulletin 1452. U.S. Government Printing Office, Washington

Whittaker AG (1978) Carbon: a new view of its high-temperature behavior. Science 200:763-764

1979 D'Amour H, Denner W, Schulz H (1979) Structure determination of α-quartz up to 68×10^8 Pa. Acta Cryst. B35:550-555

1980   Apelblat A, TamirA, Wagner M (1980) Thermodynamics of acetone-chloroform mixtures. Fluid Phase Equilibria 4:229-255

       Schuberth H (1980) Isothermal phase equilibrium behaviour of n-alcohol-water-urea systems at 60 °C. Z. physik. Chem. Leipzig 261:777-790

1981   Denbigh KG (1981) The principles of chemical equilibrium, 4th Ed., Cambridge University Press, Cambridge

       Göbl-Wunsch A, Heppke G, Hopf R (1981) Miscibility studies indicating a low-temperature smectic A phase in biaromatic liquid crystals with reentrant behaviour. Z. Naturforsch. A 36:1201-1204

       Oonk HAJ (1981) Phase Theory – the thermodynamics of heterogeneous equilibria. Elsevier, Amsterdam

1982   De Kruif CG, Blok JG (1982) The vapour pressure of benzoic acid. J. Chem. Thermodynamics 14:201-206

1983   Moerkens R, Bouwstra JA, Oonk HAJ (1983) The solid-liquid equilibrium in the system p-dichlorobenzene + p-bromochlorobenzene + p-dibromobenzene. Thermodynamic assessment of binary data and calculation of ternary equilibrium. Calphad 7:219-269

1984   Kitaigorodskii AI (1984) Mixed crystals. Springer-Verlag, Berlin

       Mulder T, Verdonk AH (1984) A behavioral analysis of the laboratory learning process redesigning a teaching unit on recrystallization. J. Chem. Education 61:451-453

       Vasyutinskii NA (1984) Stoichiometry of wustite. Izv. Akad. Nauk SSSR, Neorg. Mater.; Inorg Mater. (Engl. Transl.) 20:1324-1327

1985   Hillert M, Jansson B, Sundman B, Agren J (1985) A two-sublattice model for molten solutions with different tendency for ionization. Met. Trans. 16A:261-266

       Nordstrom DK, Munoz JL (1985) Geochemical thermodynamics. Benjamin/Cummings, Menlo Park

       Okamoto H, Massalski TB (1985) The Au-Pd (gold-palladium) system. Bull. Alloy Phase Diagrams 6:229-235

       Van Hecke GR (1985) The Equal G Analysis. A comprehensive thermodyna- mics treatment for the calculation of liquid crystalline phase diagrams. J. Phys.Chem. 89:2058-2064

1986   Hageman VBM, Oonk HAJ (1986) Liquid immiscibilityin theSiO$_2$ + MgO, SiO$_2$ + SrO, SiO$_2$ + La$_2$O$_3$, and SiO$_2$ + Y$_2$O$_3$ systems. Phys. Chem. Glasses 27:194-198

       Prausnitz JM, Lichtenthaler RN, Gomes de Azevedo E (1986) Molecular thermodynamics of fluid-phase equilibria. 2nd Ed., Prentice-Hall, Englewood Cliffs

       Schuberth H (1986) Mischphasenthermodynamik polinärer Systeme. VCH Verlagsgesellschaft, Weinheim, Germany

1987   Reid RC, Prausnitz JM, Poling BE (1987) The Properties of Gases and Liquids. McGraw-Hill, New York

1988   Angell CA (1988) Perspective on the glass transition. J. Phys. Chem. Solids 49:863-871

       Hillert M (1988) Construction of multicomponent phase diagrams using zero phase fraction (ZPF) lines and Schreinemakers' rule. Scripta Metal. 22:1085-1086

1989   Ambrose D, Walton J (1989) Vapor pressures up to their critical temperatures of normal alkanes and 1-alkanols. Pure & Appl. Chem. 61:1395-1403

       Barin I (1989) Thermochemical data of pure substances. VCH Verlagsgesellschaft, Weinheim, Germany

       Boots HMJ, De Bokx PK (1989) Theory of enthalpy-entropy compensation. J. Phys. Chem. 93:8240-8243

       Ohe S (1989) Vapor-liquid equilibrium data. Elsevier, Amsterdam

       van Duijneveldt JS, Nguyen-Ba-Chanh, Oonk HAJ (1989) Binary mixtures of naphthalene and five of its 2-derivatives. Thermodynamic analysis of solid-liquid phase diagrams. Calphad 13:83-88

       van Duijneveldt JS, Baas FSA, Oonk HAJ (1989) A program for the calculation of isobaric binary phase diagrams. Calphad 13:133-137

1990   Höhne GWH, Cammenga HK, Eysel W, Gmelin E, Hemminger W (1990) The temperature calibration of scanning calorimeters. Thermochim. Acta 160:1-12

       Preston-Thomas H (1990) The International Temperature Scale of 1990 (ITS-90). Metrologia 27:3-10

1991   Angell CA (1991) Thermodynamic aspects of the glass transition in liquids and plastic crystals. Pure & Appl. Chem. 63:1387-1392

Bottinga Y (1991) Thermodynamic properties of silicate liquids at high pressure and their bearing on igneous petrology. In: Perchuk LL, Kushiro I (eds) Physical chemistry of magmas. Springer, New York, pp 213-232

Calvet MT, Cuevas-Diarte MA, Haget Y, van der Linde PR, Oonk HAJ (1991) Binary p-dihalobenzene systems – correlation of thermochemical and phase-diagram data. Calphad 15:225-234

Klipp N, van der Linde PR, Oonk HAJ (1991) The system thianaphthene+naphthalene. Solid-liquid equilibrium.   A   case   of crossed isodimorphism. Calphad, 15:235-242

1992   Erné BH, van der Weijden AJ, van der Eerden AM, Jansen JBH, van Miltenburg JC, Oonk HAJ (1992) The system $BaCO_3$+$SrCO_3$; crystal phase transitions: DTA measurements and thermodynamic phase diagram analysis. Calphad 16:63-72

Tonkov EYu (1992) High pressure phase transformations. Gordon and Breach, Philadelphia

van der Kemp WJM, Blok JG, van Genderen ACG, van Ekeren PJ, Oonk HAJ (1992) Binary common-ion alkali halide mixtures; a uniform description of the liquid and solid state. Thermochim. Acta 196:301-315

1993   Sirota EB, King HE, Singer DM, Shao HH (1993) Rotator phases of the normal alkanes: an X-ray scattering study. J. Chem. Phys. 98:5809-5824

1994   López DO, van Braak J, Tamarit JLl, Oonk HAJ (1994) Thermodynamic phase diagram analysis of three binary systems shared by five neopentane derivatives. Calphad 18:387-396

1996   Gallis HE, Bougrioua F, Oonk HAJ, van Ekeren PJ, van Miltenburg JC (1996) Mixtures of d- and l-carvone: I. Differential scanning calorimetry and solid-liquid phase diagram. Thermochim. Acta 274:231-242

Jacobs MHG, Jellema R, Oonk HAJ (1996) TXY-CALC, a program for the calculation of thermodynamic properties and phase equilibria in ternary systems. An application to the system (Li,Na,K)Br. Calphad 20:79-88

1997   Figurski G, van Ekeren PJ, Oonk HAJ (1997) Thermodynamic analysis of vapour-liquid phase diagrams I. Binary systems of nonelectrolytes with complete miscibility; application of the PXFIT- and the LIQFIT-method. Calphad 21:381-390

1998 Hillert M (1998) Phase equilibria, phase diagrams and phase transformations: their thermodynamic basis. Cambridge University Press, Cambridge

Oonk HAJ, van der Linde PR, Huinink J, Blok JG (1998) Representation and assessment of vapour pressure data; a novel approach applied to crystalline1-bromo-4-chlorobenzene, 1-chloro-4-iodobenzene, and 1-bromo-4-iodobenzene. J. Chem. Thermodynamics 30:897-907

Perrot P (1998) A to Z of Thermodynamics. Oxford University Press, Oxford

van der Linde PR, Blok JG, Oonk HAJ (1998) Naphthalene as a reference substance for vapour pressure measurements looked upon from an unconventional point of view. J. Chem. Thermodynamics 30:909-917

1999 Calvet T, Cuevas-Diarte MA, Haget Y, Mondieig D,Kok IC, Verdonk ML, van Miltenburg JC, Oonk HAJ (1999) Isomorphism of 2-methylnaphthalene and 2-halonaphthalene as a revealer of a special interaction between methyl and halogen. J. Chem. Phys. 110:4841-4846

Rajabalee F, Métivaud V, Mondieig D, Haget Y,Cuevas-Diarte MA (1999) New insights on the crystalline forms in binary systems of n-alkanes: characterization of the solid ordered phases in the phase diagram tricosane+pentacosane. J. Mater. Res. 14:2644-2654

Pardo LC, Barrio M, Tamarit JLl, López DO, Salud J, Negrier P, Mondieig D (1999)Miscibility study in stable and metastable orientational disordered phases in a two-component system (CH3)CCl3+CCl4. Chem. Phys. Lett. 308:204-210

2000 Gemsjäger H, Königsberger E, Preis W (2000) Lippmann diagrams: theory and application to carbonate systems. Aquat. Geochem. 6:119-132

2001 Oonk HAJ (2001) Solid-state solubility and its limits.The alkali halide case. Pure & Appl. Chem. 73:807-823

2002 Hillert M, Agren J (2002) Effect of surface free energy and surface stress on phase equilibria. Acta Materialia 50:2429-2441

Rebelo LPN, Najdanovic-Visak V, Visak ZP, Nunes da Ponte M, Troncoso J, Cerdeiriña CA, L. Romaní L (2002). Two ways of looking at Prigogine and Defay's equation. Phys. Chem. Chem. Phys. 4:2251-2259

Papon P, Leblond J, Meijer PHE (2002) The Physics of Phase Transitions.Springer-Verlag, Berlin, Heidelberg

2002   van Eijck BP (2002) Crystal structure predictions for disordered halobenzenes. Phys. Chem. Chem. Phys. 4:4789-4794

2003   Valderrama JO (2003) The state of the cubic equations of state. Ind. Eng. Chem. Res. 42:1603-1618

        van Genderen ACG, Oonk HAJ (2003) The (solid+vapour) equilibrium. A view from the arc. Colloids and Surfaces A: Physicochem. Eng. Aspects 213:107-115

2004   Malakhov DV (2004) A rigorous proof of the Alkemade theorem. Calphad 28:209-211

        Chang YA, Chen S, Zhang F, Yan X, Xie F, Schmid-Fetzer R, Oates WA (2004) Phase diagram calculation: past, present and future. Progr. Mater. Sci. 49:313-345

        Marchand P, Lefèbvre L, Querniard F, Cardinaël P, Perez G, Counioux J-J, Coquerel G (2004) Diastereomeric resolution rationalized by phase diagrams. Tetrahedron: Asymmetry 15:2455-2465

2006   Jacobs MHG, Oonk HAJ (2006) The calculation of ternary miscibility gaps using the linear contributions method: problems, benchmark systems and an application to (K,Li,Na)Br. Calphad 30:185-190

        Matovic M, van Miltenburg JC, Oonk HAJ, Los JH (2006) Kinetic approach to the determination of the phase diagram of a solid solution. Calphad 30:209-215

        Los JH, Van den Heuvel M, van Enckevort WJP, Vlieg E, Oonk HAJ, Matovic M, van Miltenburg JC (2006) Models for the determination of kinetic phase diagrams and kinetic phase separation domains. Calphad 30:216-224

SUBJECT INDEX

SUBSTANCES AND SYSTEMS INDEX